Reconsidering a
Century of Flight

Reconsidering a Century of Flight

EDITED BY

ROGER D. LAUNIUS

AND

JANET R. DALY BEDNAREK

The University of North Carolina Press

Chapel Hill and London

Set in Charter and Meta Black types
by Tseng Information Systems, Inc.
The paper in this book meets the guidelines for
permanence and durability of the Committee on
Production Guidelines for Book Longevity of
the Council on Library Resources.

The publication of this book was supported by a generous
grant from the First Flight Centennial Commission, North
Carolina Department of Cultural Resources.

Library of Congress Cataloging-in-Publication Data
Reconsidering a century of flight / edited by Roger D.
Launius and Janet R. Daly Bednarek.
    p. cm.
Includes bibliographical references and index.
ISBN 0-8078-2815-7 (cloth : alk. paper) —
ISBN 0-8078-5488-3 (pbk. : alk. paper)
I. Aeronautics—United States—History. I. Bednarek, Janet R.
Daly (Janet Rose Daly), 1959–
TL521 .R335   2003
629.13′00973—dc21   2003006420

cloth 07 06 05 04 03 5 4 3 2 1
paper 07 06 05 04 03 5 4 3 2 1

For all those who follow the dreams of the Wright Brothers

CONTENTS

Part 3: Aerial Warfare

Part 4: Aviation in the American Imagination

ILLUSTRATIONS

ACKNOWLEDGMENTS

As with any thing numerous debts were incurred over the course of this book's production. We acknowledge the support and encouragement of a large number of people associated with the centennial of flight activities under way around the world, among them Tom Lambeth, Ken Mann, and Kathryn Holten and her staff at the First Flight Centennial Commission; Richard Howard and Julie Ketner Rigby and the staff of the First Flight Centennial Foundation; Brad Tillson and Madeline J. Iseli of Inventing Flight of Dayton, Ohio; and Jack Dailey and Sherry Foster of the U.S. Centennial of Flight Commission. Without these organizations, there would be no centennial of flight commemoration. Thanks must also go to Larry Tise, who was instrumental in overseeing the nuts-and-bolts organization of the conference. And, of course, David Perry of the University of North Carolina Press provided encouragement and ideas for proceeding with this volume, as did Paula Wald, Mark Simpson-Vos, and Kathleen Ketterman.

In addition, several individuals read all or part of this manuscript or otherwise offered suggestions that helped us more than they will ever know. Our thanks are extended to the staff of the NASA History Office: Stephen J. Garber, Jane Odom, M. Louise Alstork, Nadine Andreassen, Colin Fries, and John Hargenrader. In addition to these individuals, we wish to acknowledge the following individuals who aided us in a variety of ways to complete this assignment: Jeff Bingham, George W. Bradley, Andrew J. Butrica, Michael L. Ciancone, Tom D. Crouch, Virginia P. Dawson, Dwayne A. Day, Andrew Dunar, Frederick C. Durant III, Donald C. Elder, Robert H. Ferrell, Lori B. Garver, Michael H. Gorn, Trish Graboske, Charles J. Gross, John F. Guilmartin Jr., Richard P. Hallion, T. A. Heppenheimer, Frank Hoban, David A. Hounshell, Nancy M. House, Perry D. Jamieson, Dennis R. Jenkins, Stephen B. Johnson, W. D. Kay, Richard H. Kohn, Sylvia K. Kraemer, John Krige, Alan M. Ladwig, W. Henry Lambright, Elaine Liston, John M. Logsdon, John L. Loos, Howard E. McCurdy, John E. Naugle, Valerie Neal, Allan A. Needell,

Michael J. Neufeld, Frederick I. Ordway III, Anthony M. Springer, Rick W. Sturdevant, Glen E. Swanson, Stephen P. Waring, and Joni Wilson. All these people would disagree with some of the conclusions offered here, but such is both the boon and the bane of historical inquiry. We also wish to thank the authors of the individual articles for their patience and helpfulness.

Reconsidering a
Century of Flight

ROGER D. LAUNIUS & JANET R. DALY BEDNAREK

INTRODUCTION
WHITHER A CENTURY OF FLIGHT

On a cloudy autumn day in the early 1920s, airmail pilot Dean Smith was forced down on the Chicago-to-Omaha route. He cabled the superintendent of airmail: "On trip 4 westbound. Flying low. Engine quit. Only place to land on cow. Killed cow. Wrecked plane. Scared me. Smith."[1] Today it offers us what we see as a humorous account of an adventurous experience. At the time, Dean Smith (and certainly the superintendent of airmail) might have been hard pressed to find the humor in the situation. In reality, flying the airmail was dangerous, and many pilots were not as fortunate as Smith. Underneath its bare outlines, Smith's somewhat prosaic account of his deadly encounter with livestock nonetheless also offers us a compelling vision. The image of the lone pilot battling harsh weather, rudimentary technology, and immature flight operations entices our sensibilities and conjures up images of courage, self-reliance, and danger.

Perhaps that romantic vision is appropriate when considering arguably one of the most significant technological developments of the twentieth century—machines that gave individuals the ability to fly. Humans had dreams of shedding the bounds of Earth for eons. Only after 1903, with the first flight of the Wright Brothers, did this universal human goal become a reality. The path from dream to invention holds broad implications for understanding the American civilization. It prompts historians to ask what the driving forces were behind this technological change. In important ways, the invention of the airplane represented part of a process embedded in society and intended to satisfy values held by members of society, rather than the disconnected product of a single mind or of progress along one "correct" path. The story of flight, and its fruition, may provide an important example of the manifestation of curiosity, culture, intellectual puzzlement, and social attitudes toward inventiveness in general and flight in particular. For an invention like the airplane to become more than a figment of the imagination, however, the time must be ripe. A nurturing social and technical environment was present, the ideas underpinning the investigation were sufficiently well developed to allow success, a basic belief in reward for those who accomplished the feat was recognized, and, most important, a mind-set favorable to adopting the innovation existed. All existed in abundance for the Wright Brothers in 1903, and they remained an essential ingredient in the adoption of the airplane as a major technology in the United States thereafter.

At the same time, while many think of the airplane as an American invention, from the beginning there was a significant international dimension to the development, marketing, and use of the airplane. Inventors, innovators, and entrepreneurs across the globe contributed to the rapid transformation of the airplane from primitive and dangerous craft to an advanced and relatively safe means of transportation. The U.S. government also played a significant role. Not only did the government contribute to the advancement of aeronautical technology, but its involvement with the airplane also helped redefine the government's relationship with technology, a relationship that is evolving to this day. Further, government fostered the emergence of the airplane as a weapon of war. There, too, one witnesses a rather rapid development from a machine in some ways most lethal to those who used it to a machine of great lethality to those against whom it is directed. And from the beginning, the notion of flight has captured the imagination of people— engineers, adventurers, artists, and everyday humans—in ways both profound and mundane. As these themes—only a few of many that might be explored—suggest, the history of flight is complex and multifaceted. More than a history of the development of a machine, the history of flight involves cross-national influences, governmental policies and actions, and in many ways a fundamental change in warfare. All this happened within a social, political, and cultural context that helped shape and define how humans would use this new invention.

To foster the development of aircraft in America, a community of manufacturers, operators, government laboratories, and military organizations emerged early, dedicated to building and using ever more sophisticated aircraft, engines, avionics, and instruments. Related efforts also arose to support this industry: research into flight characteristics, infrastructure support to airplanes, and businesses, such as the airlines, that used aircraft as their means of generating profits. The business of flight grew increasingly complex and correspondingly expensive as the years passed and the airplane became both more capable and more valued.[2]

At first, aviation was in many ways a technology of the elite—elite exhibition fliers, military aviators, and wealthy hobbyists. Their exploits may have thrilled the masses, but for most aviation was a spectator activity, not a part of everyday life. This remained the case even after the emergence of the first airlines. Air travel in the 1920s and 1930s was limited mostly to the upper class and to those who had a good reason to fly, such as manufacturers' representatives and those involved in banking. Flying was more expensive than traveling by train and "discretionary" flying was not yet practiced. According to aviation historian Roger Bilstein, a market survey of the 2,500 airfares in 1930 revealed that "85 percent of the passengers came from major businesses and high-income residential areas."[3] Charles Solberg concluded that in 1932, the main reason people flew was speed. And, apparently, only speed could overcome the perceptions of the dangers of flight that remained. Although only twenty-five people died in air accidents that year, many remained fearful. Further, the insurance industry reinforced the image of danger: a $5,000 insurance policy for an airplane trip cost $2 while for a trip by train, the cost was 25 cents.[4]

With the introduction of the Douglas DC-2 in 1934 and the DC-3 in 1936, air travel became much more comfortable and somewhat more commonplace. The introduction of these transports of the mid and late 1930s powered a rise in the number of air passengers from 474,000 in 1932 to 1,176,858 in 1938. Other statistics are equally revealing. The number of passenger miles traveled by air in the United States increased 600 percent from 1936 to 1941. But even as late as 1939, flying travelers equaled just 7.6 percent of the long-distance train market. It would take several years more before the number of passengers traveling by air surpassed those traveling by train.[5]

The aircraft industry also remained small and in some ways a craft industry during the 1920s and 1930s. It came of age, however, with the coming of war. During World War II, the aviation industry expanded dramatically, emerging as a huge sector of the U.S. economy, employing more than one million workers and contributing more than 5 percent to the gross domestic product. By the 1940s only a few firms dominated the industry, many of which remain dominant to the present. The industry grew fat on government contracts for military aircraft. While some observers may have seen profiteering in that, its role was viewed as so significant that many claimed the industry's record of output was as important to Allied victory as the military successes on the battlefield. Production statistics were remarkable. During 1939–45, the aeronautical industry rose from forty-first to first place among heavy industry employers of American workers. In 1939 U.S. manufacturers produced only 2,141 aircraft for the military. By 1942 that number had leaped to 47,836. In 1944, the peak production year, U.S. aircraft factories turned out 96,318 aircraft. From January 1, 1940, until V-J Day on August 14, 1945, aircraft manufacturers turned out more than 300,000 military aircraft for the U.S. military and the Allies—almost 275,000 of that number after Pearl Harbor. In the peak production month of March 1944, more than 9,000 aircraft came off the assembly lines.[6]

Aerial warfare in many ways also came of age during World War II. The fleets of aircraft delivering mass destruction first envisioned by Giulio Douhet, William "Billy" Mitchell, and Hugh Trenchard, among others, became reality with the 100-plane, 500-plane, and, finally, 1,000-plane raids in both the European and Pacific theaters. When the United States first entered the war in late 1941, the air forces of its army and navy provided the only means for quickly inflicting any degree of harm on the homelands of the enemy. As the war progressed, so too did the lethality of the air weapon. Finally, the airplane became the first means of delivering the most destructive weapon yet devised, the atomic bomb. The dreams (though some might argue nightmares) of the early airpower advocates had been realized.

As a result of American victory in 1945, the nation acquired a large number of international dependencies and conquered provinces that demanded governance. The far-flung nature of this American "empire" ensured that the maturing air transportation system had to expand radically as a means of linking the realm. The Truman administration at the end of the war quickly awarded permanent international routes to the domestic carriers so prominent in the early 1940s. These were mostly the same routes the carriers had flown under military auspices during the conflict. In so doing, Harry Truman overturned a longstanding policy of giving preference to Pan American Airways as the "chosen instrument" for overseas air and began awarding several U.S. carriers regional routes. Simultaneously he created a climate of moderate and healthy, but not cutthroat, competition.[7] That competitive arena would soon also include the flagship airlines of nations in Europe and eventually Asia, as well as other parts of the world.

After the war, development of aerospace technology made the United States the undisputed (but not unchallenged) leader in this sector of the world economy, and U.S. manufacturers dominated world markets for a half-century. The aerospace industry also assured the military might of the nation. American aircraft and space systems helped to guard the nation from enemies both real and perceived, advanced the rise of quick global transportation systems, led export efforts, helped to pull along technological advances in electronics and other areas, and opened the realm above the atmosphere for communications, navigation, and early warning systems. It also served the nation's will to place an American on the moon in 1969 and to further scientific knowledge of the universe and the planet on which humans live.[8] This accomplishment, in turn, fueled dreams of technological and other heights to which Americans could aspire. All this happened within a complex of governmental action, international competition, and at

The British Aviation Comet was the first jet airliner of the postwar era, coming on line in 1953. This air-to-air view shows the two Comet prototypes and the first production machine in formation, all wearing BOAC colors. The leading aircraft is G-ALYP, the first Comet 1. (NASM photo)

least periodic public demand for demonstrations of U.S. technological superiority.

The importance of the aerospace community remains manifest down to the present. It has, of course, long been closely tied to national security, and critics have questioned its reliance upon government contracts for much of its economic health. Concerned about this relationship, in his 1961 farewell address President Dwight D. Eisenhower warned the American people of a growing military-industrial complex, which he said had the "potential for the disastrous rise of misplaced power." So closely tied to the spending for national security had the aerospace industry become by 1990 that when the Cold War ended after the collapse of the Soviet Union, the industry took a nosedive. Between 1990 and 1996, reported A. Thomas Young, former senior vice president for Lockheed-Martin, the industry declined in size—measured by employees and value of sales—by 59 percent. Regardless, it remained a major component of American culture. Even its decline found reflection in popular media. A Hollywood feature film of 1993 starring

An F-16 from the 522d Fighter Squadron, Cannon Air Force Base, New Mexico, fires an AGM-65D Maverick air-to-ground missile at a tank target during its Air-to-Ground Weapons System Evaluation Program, Combat Hammer, in 2002. (U.S. Air Force photo/Tech Sgt. Michael Ammons; VIRIN 020730-F-7709A-001)

Michael Douglas, *Falling Down*, depicted the exploits of a laid-off aerospace engineer in Los Angeles who came unhinged and exacted revenge on a political and economic system he thought should have been permanent.[9]

These issues are the concern of this collection of essays drawn from one hundred years of the history of flight. They focus on four areas central to the evolution of this important technology:

- Innovation and the Technology of Flight
- Civil Aeronautics and Government Policy
- Aerial Warfare
- Aviation in the American Imagination

Within each area we have included three essays that relate to various aspects of the subject at hand.

Perhaps the most critical component in the history of flight is the development of the technology itself. Without the evolution of more advanced flight systems, the air age would have been stunted and perhaps never have gone anywhere. The three essays in Innovation and the Technology of Flight explore critical elements through policy, technological, and international lenses. The essay by Roger E. Bilstein, the dean of serious aerospace his-

The fastest airplane ever built, the SR-71 had the capability to cruise at Mach 3+. (NASA photo)

tory, explores what he calls three sides of a coin: (1) selected features of the American and European experience during the twentieth century; (2) an appreciation of the many European contributions to America's technological milestones; and (3) commentary on some of the economic and social aspects of the history of flight. He finds that the significance of aviation technology is the sum of many parts.

In the second essay, Roger D. Launius explores the nettlesome question of the role of the U.S. government in fostering aerospace technology and finds that federal investment in both the infrastructure and the technology of flight helped to make the United States a world leader in the twentieth century. He notes that while the Wright Brothers went it alone in the invention of the airplane, within a decade after their feat on the dunes of Kitty Hawk, North Carolina, air-minded Europeans had stolen the march on the United States by investing heavily in the technology. The result was that their airplanes far outstripped American capabilities by the time of World War I. It required significant and sustained government investment to turn this around.

Finally, Hans-Joachim Braun asks the scintillating question, Was there anything distinctly "European" in the development of aviation? He explores the relationship between some innovations in aviation originating in Europe and their later application to the United States. In so doing, he finds that there was significant cross-fertilization and analyzes some of them in the context of the origins of aerodynamics, of jet propulsion, and of the European Airbus. He also notes that Europe failed to sustain a technological lead in these areas of technology, something that has chagrined political leaders since the 1950s and has prompted a significant effort to recover a central role in aerospace technology development.

In the second part of this book, Civil Aeronautics and Government Policy, three authors explore the role of government regulation and the advance of civil aeronautics. In the first essay in this part, David D. Lee takes a biographical approach in analyzing the role of Herbert Hoover as secretary of commerce in the 1920s in fostering the airline industry. Hoover's attempts to form private/public partnerships in aeronautics served well in that early era of commercial aeronautics, and some of his ideas have been retreaded in the 1990s as part of the Republican political agenda. Next, W. David Lewis also uses biography, this time of America's ace of aces Eddie Rickenbacker, to demonstrate the sometimes rocky but always symbiotic relationship between federal regulatory entities and the airlines. William M. Leary explores the theme of airline safety, a perennial concern throughout the history of the flight in the twentieth century, using as a case study the attempt to defeat the problem of icing on aircraft. The research, policy, and regulatory environments came together in a unique manner to battle icing on aircraft. That battle, as Leary points out, has not yet been won.

The third part of the book explores the theme of aerial warfare. A. Timothy Warnock writes of the primitive and dangerous nature of early military aircraft. Pioneer air power advocates dreamed of aircraft becoming new and terrifying weapons of war. They developed this vision almost despite the fact that in those early years aircraft often proved more lethal to those who flew them than to any intended targets. Technology would have to advance significantly before the dreams of the early airpower visionaries could begin to come true. Warnock's essay also explores briefly how and why the Wrights gradually ceased to be a source of aircraft to the military. Central to this was the fact that the inventors of the airplane for various reasons failed to keep abreast of the advances in the field.

John H. Morrow investigates in his essay the early development and nature of military aviation. He begins by exploring the relationship between the Wright Brothers and the development of military aviation in the United States and Europe. In part, Morrow argues, the courage and daring of the early inventors/entrepreneurs helped inspire military interest in the airplane. The same daring and courage associated with early aviators also helped produce a romantic and even mythic image of military aviation. The aces of the Great War and their exploits acquired and have maintained an aura of gallantry and individual heroism and sacrifice. To this, Morrow offers an important corrective. He notes that military aviation in World War I was far more than the collective stories of its most celebrated pilots. Rather, it involved the mass mobilization of governments, industry, and the military itself. By the end of the war, mass aerial attacks, not individual duels, were the norm. Rather than glorious and romantic, aerial warfare was brutal, gory, and deadly.

Finally, Tami Davis Biddle discusses the enormously flexible and overwhelmingly destructive power of airplanes in the arsenals of the world and how their power marked a profound and permanent change in the waging of war. She writes that "airplanes quickly became key tools in the prosecution of twentieth-century conflict. In their most dramatic role, they delivered bombs onto the homeland territory of combatant states. What evolved into the theory of strategic bombing began with a simple assumption: aircraft, specifically long-range or "strategic" bombers, can avoid an enemy's army and navy and proceed directly to its "vital centers," where they can cause enough destruction and disruption to induce surrender on terms favorable to the attacker. It was a simple idea, but it would prove to be a powerful and tenacious one."

Biddle traces the evolution of aerial warfare from the earliest thought on strategic bombardment to the Gulf War. She find that while the technology has evolved significantly since the first military pilots took to the air, the core ideas behind the use of air power have remained largely the same. Those ideas were the product of a complex social and political mix of ideas and opinions, metaphors, and wishes that informed the military mind throughout the twentieth century.

The final part of this book provides the setting for three essays on aviation in the American imagination. In the essay opening this part, David T. Courtwright focuses attention on the rhetorical application of Frederick Jackson Turner's frontier imagery to flight in America. The sky as frontier, he argues, has been used as a gimmick without clear definition and rigorous analysis. As such it is essentially a cliché. But Courtwright finds that one might properly evoke an aviation frontier by analyzing it from an anthropological perspective. He asserts that the sky "underwent a settlement process analogous to that of the nineteenth-century mining and ranching frontiers. Both began with a sprinkling of mostly male adventurers and ended with a much larger and more or less demographically normal slice of the general population. By the last two decades of the twentieth century, . . . ordinary people could fly in relative comfort and safety and at low cost, but without much excitement. . . . [T]he sky had become . . . Kansas without the Indians."

In his essay Courtwright explores how the aviation frontier became settled, the routinization process so often discussed in sociological literature. In the end, flying became just another form of mass transportation little different from a crowded commuter bus or train except that it operates at 30,000 feet above the surface of the Earth.

Then Anne Collins Goodyear reviews artistic works with an aviation theme to investigate the effect of the ability to fly on the development of twentieth-century art. The substance and symbolism of flight influenced artists not only by giving a new perspective, but also by inspiring a new "understanding of materials and mythologies—an appreciation for the meaning of weight and gravity and for the mystical and political dimensions of human flight that became a reality nearly a century ago." Finally, Dominick A. Pisano reviews the significance of Charles Lindbergh's airplane, *The Spirit of St. Louis*. As a cultural icon, this aircraft holds a special place in the history, and especially the mythology, of the twentieth century. Pisano suggests that the *Spirit of St. Louis* has become idealized as a totem or symbol of American technological prowess. Unraveling that symbolism will be the task of many historians over a long period.

From the Wrights to the present the first one hundred years of flight have been exhilarating and frustrating, elusive and obvious, awesome and awful. These years have witnessed enormous changes in the manner in which humanity has interacted, enabling travel over great distances with relative ease on short schedules. Another issue, suggested but not developed in a focused way by these essays, though, is the question of progress. Aviation technology certainly changed during the course of the twentieth century. The many enthusiasts drawn to the technology emphasized the progress—in particular, higher, farther, faster—they saw as inherent in those changes. The desire for advancement toward higher, farther, and faster was definitely established early, and clear milestones mark the evolution of aviation technology toward those goals. However, one could also argue that not all the changes represented progress. Do the many changes (technological, political, strategic, etc.) that made aircraft more lethal, discussed in Tami Biddle's essay, represent progress? Has the long-held notion that somehow air combat is cleaner and more efficient caused the downplaying or even dismissal of the real horrors associated with the airplane and combat, a point developed by John Morrow's essay? And as Dom Pisano's essay suggests, we need to examine closely the meanings we give to the people, events, and artifacts

involved with the history of flight. While Lindbergh's flight certainly captured the public's imagination and helped stimulate the aviation industry in
the United States in many ways, one could also argue that it did so at least
in part because it reaffirmed the importance of humans, of the individual in
the face of rapid industrialization and modernization. It seemed central to
the Lindbergh mythology that he made the flight alone. He had mastered
a technology in order to accomplish an individual feat. Unlike the character in Charlie Chaplin's classic *Modern Times*, who was swallowed up and
became a cog in the machine, Lindbergh's flight reasserted and reaffirmed
the importance and power of the individual. Despite the mechanization of
modern society, the individual human was still capable of greatness. Thus,
the notion of progress when applied to Lindbergh's flight is far more complex than might be suggested at first glance. Did the ability to fly represent
technological progress or did it, or at least in the reception it received, represent an unease with the growing role of technology in American society?
Again, these essays suggest rather than fully develop a discussion on the
issue of progress in relation to the history of flight, and they certainly do
not provide any definitive answers. A definitive answer, should one ever be
developed, would undoubtedly have great complexity and nuance. The debate is still wide open. Perhaps, though, in addition to what they accomplish
in the context of this collection, these essays can also, as Pisano's does, remind readers of the many themes and issues in the history of flight still to
be explored.

The essays presented in this collection explore several aspects of what it
has meant to fly in the twentieth century. Virtually all of them were originally given at a symposium on the history of flight at North Carolina State
University, Raleigh, in October 2001. The symposium, "They Taught the
World to Fly: The Wright Brothers and the Age of Flight," was sponsored
by the First Flight Centennial Commission and the First Flight Centennial
Foundation of North Carolina and coordinated by Dr. Larry Tise. It featured
more than two hundred participants speaking on all manner of aerospace
historical topics. From these presentations we have drawn a selection of outstanding essays for inclusion in this volume. It represents some of the best
recent work on the history of flight in America and is a fitting reflection on
a century of flight commemorated in 2003.

NOTES

1. This story is delightfully told in Dean Smith, *By the Seat of My Pants* (Boston: Little,
Brown, 1961), 134–35.

2. An excellent general history of flight is Roger E. Bilstein, *Flight in America: From the Wrights to the Astronauts*, 3d ed. (Baltimore, Md.: Johns Hopkins University Press, 2001).

3. Roger E. Bilstein, "Air Travel and the Traveling Public: The American Experience, 1920–1970," in William F. Trimble, ed., *From Airships to Airbus: The History of Civil and Commercial Aviation*, vol. 2 (Washington, D.C.: Smithsonian Institution Press, 1995).

4. Carl Solberg, *Conquest of the Skies: A History of Commercial Aviation in America* (Boston: Little, Brown, 1979).

5. U.S. Department of Commerce statistics, 1932–41. On the DC-3 and the Boeing 247, see F. Robert Van der Linden, *The Boeing 247: The First Modern Airliner* (Seattle: University of Washington Press, 1991); Henry M. Holden, *The Douglas DC-3* (Blue Ridge Summit, Pa.: PNAero, 1991); Arthur Pearcy, *Sixty Glorious Years: A Tribute to the Douglas DC-3* (Osceola, Wis.: Motorbooks International, 1995).

6. See Roger E. Bilstein, *The American Aerospace Industry: From Workshop to Global Enterprise* (New York: Twayne, 1996).

7. The story of Pan Am's special relationship with the prewar federal government has been detailed in Marilyn Bender and Selig Altschul, *The Chosen Instrument* (New York: Simon and Schuster, 1982). The change in policy is clearly described in Frederick C. Thayer Jr., *Air Transport Policy and National Security: A Political, Economic, and Military Analysis* (Chapel Hill: University of North Carolina Press, 1965), 68–75; John R. M. Wilson, *Turbulence Aloft: The Civil Aeronautics Administration amid Wars and Rumors of Wars, 1938–1953* (Washington, D.C.: Federal Aviation Administration, 1979), 196–204; Nawal K. Taneja, *The Commercial Airline Industry: Managerial Practices and Regulatory Policies* (Lexington, Mass.: D. C. Heath, 1976), 170–87.

8. This is the thesis of John B. Rae, *Climb to Greatness: The American Aircraft Industry, 1920–1960* (Cambridge, Mass.: MIT Press, 1968).

9. Roger D. Launius, "End of a Forty Year War: Demobilization in the West Coast Aerospace Industry after the Cold War," *Journal of the West* 36 (July 1997): 85–96.

Part I

INNOVATION AND THE TECHNOLOGY OF FLIGHT

CHAPTER 1
THE TECHNOLOGY OF FLIGHT
THREE SIDES OF A COIN

The history of America's aviation technology combines many strands of Yankee ingenuity, federal partnership, regulatory influences, and changes in the business climate. International factors also played important roles: technological legacies from abroad, accelerated development spurred by wartime emergencies, and a variety of overseas events. As a technological phenomenon, aviation became a significant economic component in both domestic and foreign arenas. Rounding out this technological picture, the evolution of aviation left its imprint on popular culture in subtle but myriad ways, cementing its presence as an integral feature of global civilization in the twentieth century. Clearly, the significance of aviation technology is the sum of many parts. In the essay that follows, these multiple factors are organized around three themes: selected features of the American and European experience during the twentieth century; acknowledgment of European contributions to many of America's technological milestones; and commentary on some of the economic and social aspects of the history of flight. In other words, the themes represent three sides of a coin.

The Early Years

Late in the nineteenth century, when the Wrights first started to think about the possibilities of human flight, they went about the venture in their typical, methodical way. Among their first decisions, they contacted the Smithsonian Institution in Washington, D.C., to acquire extant literature on the study of aeronautics. The Smithsonian, chartered by Congress in 1846, had been established with funds from a bequest established by James Smithson, the illegitimate son of a British aristocrat. Had it not been for this European connection, and the rich legacy of European experimentation in the realm of aeronautics, the history of flight might well have gone off in some different direction.

The information from the Smithsonian included considerable references to European literature on the subject of aeronautics. With a long history of ballooning, experimentation with powered and maneuverable lighter-than-air vehicles, gliders, and active scientific groups, European aeronautical research led the world. The Aero Club de France announced its formation in 1898; the Aero Club of Britain followed in 1901. French terms for the parts of airplanes pervade the early speculations of the Wright brothers: empennage, aileron, pitot tube, canard, longeron, nacelle, fuselage, and other European nomenclature (including aeronautics and aviation) persist even to the present. Consequently, the Wrights first looked to European experimentation for promising avenues to follow when they began their own quest for human-controlled, powered flight. They read accounts of early piloted gliders built by the British engineer Sir George Cayley; examined the elaborate tables of lift and aerodynamic data accumulated in dozens of flights by the German pioneer Otto Lilienthal; and studied other information about the dynamics of flight amassed by numerous Europeans. Much of the significance of their eventual success lay in the fact that they found these records to have frequent errors and erroneous assumptions. But as starting points, the European legacies cannot be ignored. Note also that the brothers achieved powered flight with a four-cycle gasoline engine, an innovation from Germany, where Karl Benz and Gottlieb Daimler demonstrated it in early automobiles about 1885.[1]

Because the Wright Brothers had used a biplane configuration for their successful flights, many early builders followed that same pattern. Other aspiring fliers used a monoplane design made famous in 1909 when the French pioneer, Louis Blériot, made his historic flight across the English Channel. An American firm in New York City sold numerous Blériot-licensed planes to aspiring fliers all over the United States, and numerous backyard builders constructed flying machines with the monoplane in mind. Blériot's monoplanes undeniably had an important role in the lively aviation scene in the United States.

Continuing the European influence, in the years before World War I a number of European émigrés came to the United States and enhanced America's aviation progress in a variety of ways. English immigrants included Douglas Thomas, who joined the Glenn Curtiss organization and helped design the famous JN-4 "Jenny" trainer at the time when the United States entered World War I and needed to instruct legions of new military pilots. After scores of aspiring American fliers won their Army Air Service wings in this ubiquitous trainer, dozens of war-surplus Jennies populated the skies as part of the barnstorming era. Barnstorming in an old JN-4 not only helped launch the aviation careers of many postwar fliers—like Charles Lindbergh, for example—but also became a symbol of future careers for uncounted youngsters whose first encounter with an airplane was with a Curtiss Jenny.[2]

It is instructive to track the imagery of flight through the use of aeronautical themes in avant-garde literature and avant-garde painting. As analyzed by historian Robert Wohl in *A Passion for Wings: Aviation and the Human Imagination, 1908–1918*, avant-garde sources delineate revealing awareness relative of the symbolism of flight and its relevance to the human experience. From the early years of aviation, airplanes often appeared as aesthetic objects with unique beauty. For many writers, poets, and artists, they symbolized humanity's ascent from Earth. Aircraft embodied lofty new perspectives of human activity, achieving a kind of liberation from the constraints of day-to-day existence. At the same time, airplanes appeared as threatening machines that would enable malign influences to dominate the skies. This feature led patriots from various nations to sermonize in favor of aerial superiority for domination as well as protection.[3]

Certain plateaus of aviation technology can also be identified by the appearance of aeronautical imagery in day-to-day products and lifestyles. These images serve as archeological waypoints of aeronautical awareness in popular culture and daily life. For example, it is possible to point to the illustration of an aircraft on the cover of a popular turn-of-the-century mass-circulation magazine or the use of aeronautical themes as an advertising gimmick to attract consumers to an otherwise down-to-earth product, such as bread or hand soap. Similarly, the introduction of aviation into the vernacular literature of popular plays or popular novels is worth noting as evidence that aeronautical technology had achieved recognition with a significant segment of the general population, even if many people had yet to see a flying machine drone through the skies above them.

Consider, for example, George Bernard Shaw's play, *Misalliance*, produced in 1910, in which a group of suburbanites at a house party discuss the new phenomenon of aviation and its possible implications in modern life. Offstage, the sound of an engine intrudes—it is an airplane—and a crash ensues, which turns out to be an errant aviator whose craft has suddenly plunged into the garden. In the context of the play, it becomes a moment to consider the complexities and problems of a technological age. Shaw wrote the script during 1909 and was a keen observer of public response to Blériot's flight across the English Channel as well as several highly publicized aerial events in Britain the same year. He inserted the aviation sequence partly as a ploy to attract more playgoers. Aviation also plays a dramatic role in Scottish author John Buchan's novel, *The Thirty-Nine Steps*, written in the winter of 1914. When intimations of an impending conflict already shadowed Europe, the brief novel is considered one of the first true espionage adventures. In it, the protagonist, Richard Hannay, becomes aware of a dark plot to trigger a war and compromise Great Britain. Hannay becomes the quarry in a dramatic manhunt ranging into Scotland, where speeding automobiles and a clattering airplane pursue him. The latter becomes a particularly menacing foe because of its mobility and its altitude, which allows Hannay's enemies to scan wide expanses of the Scottish countryside, keeping him constantly on the run.[4]

In a variety of other ways, images of aviation technology made an appearance in everyday life. In America, mundane but revealing artifacts include an issue of *Century Magazine* from the turn of century, which depicted an adventurous young woman swooping across the cover in a glider. Aeronautical themes turned up in popular ditties like "Come Fly with Me, Josephine, in My Flying Machine." Also in the pre–World War I era, the purveyors of Armour cleansing products touted "White Flyer Laundry Soap—Makes Dirt Fly" that featured a Blériot monoplane logo. In 1910, perhaps as an enticement to parents, the folks at Butter-Nut Bread offered young consumers a small paper glider with a 4½-inch wingspan to play with. These halcyon prewar years also witnessed numerous advertisements for model airplane kits and a spate of juvenile novels about "aeroplane" adventures.[5] Although military themes occasionally intruded, there seemed to be an air of innocence about the whole thing.

World War I itself marked a stark change in public attitudes about aviation. Early nonsense about knights of the air aboard flying steeds, charging straight for each other in a sort of aerial joust, eventually gave way to the realities of wartime combat. Military aviators dealt with numbing cold, gut-wrenching fear, and high casualty rates. The public might lionize particular aces, but the fliers themselves knew that survival depended on staying out of swirling dogfights in order to pick out a specific target, close on it out of a blinding sun, and attack from the rear before the victim knew what was going on. The war also introduced high-altitude bombing raids, with crews dependent on oxygen while attacking anonymous targets far below. But the targets were often sited within large population centers, bringing sudden death and destruction to civilians far removed from traditional battle lines on the ground. In this sense, air warfare gave special meaning to the term "total war." German attacks on London, which killed several hundred people and wounded thousands more, stirred up such fear and consternation that the British created the Royal Air Force, the world's first independent air service, in response to the enemy's aerial threat.

The wartime experience played a key role in pushing the United States into the modern era of aviation. In terms of performance, prewar speeds of 50 to 60 miles per hour or so soon doubled. More important, an infrastructure of suppliers of engines, propellers, magnetos, flight instruments, and so on became part of the technological landscape. Manufacturers learned that the mass production of aircraft required approaches different from those in the mass production of automobiles. Military leaders took the first steps in developing a new literature of tactics, strategy, and doctrine.[6] Although the term "Air Age" became more widespread with the advent of World War II, public and political attitudes in support of aviation gained considerable momentum in the post–World War I era.

The Interwar Period

During the 1920s and 1930s, technological progress and the adoption of aviation technology as an integral feature of modern life also required some sort of demonstration of its utility. One of the more practical milestones was the legislation of the U.S. Congress implementing a workable airmail service.

Between 1918 and 1925, the U.S. Air Mail Service operated as an arm of the Post Office Department. The department assembled planes and pilots, planned effective airmail routes, installed a series of beacons to guide fliers on night schedules, and developed a clientele. In these pioneering operations, the department relied on a European airplane that significantly shaped America's postwar aeronautical experience—the rugged old de Haviland DH-4. An open-cockpit, two-seat British design built in the United States during World War I (and the only American-built plane to see much aerial combat in the war), the sturdy DH-4 forged the first successful airmail routes in America, churning along at about 100 mph. Modified to carry 400 pounds of mail sacks in the forward cockpit section, the DH-4s expanded airmail service across the entire nation and inaugurated the first transcontinental schedules in 1920, including the remarkable journey of Jack Knight across the Midwest during a frigid night in February.

By 1924 a series of airmail beacons and designated fields regularly delivered coast-to-coast mail in about twenty-four hours—three days faster than rail service. Airmail quickly made an impact on the way in which American citizens corresponded with one another and demonstrated an even stronger influence on many aspects of American commerce. Airmail service increased the tempo of financial activities by carrying cash, checks, and other financial paper, along with aerial express of many time-sensitive commodities such as advertising copy, electrical components, flowers, motion-picture film, pharmaceuticals, and similar lightweight items. Airmail aided the rapid distribution of business correspondence in an era known for the rapid evolution

The predecessor of this Douglas M-3 mail plane was first manufactured in about 1926 as a replacement for the ancient DH-4, which was widely used at that time for carrying the mail. The aircraft was powered by the World War I Liberty engine of about 400 HP. (NASA photo no. L-5470)

of business as a national phenomenon. As the pace and volume of business activities increased, the role of airmail became essential in the day-to-day conduct of American commerce. The lighted transcontinental airway constituted a backbone of operations with links that reached into every region of the country. "Of all American contributions to technique of air transport operation," declared aviation expert Edward Pearson Warner, "this was the greatest." As late as 1930, describing the ambitious system to European friends, Warner found them "politely incredulous," taking his story as "a manifestation of American bluff."[7]

During the same period, several aviation firsts mesmerized Americans as well as people overseas. In May 1919, an American seaplane crossed the Atlantic in stages by flying from Newfoundland to the Azores and then on to Portugal; the British accomplished the first nonstop transatlantic flight, from Newfoundland to Ireland, the following month. The promotion of aviation technology often occurred in a context of politics and national self-interest. Several European countries also launched dramatic long-distance flights, partly to demonstrate national aeronautic prowess, explore the potential of airplanes for long-distance communications, and to establish closer links to far-flung colonial outposts. During 1919–20, one adventurous British crew hopscotched their way from London to Australia in twenty-eight days; another crew made it from London, England, to Johannesburg, South Africa, in about three weeks. Scheduled routes of British Imperial Airways followed. The Dutch pioneered airmail routes through the Middle East and on to the Dutch East Indies. The French established a series of mail routes across the Mediterranean to their territories in North Africa and western Africa and then pushed service across the south Atlantic to South America. Germany exploited aviation as an antidote to the burden of Versailles, filling the skies with novice glider pilots and nationalistic fervor and extending an aeronautical presence as far afield as Latin America. Not to be left out of such technological demonstrations, the U.S. Army mounted a round-the-world flight in 1924, succeeding in the face of daunting geographic challenges and difficult weather. Such early efforts reflected a mixture of pioneering spirit spiced with practical concerns of colonialism and empire building. In the same way that the British and Dutch vied with each other in Asia and the Pacific, the presence of European airline ventures in Latin America clashed with American interests. These factors had a direct relationship to the organization of Pan American Airways in the late 1920s and substantive federal collaboration with Pan Am's expansion in the following decades.[8]

Collectively, such activities convincingly demonstrated the potential of global aviation to vault age-old barriers of formidable mountain ranges, parched deserts, and intimidating ocean expanses. The phenomenon of flight fundamentally altered the way in which people considered the world. Mark Sullivan, a widely read popular journalist of the era—and a keen observer of human nature and cultural change—took special note of aviation achievements. In his monumental six-volume study, *Our Times: The United States, 1900–1925*, he commented that of all the things that had influenced Americans during the first quarter of the twentieth century, "by far the greatest was the sight of a human being in an airplane."[9] And this was even before consideration of the extensive impact of aviation stemming from the Airmail Act of 1925, the Air Commerce Act in the following year, and the electrifying nonstop solo flight of Charles Lindbergh from New York to Paris the year after that.

In 1925, responding to pressure from the business community, the government turned over the airmail business to private contractors. The formation of new airlines to carry the mail represented what became the core of major air transport corporations of the future—American Air Lines, Eastern, TWA, United, and others, including the intercontinental network of Pan American World Airways. Passenger services eventually followed, and the convenience of rapid air travel throughout the United States as well as throughout the world became a facet of contemporary life.

At about the same time, the Air Commerce Act of 1926 helped sustain the proliferation of airlines and the spread of aviation services. Among other things, the newly established Bureau of Air Commerce began the process of requiring pilots to meet standard requirements of training and aviation skills before they could receive a license. Airplanes had to meet federal standards for manufacture and to pass periodic checks for airworthiness. Mechanics had to demonstrate specific levels of proficiency before winning a license to work on aircraft that carried private fliers or carried dozens of trusting passengers in air transports. All this meant that bankers began to feel less nervous about extending loans to new concerns determined to build or operate flying machines as public conveyances. And one of the reasons that bankers were less nervous is that insurance companies now felt confident about insuring planes that met standardized requirements for safety, were flown by certified pilots, serviced by certified mechanics, and periodically inspected to assure their continued airworthiness.

The notable progress in flying activity stemmed from a growing infrastructure and reflected significant advances in aeronautical technology. Here again, the federal government played a central role in elaborating the infrastructure in the course of pursuing practical and theoretical research that influenced civil and as well as military aviation. The principal agency in this role is known today as the National Aeronautics and Space Administration, although its genealogy originated in the troubled years of World War I. The heritage of the National Aeronautics and Space Administration (NASA) dates back to the National Advisory Committee for Aeronautics (NACA), formed in 1915 largely as a response to the alarming progress of military aeronautics in Europe during World War I. In organizing the NACA, American officials unabashedly followed the format (and the title) of the British Advisory Committee for Aeronautics in Britain, including the ringing phrase of its famous directive, "the scientific study of the problems of flight, with a view to their practical solution," which was borrowed directly from the original charter for the British institution.[10]

The decades of the twenties and thirties represent an era of aeronautical headliners like Charles Lindbergh, Amelia Earhart, and Douglas "Wrong Way" Corrigan. Behind the headlines of record flights, a growing number of "aviation professionals" played key roles in the evolution of aviation technology and helped erect the architecture of an essential infrastructure. Herbert Hoover, as secretary of commerce, was one such individual, presiding over early federal departments to oversee and regulate the young aviation business. Jerome Hunsaker was another, playing a central role in the NACA, creating one of the country's first college degree programs in aeronautical engineering at the Massachusetts Institute of Technology and teaching future aerospace leaders like Donald Douglas.

Another of Hunsaker's protégés was Edward Pearson Warner, who also taught students at MIT, held influential positions with the NACA, wrote aeronautical texts, helped frame significant federal laws, and took a leading role in establishing major institutional entities like the Institute of Aeronautical Sciences (which became the American Institute of Aeronautics and Astronautics) as well as the influential periodical *Aviation*, which eventually morphed into the authoritative periodical *Aviation Week and Space Technology*. Individuals like these sat on key committees, ran agencies, and shaped the course of aviation development as it reached maturity.[11]

The record of subsequent aviation progress in America abounds with examples of useful technology, institutional concepts, and scientific data directly attributable to European sources. European émigrés often brought their unique qualifications and perspectives directly to the New World.

The pace of European interaction with American aviation technology seemed to pick up after World War I. Examples of overseas influences included obvious areas such as research and engineering as well as apparently obscure areas with long-term significance. Consider, for example, the role of two brothers, William Thomas and Oliver Thomas, who received technical training from the London Central Technical College in Great Britain. They arrived in the United States prior to World War I, became employees of Glenn Curtiss and others, and then launched their own Thomas-Morse Aircraft Corporation, which became one the principal builders during America's involvement in World War I. The brothers went into other businesses, but their original firm evolved into Convair, builder of the B-24 bomber during World War II, and eventually General Dynamics, a major Cold War enterprise.

The career of William Thomas, meanwhile, took a significant turn during the 1920s when he became engrossed in the activities of flying scale-model airplanes. For many years, he continued to be active in promoting the hobby of flying scale-model aircraft, an increasingly popular pastime that finally led to the organization of the American Academy of Model Aeronautics in 1936. The academy's programs became a sort of seedbed for hundreds of future aviation engineers and professionals who began their careers as eager constructors and fliers of scale-model aircraft. Neil Armstrong, the first person to set foot on the moon, later recalled his fascination with model aircraft in his youth and acknowledged the academy as one of the steps leading to his career as a test pilot and pioneer astronaut.[12]

In the aftermath of World War I, political turmoil in Europe caused an increase in individuals seeking more favorable employment opportunities in the United States. In some cases, American organizations actively sought them out and enticed them to cross the Atlantic. As the young NACA began to expand its research activities, it became apparent that more expertise in the field of theoretical aeronautical research was needed. Consequently, the United States made a special effort to recruit a brilliant young physicist, Max Munk, from Göttingen, Germany, a center of advanced European aeronautical investigation. Because the United States did not sign the Versailles Treaty ending World War I, getting Munk into the United States took some doing, since America technically remained at war with Germany in 1919. Jerome C. Hunsaker was a central figure in getting Munk to the United States. It required high-level diplomacy and a special executive order signed by President Woodrow Wilson to get Munk to the United States in 1920. Once at the NACA's Langley Memorial Aeronautical Laboratory at Hampton, Virginia, the temperamental Munk often frustrated his colleagues, but his legacy included invaluable aerodynamic work resulting from the variable density tunnel which he helped design and build. Moreover, Munk contributed the crucial line of theoretical research that helped make NACA Langley into one of the world's premier institutions in flight research. The NACA's debt to European sources for advanced aeronautical trends was also reflected in the agency's European office in Paris, which kept a close watch on foreign advances up to the outbreak of World War II.[13]

Imaginative aeronautical investigation in Europe often paralleled the NACA's most promising programs. The renowned NACA engine cowling of the late twenties appeared at the time that the British unveiled a similar design, named the Townend ring, after its developer, Hubert Townend. In fact, as one veteran Langley engineer wrote in 1931, "It is regretted that the Laboratory, in its report on cowlings, did not mention the work of Townend and give him credit." In certain instances, the NACA successfully folded European research into its own investigations. NACA's outstanding work in the understanding of laminar flow over aircraft wings and its relation to high-speed flight clearly profited from a visit to England by a Langley scientist, Eastman Jacobs. In 1935 he had stopped off to visit friends at Cambridge University, where they shared insights about their work in progress. Returning home, the NACA researcher enlisted colleagues to help in further analysis, which led to the publication of basic research papers on the phenomenon of laminar flow and its role in the design of advanced aircraft. The famed

North American P-51 Mustang of World War II became one of the first high-performance military planes to benefit from this research.[14]

The expanding library of aeronautical data from the NACA and other agencies clearly contributed to the evolution of modern airliners of the late 1930s. These remarkable flying machines embodied high strength-to-weight ratios that gave them enviable performances in range and payload. Much of this technology evolved from the technique of metal airframes and the use of stressed-skin construction, in which the skin of the airplane itself, attached to spars, formers, and stiffeners, became part of the load-bearing structure, all of which saved weight and contributed to more efficient aircraft. Although some American builders employed stressed-skin concepts in wooden airframe designs, German innovators like Hugo Junkers and Adolph Rohrbach strongly influenced the trend toward all-metal stressed-skin aircraft. During a visit to America, Rohrbach gave a detailed analysis of his stressed-skin construction techniques in a paper he presented during a meeting of the Society of Automotive Engineers in 1927. His paper helped reshape the style of aeronautical technology in the United States. Other American designers and engineers refined additional key elements such as controllable-pitch propellers, retractable landing gear, reliable engines, and high-octane fuel. The NACA's engine cowling, anti-icing research, reversible-pitch propellers, and studies dedicated to the reduction of aerodynamic drag represented other aeronautical dividends. Moreover, designers of a new generation of airliners paid more attention to passenger comfort by integrating heating and fresh-air ducts into the passenger cabin and making the effort to avoid nauseating fumes from hydraulics, lubricants, and aviation fuel. Seats were engineered to reduce vibration, and soundproofing added to the comfort of passengers. In many ways, the integration of these advances into the design of the Boeing 247 and the DC-3 symbolized the notable degree of sophistication in research and design achieved by the late 1930s.[15]

Built to carry twenty to thirty passengers at about 180 miles per hour, the DC-3 possessed the operational economy and reliability that made it a major factor in establishing air travel as a modern phenomenon. For years, fear of flying led the list of reasons why travelers preferred trains or busses. Throughout the twenties and thirties, wives and relatives habitually discouraged airline travel for businessmen. "Some of us kept it a secret, not only from wives, but creditors as well," one airline patron recalled. "But every one of us was proud to make a business appointment in a distant city, saying that we would fly to keep it. It gave us prestige."[16] Better equipment like the DC-3 helped dispel these concerns. One of the most substantial examples of this new respectability appeared in ads sponsored by the Air Transport Associa-

The Douglas DC-3 was one of the breakthrough aircraft of the twentieth century, coming into widespread use during World War II with some still flying fifty years later. (Photo courtesy of Northwest Airlines)

tion (ATA), which appeared in the *Saturday Evening Post* during November 1940. The Equitable Life Assurance Society had determined that air travel had reached a level of safety and reliability that justified a reduction in rates and across-the-board availability to all air travelers. Aerial insurance at the cost of one dollar per unit of $5,000 coverage now dropped to twenty-five cents per unit, the same rate charged for rail travel. For the airline industry, the ATA proclaimed, this represented a true benchmark of progress. In another sign of the times, souvenir postcards from many cities that had proudly boasted of rail passenger facilities now began to depict airline terminals as hallmarks of civic progress.[17] Nevertheless, airline travel remained a pricey venture, especially at a time when the effect of the 1930 Depression still lingered. The democratization of air travel did not occur until the latter years of the 1950s and early 1960s.

During the interwar years, images of flight continued to evolve. Concern about aerial weapons dealing death and destruction did not disappear, but in the peacetime environment in America during the 1920s and much of the 1930s, aeronautical imagery took on a much more benign profile. In popular culture, references to aeronautical technology in contemporary life became increasingly commonplace. The success of the U.S. Air Mail service caught public imagination. In 1925 the Paramount film company released an aerial epic titled *The Air Mail* with Warner Baxter in a starring role along with none other than Douglas Fairbanks Jr. A number of other producers made films with aeronautical themes, often casting an airplane as a useful technological phenomenon of the contemporary world. Aeronautical imagery also appeared in other formats of mass culture, such as the newspaper comic strip *Ace Drummond*, which ran in more than a hundred papers. Wartime ace Eddie Rickenbacker wrote much of the strip's continuity, based on his early flying experiences. This comic strip established a precedent for several nationally syndicated cartoon sagas that followed in the thirties: barnstorming swashbucklers like Roy Crane's *Captain Easy*, Zack Mosley's *Smilin' Jack*, early panels for Milton Caniff's *Terry and the Pirates*, and others.

Even before Charles Lindbergh completed his epic crossing of the Atlantic, many widely marketed periodicals sensed the popular fascination with the technology of flight and began to incorporate aviation information as standard features in every issue. The *Literary Digest* passed along aeronautical notes in its "Motoring and Aviation" column in the early twenties; *World's Work* added an "Aviation Bureau" to its regular format; *American City* launched an "Airports and Aviation" column edited by aviation consultant Archibald Black, author of books on how to build and operate modern aerodromes. During the late 1920s, *Cosmopolitan* hired Amelia Earhart as its aviation consultant to assist budding women aviators in keeping up with the latest in flying news. By 1929 bookstore shelves featured a string of autobiographical accounts by record-setting pilots, a welter of technical books covering everything from airports to handbooks for aviation mechanics, and numerous aeronautical titles for juvenile girls and boys alike. As one writer remarked in the trade journal *Publisher's Weekly*, aviation had become the hottest subject in America, and bookstores that failed to stock aviation-related titles were missing a "phenomenal opportunity."[18]

With the subject of aviation attracting such a diverse readership, advertisers perceived aerial themes as a useful means of catching the attention of consumers. Automotive manufacturers, in particular, appreciated the aura of speed and advanced technology implicit in aircraft. As the builders of the Franklin line of cars boasted in 1929, "The AIRPLANE FEEL of the Franklin opens the road to new motoring thrills." The advertisement proceeded to explain that drivers could "smoothly, powerfully slip through space . . . just like craft of the air." Other advertisers disregarded any sort of comparisons but simply used the appeal of aviation to call attention to their products. This seems the only way to explain the use of aeronautical imagery for Lydia

Pinkham's Vegetable Compound, endorsed by grateful women across the nation who relied on its beneficent qualities for the "Change of Life," listlessness, and other ailments. It seems that Mrs. Pinkham's granddaughter, "a tall, healthy young woman who has not bobbed her hair" (and advertising manager for the Pinkham Compound empire), acted on a whim and chartered a plane to fly from California to Boston. Interspersed with glowing testimonials, a thirty-two-page booklet published in 1926 described the journey and announced a related essay contest sponsored by Pinkham's.[19]

To many contemporary observers, modern aeronautics represented a positive trend, symbolizing the best of American ingenuity and technology. Airplanes came to embody the best of the streamline style in their lightness, economy of design, and use of modern materials like aluminum. As social critic Lewis Mumford noted in *Technics and Civilization* (1934), aeronautical requirements set the standards for refined and exact engineering. In his script for a pioneering documentary, *The City* (1938), Mumford dramatically included a sequence that filmed a DC-3. It taxied into position for takeoff and then soared into the skies accompanied by the crescendo of Aaron Copeland's score. The script's dialogue remarked how the swift airplane symbolized the essence of aesthetic logic and design.[20] Increasingly, aeronautical imagery became linked with consumer goods as a measure of modern design, sophistication, and quality. During 1939 and 1941, the pages of the *Saturday Evening Post* carried a variety of ads for cigarettes, automotive lubricants, typewriters, shirts, and rain coats in which airline themes were used to bolster the reputation and quality of the product.[21]

In order for aviation technology to reach these levels of sophistication and reliability, aircraft and their systems required knowledgeable designers and engineers. The thousands of new aeronautical engineers who took up their profession during the 1920s and 1930s and who pushed the performance envelope of postwar American designs owed much to a legacy of European training and pedagogy. In 1913, the originator of aeronautical engineering courses at the University of Michigan, one of the early centers of instruction in aeronautical engineering, was Felix Pavlovsky, a native of Poland who received his training in Europe before arriving in America. On the eve of World War I, the Massachusetts Institute of Technology turned to Jerome Hunsaker, who discovered that no suitable texts were available. Consequently, he translated a work by Alexandre-Gustave Eiffel, builder of the Eiffel Tower in Paris, who had pursued a subsequent interest in aeronautics by dropping models from the top of the tower to test promising configurations.

In the United States, the wealthy Guggenheim family looked around for a promising postwar technology to assist. The Daniel Guggenheim Fund for the Promotion of Aeronautics began its activities during the mid-twenties. Its program that led to successful "blind flying" (instrument flying) represented only one of its several significant legacies. Another of its principal efforts involved financing to establish a series of pioneering aeronautical engineering programs at selected universities across the country. During the late 1920s, leaders of the Guggenheim Fund decided that the west coast of America, the home of builders like Boeing, Douglas, Lockheed, and others, needed a first-rate center of aero-engineering instruction and a leading teacher/scientist to give it momentum. The Guggenheim Fund eventually established a highly acclaimed aeronautical engineering program at the California Institute of Technology and in 1930 finally lured Theodore von Kármán (born in Hungary) from Germany to give it direction. The distinguished von Kármán not only made Caltech a center of research but also presided over the education of dozens of engineers who became leading teachers and engineers themselves. Moreover, von Kármán became a key consultant to the U.S. government for numerous national security ventures that shaped the future of the postwar air force and air doctrine during the Cold War era.[22]

World War II

During the late 1930s, the use of bombers to attack cities in China and in Spain during the Spanish Civil War demonstrated the deadly potential of military aviation. In 1939 the outbreak of World War II in Europe underscored the massive, cataclysmic reality of modern air power. The Battle of Britain, waged during the summer of 1940, stands out as a landmark in aeronautical technology. German bombers spread destruction throughout London and other British population centers. During this desperately fought aerial conflict, the Royal Air Force finally turned back aggressive bombing attacks and fighter offensives of the German Luftwaffe. Had the Germans prevailed, and an invasion of Great Britain succeeded because the German Air Force controlled the skies, then the outcome of World War II would have been markedly different. As it was, Britain's survival meant that a key ally remained in the fight against totalitarian forces in Europe, the Mediterranean, and the Pacific theaters of operations. British factories continued to turn out military equipment for its own and Commonwealth forces; the RAF continued to operate from British bases to harass German forces on the continent. Most important, Britain's survival gave the United States an essential location in which to build numerous air bases to support the U.S. Army Air Forces in its own aerial assaults on Adolf Hitler's military machine. Finally, Britain provided the essential seaports, railways, roadways, and communications essential to a staging area for the hundreds of thousands of troops and mountains of equipment needed to launch D-Day on June 6, 1944. Without the positive outcome of the Battle of Britain, Hitler might well have prevailed in Europe. The technology of air power determined one of the most significant engagements of the twentieth century and helped turn the tide of history.[23]

The United States officially entered World War II in the aftermath of the dramatic assault delivered by Japanese naval air power at Pearl Harbor on December 7, 1941. Aviation technology subsequently bought the United States valuable time to mobilize against Japanese offensives in the Pacific theater of operations. In April 1942, during the battle of the Coral Sea, U.S. naval fighters and attack planes, operating from American aircraft carriers, engaged enemy planes from Japanese aircraft carriers. American forces finally carried the day, marking it as one of the first times that a major naval engagement had been decided by airplanes. No ships from either side spotted one another or traded gunfire. The following month, American planes operating from aircraft carriers intercepted and routed a strong Japanese naval task force in the battle of Midway. Carrier-based air power in these two engagements delivered a telling blow to enemy carriers and aircraft from which the Japanese never quite recovered. Momentum shifted to American and allied forces in the Pacific.[24]

In the meantime, America's industrial war machine began to gear up. During the late 1930s, as Europe began to prepare for a possible conflict, production orders from French and British sources had already affected aircraft manufacturers in the United States. Orders from abroad represented $400 million of a backlog totaling $680 million. After the fall of France in 1940, Britain picked up French orders; by that autumn, the RAF had American contracts for 14,000 planes and 25,000 engines worth $1.5 billion. During official postwar assessments, planners concluded that the French and subsequent British orders had increased aircraft production in the United States far beyond domestic sales.[25]

The evolution of the P-51 embodied several strands of this European influence. The initial impetus came not from the Army Air Corps but from the British, who, during the spring of 1940, arrived in America in search of more fighter planes for the Royal Air Force. They finally wound up at North American's plant on the West Coast, where they convinced the American firm to launch a new design. One of the principal engineers was Edgar Schmued, a German trained in Europe who had taken a series of jobs in Latin America and the United States before finally joining North American. After production versions of the P-51 reached the RAF, a savvy test pilot for the Rolls-

Royce firm, Ronald Harker, took one of the new planes up for a spin. He liked the Mustang but recommended a Rolls-Royce power plant as a replacement for the American equipment. With its British engine, the Mustang turned into a thoroughbred of the skies. License-built versions of the Rolls-Royce engine also powered later planes manufactured in the United States. And so, this quintessential American fighter, which played such a key role in securing air superiority in Europe during the war, would not have existed except for the insistence of the RAF.[26]

Other foreign legacies affected American aeronautical experience in the war. Certainly one of the most famous of the Russian émigrés who came to the United States in the aftermath of the Russian Revolution was Igor Sikorsky. After arriving in America in 1919, Sikorsky—and a number of émigrés who joined his firm—developed a series of pioneering flying boats. In the drive to raise money for the fledgling Sikorsky firm, other émigrés contributed money, led by the renowned pianist and composer Sergey Rachmaninoff. The Sikorsky firm's products included the S-42 amphibian aircraft, which played a key role in the evolution of Pan American Airways routes in the Caribbean and also in the western Pacific. By 1939 Sikorsky had also developed and flown a helicopter, which used a configuration that became the basis of subsequent practical rotary-winged aircraft and represented the first helicopters to enter service during the war. Another Russian émigré, Alexander de Seversky, founded the manufacturing firm that became known as Republic Aviation. With Alexander Kartveli, a fellow Russian émigré, as chief designer, Republic built the P-47 Thunderbolt fighter. Along with the P-51, the Thunderbolt played a premier role in escorting Allied bombers to key enemy targets in Europe and in taking the measure of the German Luftwaffe.

During the war, American fighter planes reaped considerable benefit from the distinctive bubble-shaped canopies of planes like the Mustang and Thunderbolt, allowing pilots far better 360-degree views of hostile skies and offering greater advantage in sighting enemy aircraft. Plexiglas canopies evolved from material developed by the American division of a German firm, Rohm and Haas, after the American group successfully worked out details for mass production and forming of the Plexiglas material. There were other, seemingly inconsequential, European legacies that collectively enhanced the performance of American combat planes in winning the war in the air. These included improved lubricants and hydraulic fluids—also perfected by Rohm and Haas—that did not become sluggish or freeze in the extremely cold temperatures encountered at high altitudes where the U.S. Air Force carried out its combat missions over enemy-held territory.[27]

Like several other countries, the United States investigated gas turbine technology—the jet engine—during the 1930s. Nevertheless, the first practical jet engine design—built by Frank Whittle—roared to life in Britain in 1937. Germany was not far behind and succeeded in flying the first jet plane, the Heinkel He-178, two years later in 1939. During the dark early days of World War II, the British decided to share their jet engine technology with the United States in case a German invasion of Britain might succeed. With detailed plans handcuffed to his wrist, an Air Force officer boarded a special plane to fly across the Atlantic to America. Later, the British sent a complete engine, along with Frank Whittle to explain how the thing worked. America's first operational jet fighter, the Lockheed P-80 Shooting Star, used an American copy of this engine, and its British cousins powered a number of important early U.S military jets such as the Grumman F9F Panther. These planes saw extensive duty during the Korean War and populated U.S. Air Force and Navy squadrons around the world during the early period of the Cold War. Later generations of American jets benefited form additional sources of German wartime technology, which the Americans eagerly scooped up after Germany's defeat. Aeronautical treasures included more powerful axial-flow jet engines, swept-wing configurations, and reams of additional aeronautical data that substantially contributed to swift American progress in jet-powered military planes as well as a new generation of jet airliners in the postwar decades.[28]

Additional legacies came from Great Britain, producers of the Martin-Baker ejection seat, which still equips many of the U.S. Navy's front-line combat aircraft. British innovations also bequeathed the canted decks of modern American nuclear aircraft carriers, along with the Harrier "jump-jet" that equips many U.S. Marine flying squadrons and the Navy's latest advanced jet trainer, the T-45, a design adapted from its British counterpart, the British Aerospace Hawk aircraft.

American bombers of World War II constituted a key element in the eventual Allied victory. The sequence of events in which America evolved this potent aviation technology is notable. Considerable credit must go to a determined group of U.S. Army Air Corps officers during the Depression years of the 1930s. Despite slim budgets and little congressional interest in their theories about the potential threat of aerial warfare in the coming decades, they promoted projects that eventually led to the construction of two remarkable prewar experimental bombers that pointed the way to the future. One was the Boeing XB-15; the second was the Douglas XB-19; both evolved in the middle of an era remembered for its pacifism and isolationist attitudes. Design and engineering on the Boeing XB-15 began in 1934, when the

Army Air Corps considered long-range aircraft to intercept enemy fleets at sea. By the time of the plane's first flight, in 1937, it held the title of the largest aircraft in the world. Nevertheless, it became upstaged by a smaller version, known as the B-17, which became one of the most significant aerial weapons of World War II. The Douglas XB-19, begun in 1935, had an even longer wingspan (212 feet, compared to the XB-15's 149 feet) and sported the first tricycle landing gear used on a large bomber. Although neither plane achieved combat status—both became designated transports during the war—they pioneered many techniques for construction of unusually large aircraft and associated electrical and other systems. And they led the way to the biggest bomber of World War II—the Boeing B-29.

The order for the first XB-29 went out in August 1940, before America entered the war. Prior to the plane's first flight in September 1942, Boeing had signed contracts to deliver more 1,500 copies of the big bomber, which weighed twice as much as the B-17. Except for the atom bomb—which the B-29 delivered in 1945—the aircraft arguably represented the largest and most complex weapon development program of the war. In addition to its pressurized fuselage sections, the plane featured remote-controlled machine guns along with an unprecedented maze of other control systems and instrumentation. Government and contractor officials alike faced a daunting process of constructing cavernous new assembly buildings, acquiring innumerable machine tools, training thousands of unskilled workers, and assuring a precise fit for tens of thousands of parts produced in locations from coast to coast. The training films developed for production workers as well as maintenance personnel, accompanied by a basic 2,000-page maintenance manual, plus separate documents for specific systems, resulted in a small production and printing industry in itself. "The B-29," wrote one admiring journalist during the war, "is the most *organizational* plane ever built." In all this, the myriad production elements of the B-29 required advanced management techniques that presaged systems engineering associated with the Cold War era.[29]

Given the globe-girdling dimensions of World War II, the need for rapid transport of key personnel, cargo, and correspondence spurred the use of aircraft to do the job. By 1942 various efforts coalesced into the Air Transport Command (ATC). By the end of the war, the ATC represented a remarkable airline system, involving more than 300,000 civil and military personnel and fielding some 3,700 aircraft that flew anywhere in the world. The ATC's activities changed intercontinental air travel from a state of high-risk adventure to a matter of daily routine. The ATC built airfields on every continent, installed radio installations along its far-flung routes, accumulated priceless meteorological and navigational data, and generated a considerable pool of skilled airway professionals. At its peak of operations, ATC aircraft crossed the Atlantic at an average rate of one plane every thirteen minutes. In the process, the time required to travel from continent to continent shrank dramatically from a matter of weeks to a matter of days, or, within a military theater of operations, to a few hours.[30] Often unheralded as a wartime institution, the ATC left an invaluable legacy of postwar aviation technology in worldwide airline operations.

The Cold War Years

In the aftermath of World War II, global leaders contemplated the radioactive ashes of Hiroshima and Nagasaki, the scorched ruins of dozens of German cities, and the numbing statistics of military personnel and civilians killed in cataclysmic air raids and realized that aviation technology had transformed the nature of warfare. For the United States as well as other nations, air power emerged as a major force in postwar national security and foreign relations.[31]

During the war years, the air force actually functioned under the name of the U.S. Army Air Forces — a component of the U.S. Army analogous to units for artillery, tanks, and so on. Given its destructive power, complexity, and global capability, military aviation seemed destined to emerge as a separate component of the U.S. armed services. Accordingly, in the National Defense Act of 1947, the United States Air Force emerged as a discrete element, along with its own Secretary of the Air Force within the Department of Defense. The Strategic Air Command, with its ability to deliver nuclear bombs anywhere on the surface of the globe, also emerged as a premier factor in Cold War confrontations with the Union of Soviet Socialist Republics (USSR). The Soviets in turn soon fielded their own intercontinental nuclear bombers, threatening strategic targets and population centers in Europe and North America. In technological response, America spent billions in constructing defensive ramparts such as the Distant Early Warning system, a line of radar stations stretching across the inhospitable frozen wastes of the Arctic region. The intent was to observe enemy aircraft flying from Europe and Asia up over the Arctic in a "great circle" route and down across North America. But officials realized that the radar defenses were not perfect. Hostile Soviet bombers might slip through the radar net at low altitudes to strike American cities.

To plug the gaps, the Air Force organized the Ground Observer Corps, relying on volunteers with binoculars and sharp eyes to glimpse marauding Soviet planes on northern horizons and then alert military officials. All this brought home the dangers of living in a Cold War environment. Not just military personnel, but civilians also, served as integral links to defend the nation and to save the lives of their families and fellow Americans. To the adults and adolescent children who populated the Ground Observer Corps defenses, it became a sobering responsibility. For generations, the Atlantic and Pacific oceans had been comforting barriers against invasion. In the age of air power, the United States (and North America) no longer existed as a continental "island" safe from direct attack because of its geographic position.[32]

And, in case everything failed, American cities became studded with bright yellow signs emblazoned with emergency instructions: "nuclear fallout shelter this way" with a large black arrow pointing the direction to presumed safety. Thousands of frightened citizens constructed their own fallout shelters or purchased prefabricated models from one of the new nuclear shelter businesses that sprang up in the early Cold War era. On Sunday mornings, clergy often delivered agonizing sermons about moral dilemmas involved in helping—or not helping—neighbors who might be exposed in the wake of a nuclear attack. Many Americans equipped their shelters with shotguns and plenty of ammunition, not to defend against a Soviet attack, but to ward off hapless acquaintances seeking protection in the wake of radioactive fallout. Grade school teachers regularly instructed their charges on techniques called "duck and cover" to survive nuclear explosions while at school: drop to the floor, scurry under your desk, and cover your face and eyes with your arms. Stay away from windows where shattering glass could kill you. Don't drink the water, but use the juice from canned fruits and vegetables. Don't go home until instructed—it might not be there any more. Also, mommy and daddy might be gone as well. Little wonder that pediatricians reported a sharp rise in gastrointestinal complaints and anxiety attacks among the nation's schoolchildren.[33]

Air power in an air age world inevitably embraced positive and negative elements. Academics launched bravely into this new world. As early as 1944, the Brookings Institution released a two-volume study under the general title *America Faces the Air Age*. Four years later, the Finletter Committee, appointed by President Harry Truman to probe swirling controversies about the relationships between aviation and national security, released its report. Its tendentious and cautionary title, *Survival in the Air Age*, underscored the fundamental realities of aviation technology the postwar era. Somewhat less dramatic was a new journal, *Air Affairs*, which released its first issue in 1946. Although its articles often addressed the Armageddon-like prospects of nuclear bombing, other pieces addressed the political implications of living in a world shrunk in time by airlines or discussed the societal implications of expanding aviation activities. William Fielding Ogburn, the distinguished anthropologist and sociologist at the University of Chicago, used the pages of *Air Affairs* for his first speculations about aviation and society in an article so titled, which he later expanded into *The Social Effects of Aviation*. Published in 1946, Ogburn's survey book accurately commented on the likely impact of aviation on business travel, agriculture, and other activities.[34]

Although none of these publications predicted anything like the Berlin airlift of 1948–49, the significance of this event for avoiding direct military confrontation is worth remembering. In the spring of 1948, following a series of Cold War diplomatic incidents, the Soviets cut off surface access to the occupied city of Berlin, situated in the Soviet occupation zone after World War II. They were unable, however, to prevent air traffic to the city because of an agreement between postwar occupation nations to provide air corridors through East Germany. To reach their designated occupation areas in western Berlin, British, French, and American allies mounted an emergency airlift to supply food, clothing, and fuel for 2.5 million people. Contending with Soviet harassment, poor facilities, and lousy weather during the winter of 1948–49, the western allies turned a potential diplomatic disaster into a rousing success.[35] The Berlin airlift demonstrated the effectiveness of aviation technology in postwar diplomacy, providing an alternative to an explosive confrontation that might have triggered armed conflict.

In the following decades, aeronautical technology in the form of air power continued to play a decisive role in Cold War confrontations as well as in hot wars. Already during the latter stages of World War II, Germany's Luftwaffe had caused considerable consternation when Messerschmitt Me-262 twin jet fighters intercepted Allied bomber formations, carried out high-speed attacks, and streaked away at speeds that left the best Allied fighters lagging far behind. Using captured German documents, hardware, and experts, American engineers continued to build on the expertise shared with us by the British in the research and development of jet-propelled aircraft like the Lockheed P-80. The following generation of swept-wing jet fighters and bombers not only relied on axial-flow jet engines with direct lineage to wartime German technology but also utilized German aerodynamic information during the evolution of American swept-wing designs such as the 600-mile-per-hour North American F-86 Sabre and the 500-mile-per-hour Boeing B-47 Stratojet.

The new jets, requiring longer runways and complex support systems, launched an expensive round of airfield construction and improvements around the world to enhance American security commitments during the Cold War era. Jet technologies quickly made an impact on the U.S. Navy, with subsequent development of increasingly advanced aircraft and huge new aircraft carriers, including nuclear-powered vessels. The elaboration of this growing panoply of postwar air power technology also absorbed billions of tax dollars, making military aviation a major factor in budget debates and setting off sharp inter-service rivalries between the U.S. Navy and the U.S. Air Force.[36] Succeeding generations of combat planes attained speeds more than twice the speed of sound, appeared with variable-sweep wings, came packed with exotic electronics, mounted smart missiles with their own guidance systems, and eventually became "stealthy" to avoid detection by enemy radar.

Research, design, fabrication of test articles, and manufacturing of these new combat aircraft also demanded new levels of technological sophistication and an increasingly capable infrastructure. To save weight and preserve strength, use of titanium components and other light-alloy forgings to unprecedented extents became standard, but not before an extensive research and development program found ways to produce the materials, build usable machine tools, and perfect complex numerical control formulas to run the new machine tools. Similar advances characterized successive generations of avionics. Another wave of innovation coursed through the airframe industry as composite materials became more widespread after the 1970s. In order to avoid excessively long periods before planes could enter operational service, "concurrency" became the new watchword, a planning/manufacturing process in which tooling, avionics, weapons, and other paraphernalia were designed and planned for production before many details of the plane itself had been frozen. New military doctrine and strategy might also enter the equation with some types of new equipment. This fed back into the design/manufacturing process, and so on. The systems-management approach, as it evolved in the United States in the 1950s, provided a means to keep track of all these events and effectively integrate them. The astute British aerospace expert Bill Gunston singled out the systems-management approach as a particularly American phenomenon: "The Americans are very good at identifying abstract things and naming them," he wrote. "Such a concept has [had] tremendous and far-reaching implications in practice."[37]

The escalating costs of billion-dollar aircraft projects eventually drove manufacturers and the Department of Defense to undertake new methods in order to pursue development of many advanced combat aircraft. Contractor teams formed to share the risks as well as potential profits involved in design and prototype competitions. The General Dynamics F-16 jet fighter (later built by Lockheed Martin), which entered service during the 1970s, required a series of international agreements before it entered production. The diplomatic negotiations embraced several European countries that committed funding for its development and who manufactured major components of the aircraft. Internationalization of the aerospace manufacturing industry characterized other military projects as well as commercial air transport programs.[38]

Considering the dramatic progress of aeronautical technology during World War II, and given the remarkable evolution of the ATC, air transportation seemed destined to play a major role in the postwar era. The phrase "air age world" not only implied the possibility of armed aerial armadas but also promoted the more positive image of accessible, reliable airline travel all over the globe. Visionary individuals realized that the orderly progress of international transport aviation required a rational structure to facilitate worldwide airline operations. There had to be agreement on a standard language used by pilots and traffic controllers, consistency in airport arrival and departure procedures, commonality in radio equipment and frequencies, standardized runway marking and night lighting, up-to-date meteorological information, and a host of other questions to be resolved. The International Civil Aviation Congress, convened in Chicago late in 1944, represented an important step in addressing the issue.[39]

Delegates from fifty-two countries attended, overcoming geographical and political barriers implicit in a world still at war. Despite a number of rancorous debates over landing rights and other topics, the delegates managed to agree on a new organization—the International Civil Aviation Organization, or ICAO, a specialized agency of the United Nations. Officially launched in 1947, the ICAO sited its headquarters in Montreal, Canada. English became the official language; an American, Edward Pearson Warner, became the first president, and he continually won reelection until his retirement in 1957. With a membership of 183 nations in 2001, the ICAO continues to provide essential service to the technology of international airline operations. In addition to standards for airworthiness of passenger planes, airfields, and airways, the ICAO presides over telecommunications links, adjudicates legal hassles over international tax codes, cooperates with the Universal Postal Union on questions involving international mail and cargo procedures, provides statistics and data, and more. It has become a model for effective international collaboration in dealing with modern technology.[40]

The ICAO became one of the prime factors in the rapid evolution of international travel during the postwar era, when piston-engine transports forged a network of airways across oceans and continents. The revolution represented by air travel first showed up in large markets such as the United States. Larger, pressurized aircraft like the DC-6, Boeing Stratocruiser, and the elegant, triple-tailed Lockheed Constellation followed modern, four-engine transports like the Douglas DC-4. With accommodations for about sixty to eighty passengers and cruising speeds of 300–350 miles per hour, these airliners could speed coast to coast in a matter of hours, at altitudes where travelers experienced much less discomfort from turbulence and where airliners could vault the sorts of bad weather that used to keep passengers grounded. Business travelers began to flock to airline terminals, forsaking traditional conveyances like railways with Pullman accommodations for longer trips. In 1951 airline passengers exceeded rail Pullman passengers for the first time. Costs came down; rising incomes and the growing practice of traveling on credit encouraged a greater cross section of the American public to travel by air. Vacationers began to join the queues at airline ticket counters. By 1957 airline passenger miles surpassed both train and bus passenger miles. That same year, the acclaimed social critic Max Lerner referred to the air-travel phenomenon in his new book *America As a Civilization*. "The new Air Age, whose impact is just beginning to be felt, has further heightened the mobility of Americans," he observed. Holiday travel took on a new look. Lerner predicted that airliners would continue to alter traditional patterns of American travel and would certainly bring a greater democratization of travel. In Lerner's words, "What the Air Age has done has been to make the faraway vacation possible for the boss's secretary as well as the boss."[41]

The democratization phenomenon reshaped international travel as well. Before World War II, transoceanic leisure travel by steamship consumed many days of time; only the elite could afford the fares and several weeks required to arrive in Europe, see the sights, and sail home again. Postwar flights across the Atlantic in less than twenty-four hours permitted a ten-day junket to fit both pocketbook and available vacation time for a remarkable cross section of holiday seekers. The arrival of jets in the 1960s enlarged the global range of travelers. The British inaugurated the jet era in 1952 when the de Havilland Comet went into service. A series of tragic accidents grounded this airliner for several crucial years, giving American manufacturers an opportunity to adapt technologies learned from large jet bombers and jet tankers. By 1958 the classic Boeing 707 entered international service on Pan Am's transatlantic routes, joined by an improved Comet, and followed by the Douglas DC-8 in 1959. Jet airliners boasted speeds of close to

The Boeing 747 wide-body has been one of the most successful passenger airliners of the last half of the twentieth century, being used by airlines around the world. (Photo courtesy of Virgin Atlantic Airlines)

600 miles per hour and could carry one hundred or more passengers. Their comfort, convenience, and affordable fares marked a new era of passenger flying, especially in terms of international travel for the American public. Daniel Boorstin, a keen observer of American socioeconomic history, wrote: "The United States was the first nation in history so many of whose citizens could go so far simply in quest of fun and culture. The size of this phenomenon made international travel, for the first, time, a major element in world trade, a new problem for the American economy and for American balance of payments, and a new opportunity for the destination countries."[42]

In time, a variety of other jets populated the skies, including a new generation of jumbo jets like the Boeing 747, introduced into service in 1970 (with about 380–400 seats) and the proposed Airbus A380 (about 550 seats), announced in 2001. The number of passengers from all over the world also proliferated, completing the transition of air travel from the exceptional to the usual mode of transportation.

In the postwar years, the activities of personal planes populating American skies became known as general aviation, a broad term that covered all planes not in the military services or airlines. The use of helicopters, or rotorcraft, also became widespread, in applications such as passenger shuttle missions for offshore oilrigs and emergency ambulance service. General aviation planes ran the gamut, from diminutive single-seaters and gliders to converted military aircraft and retired airliners. For the most part, however, general aviation aircraft came from a specialized group of light plane manufacturers like Beechcraft, Cessna, Piper, and others. The phenomenon began to emerge during the 1920s, when experiments in dusting cotton fields to control boll weevils evolved into agricultural aviation for dusting and spraying to control insects. Before long, planes spread herbicides, distributed seeds, and applied fertilizers. During the same decade, businesspeople began to fly their own planes in order to save time and to reach out-of-the-way locations. As American businesses continued to decentralize their operations, the practice of business flying continued to grow, either in personally flown aircraft or in larger executive planes to carry several passengers and operated by professional corporate pilots. From only a few hundred planes before the war, the number of private planes in the United States numbered about 37,000 in 1945. Hopeful predictions for low-cost general aviation designs did not materialize in the postwar era, and many thousands of aspiring private pilots were left sitting on the sidelines.[43]

Nonetheless, the general aviation fleet burgeoned after World War II, taking advantage of new labor sources, lower overhead, proximity to markets and raw materials, and similar enticements. Expansion of the Interstate Highway System, with its obvious advantages for trucking, further accelerated decentralization of production and distribution. But many of the new business centers and industrial operations wound up in regions with shrinking railway passenger routes and infrequent or nonexistent scheduled airline service. For example, in the late 1990s, the United States counted some 5,000 public airports, but only 600 listed service by a scheduled airline. Faced with days of driving to reach many municipalities, businesspeople turned to aviation technology—general aviation activities literally took off. By the mid-1960s, the American general aviation fleet numbered about 100,000 aircraft. By the turn of the century, the U.S. general aviation fleet numbered more than 200,000 planes (out of an estimated worldwide fleet of 260,000), compared to about 6,000 planes in the U.S. armed services and 6,000 aircraft in scheduled airline service. The vast majority of general aviation planes was engaged in business flying and included single-engine piston planes for four passengers as well as luxurious corporate jets with spacious accommodations for four to twelve business fliers. About 65 percent of all general aviation flights took place for business or commercial purposes. Some of the larger corporate jets could conduct transoceanic flights. By the start of the twenty-first century, manufacturers like Boeing and Airbus offered conversions of 100-passenger airliners for executive travel. American presidents and other chiefs of state jetted around the world in specially outfitted Boeing 747 transports.

Contemporary Developments   Continuity with the Past

Symbols and icons of aviation technology continued to maintain a high profile in contemporary life. During World War II, Harley J. Earle, the chief stylist for General Motors, became enamored of the twin tails of the Lockheed P-38 fighter. In Earle's postwar designs, upswept rear-fender embellishment represented a hallmark of postwar Cadillac cars, a silhouette soon copied by other auto builders. Harley also loved the clean, aerodynamic look of new jet planes, and General Motors automobiles adopted trendy, curved windshields and grille treatments that echoed jet plane air intakes. Inevitably, rival designers introduced these elements. The 1956 Plymouth Belvedere displayed fin-like rear fenders blended into tall taillights. It had a wrap-around windshield and side trim to suggest speed lines. The whole approach was billed as "Flight-Sweep," with ad copy touting the car's "long, low aerodynamic lines." The ad layout included a photo and endorsement by the chief design engineer for the Douglas DC-8 transport. Although airline toys began to show up in the 1930s, a symbolic pop-culture event occurred in 1964, when the makers of Barbie dolls introduced boyfriend Ken as an airline pilot. About the same time, the folks from Barbie offered a scaled-down jet airliner "playhouse" for aspiring young stewardesses, complete with a galley. By the late 1990s, gender equality dictated the debut of Barbie as a pilot herself. In recent decades, Macy's ad agency still utilized aviation imagery to emphasize high style, state-of-the-art design and quality and as a gimmick to catch the consumer's eye.[44]

There were troubling and problematic facets to the growing role of aviation technology in contemporary life. During the 1960s, the high profile of international airline operations turned commercial airliners into targets of callous terrorism; deadly incidents continued through the 1980s. As a factor in national security, aeronautical assets also triggered international confrontations. In 1960 the Soviet Union downed a Lockheed U-2 spy plane, scuttling a scheduled summit between President Dwight Eisenhower and Premier Nikita Khrushchev and heating up Cold War rhetoric. Early in 2001 a U.S. Navy intelligence plane experienced a midair collision with a fighter plane from the People's Republic of China. The Chinese pilot died; the American plane made a forced landing at a Chinese airfield, setting the stage for several weeks of tense negotiations before American officials could recover the damaged aircraft and its crew. In a different context, the competition between the American aircraft manufacturer Boeing and its European counterpart, Airbus Industrie, involves economic factors tied to global markets representing tens of billions of dollars as well as demonstrating global leadership in a highly visible technological enterprise.

Other aspects of current aeronautical technology represent both change and continuity. During the late 1990s, the U.S. Department of Defense launched development of a Joint Strike Fighter (JSF) program. In addition to stealth characteristics, several features of the project made it unique. For one thing, all three of the major American armed services—the U.S. Navy, Air Force, and Marine Corps—actively participated in setting out requirements for the new combat plane. For another, the plane was designed from the start to operate from the decks of aircraft carriers as well as from land bases. Moreover, the Marine Corps version of the JSF would have the capabilities for vertical takeoff and landing (VTOL). A particularly unusual aspect of the program was the key role played by Great Britain's RAF, which committed tens of millions of dollars to the JSF program and became a major partner in the plane's evolution.[45]

All this underscores a major theme in recent aviation technology, the multiple partners required to fund increasingly costly new aircraft. Partnerships may take the form of partnerships between major contractors (as in the case of the Advanced Tactical Fighter) as well as partnerships between foreign entities. The Europeans have become especially skilled at launching multinational projects, but the intimate participation of the RAF in the American effort for a new combat plane such as the JSF, especially so early in the program, is quite remarkable.[46] However, looking back at the history of American aircraft development, the consistent element of European influence snaps into focus in so many instances that what we see at present is really not much more than a reflection of things past.

NOTES

1. The Wrights' comprehensive search for information is recounted in Tom D. Crouch, *The Bishop's Boys: A Life of Wilbur and Orville Wright* (New York: W. W. Norton, 1989). See also Peter L. Jakab, *Visions of a Flying Machine: The Wright Brothers and the Process of Invention* (Washington, D.C.: Smithsonian Institution Press, 1990). The considerable legacy of European aviation activities is evident in Charles Harvard Gibbs-Smith, *Aviation: An Historical Survey from Its Origins to the End of World War II* (London: Her Majesty's Stationery Office, 1970).

2. Grover Loening, a pioneer American aviator and builder, describes his association with Blériot designs in his autobiographical book, *Our Wings Grow Faster* (New York: Doubleday, Doran, 1935), 21–30, 48–55. See also Tom D. Crouch, *Blériot IX: The Story of a Classic Aircraft* (Washington, D.C.: Smithsonian Institution Press, 1982). On Douglas

Thomas, see Louis Casey, *Curtiss: The Hammondsport Era, 1907–1915* (New York: Crown, 1981), 176–77; Cecil R. Roseberry, *Glenn Curtiss: Pioneer of Flight* (Garden City, N.Y.: Doubleday, 1972), 365–66.

3. Examples can be found in studies of European as well as American responses to flight. For the European response, see Robert Wohl, *A Passion for Wings: Aviation and the Western Imagination, 1908–1918* (New Haven, Conn.: Yale University Press, 1994), esp. 137, 153, 255–56, 271–72. For the United States, see Joseph J. Corn, *The Winged Gospel: America's Romance with Aviation, 1900–1950* (New York: Oxford University Press, 1983), esp. 29–50.

4. Robert G. Everding, "Bernard Shaw, *Misalliance*, and the Birth of British Aviation," in Stanley Weintraub, ed., *Shaw: The Annual of Bernard Shaw Studies* (State College: Pennsylvania State University Press, 1988), 69–76; John Buchan, *The Thirty-Nine Steps* (Ware, U.K.: Wordsworth Editions, 1993). Regrettably, *Misalliance* closed in London after eleven performances. *The Thirty-Nine Steps* first appeared as an Alfred Hitchcock film in 1935; other adaptations followed.

5. I have scanned sheet music, juvenile books, and other printed ephemera in the Bella Landauer Collection and National Air and Space Museum, as well as juvenilia in the Ross-Barrett Collection of the Denver Public Library. The paper glider is in the author's collection, along with a reproduction of the Armour advertisement. See also the discussion in Roger E. Bilstein, *Flight in America: From the Wrights to the Astronauts*, 3d ed. (Baltimore, Md.: Johns Hopkins University Press, 2001), 19–20.

6. A vast literature on World War I exists. For an outstanding introduction to the subject, see Dominick Pisano et al., *Legend, Memory and the Great War in the Air* (Seattle: University of Washington Press, 1992). The German air raids over England and London are memorably described in Raymond Fredette, *The Sky on Fire: The First Battle of Britain, 1917–1918* (Washington, D.C.: Smithsonian Institution Press, 1991). See also the insightful commentary on aerial engagements in Lee Kennett, *The First Air War, 1914–1918* (New York: Free Press, 1991). The RAF was formalized in 1918 by joining the Royal Flying Corps and the Naval Air Service.

7. Roger E. Bilstein, *Flight Patterns: Trends of Aeronautical Development in the United States, 1918–1929* (Athens: University of Georgia Press, 1983), 29–56. Quote is from Edward Pearson Warner, *The Early History of Air Transportation* (York, Pa.: Maple Press, 1937), 27. See also William M. Leary, *Aerial Pioneers: The U.S. Air Mail Service, 1918–1927* (Washington, D.C.: Smithsonian Institution Press, 1985).

8. For a convenient record of the flurry of long-distance and pioneering commercial flights between the wars, see John W. R. Taylor et al., *Air Fact and Feats* (New York: Bantam Books, 1979), 101–18; Carroll Glines, *Around the World in 175 Days: The First Round-the-World Flight* (Washington, D.C.: Smithsonian Institution Press, 2002). The fascinating story of British and Dutch rivalry is told by Mark Dierikx, "Struggle for Prominence: Clashing Dutch and British Interests on the Colonial Air Routes, 1918–41," *Journal of Contemporary History* 26 (1999): 333–51. Germany's nationalistic approach to aviation is the subject of Peter Fritzsche, *A Nation of Fliers: German Aviation and the Popular Imagination* (Cambridge, Mass.: Harvard University Press, 1992). On American activities in the 1920s, see Bilstein, *Flight Patterns*, 164–73, and especially Wesley Phillips Newton,

The Perilous Sky: U.S. Aviation Diplomacy and Latin America, 1919–1931 (Coral Gables, Fla.: University of Miami Press, 1978).

9. Mark Sullivan, *Our Times: The United States, 1900–1925*, vol. 2 (New York: Charles Scribner's Sons, 1927), 556–57.

10. Bilstein, *Flight Patterns*, 36–53, 127–32, 138–44; Nick A. Komons, *Bonfires to Beacons: Federal Civil Aviation Policy under the Air Commerce Act, 1926–1938* (Washington, D.C.: U.S. Government Printing Office, 1978). On the NACA, see Roger E. Bilstein and Frank Anderson, *Orders of Magnitude: A History of NACA and NASA, 1915–1990*, NASA Special Publication 4406 (Washington, D.C.: U.S. Government Printing Office, 1989). The quote is cited in Roger E. Bilstein, *The Enterprise of Flight: The American Aviation and Aerospace Industry* (Washington, D.C.: Smithsonian Institution Press, 2001), 14.

11. For insights on the unsung aviation professionals of the time, see William M. Leary, ed., *Aviation's Golden Age: Portraits from the 1920s and 1930s* (Iowa City: University of Iowa Press, 1989). See also William F. Trimble, *Jerome C. Hunsaker and the Rise of American Aeronautics* (Washington, D.C.: Smithsonian Institution Press, 2002).

12. European influences are cataloged in Roger E. Bilstein, "American Aviation Technology: An International Heritage," in Peter Galison and Alex Roland, eds., *Atmospheric Flight in the Twentieth Century* (Boston, Mass.: Kluwer Academic, 2000), 207–22.

13. The story of Max Munk is told in James R. Hansen, *Engineer in Charge: A History of the Langley Aeronautical Laboratory, 1917–1958*, NASA Special Publication 4305 (Washington, D.C.: U. S. Government Printing Office, 1987), 72–95, 119–22, 522–25.

14. The NACA's early years are summarized in Bilstein and Anderson, *Orders of Magnitude*, 16–121. For a more detailed coverage, see Hansen, *Engineer in Charge*; Alex Roland, *Model Research: The National Advisory Committee for Aeronautics, 1915–1958*, NASA Special Publication 4103, 2 vols. (Washington, D.C.: U.S. Government Printing Office, 1985). See especially Roland's commentary on the cowling issue in vol. I, 113–17; quote is from 352, fn. 44.

15. For a comprehensive analysis of airframe development and ancillary technologies, see Ronald Miller and David Sawers, *The Technical Development of Modern Aviation* (New York: Praeger, 1970), 53–86. This book also notes the special role of the DC-3. Although the Boeing 247 featured virtually all the advanced features of the era, seating for only 10 passengers compromised its profitable operation. This pioneering transport is skillfully assessed by F. Robert van der Linden, *The Boeing 247: The First Modern Airliner* (Seattle: University of Washington Press, 1991). For a catalog of beneficial NACA and American research efforts, see Pamela Mack, ed., *From Engineering Science to Big Science: The NACA and NASA Collier Trophy Research Project Winners*, NASA Special Publication 4219 (Washington, D.C.: U.S. Government Printing Office, 1998). See also Roger D. Launius, ed., *Innovation and the Development of Flight* (College Station: Texas A&M University Press, 1999).

16. John H. Frederick, *Commercial Air Transportation* (Chicago: Richard D. Irwin, 1942), 325–26; quote is from Paul Peter Willis, *Your Future in the Air* (New York: Prentice-Hall, 1940), 33. Willis was a marketing and public-relations executive for American Air Lines during the 1930s.

17. Advertisement, *Saturday Evening Post*, Nov. 16, 1940, 83. Samples of postcards featuring airline terminals are in the author's collection.

18. Bilstein, *Flight in America*, 310–11; Bilstein, *Flight Patterns*, 147–53.

19. The Franklin auto ad appeared in a 1929 number of the magazine *Country Life*, reproduced as a full-page (unnumbered) plate in Robert Atwan, *Edsels, Luckies, and Frigidaires* (New York: Dell, 1979). "Lydia E. Pinkham: Pinkham Pioneers," a 1926 advertisement booklet, is in the author's collection. To many in the jazz age, bobbed hair smacked of radical political and social ideas.

20. Lewis Mumford, *Technics and Civilization* (New York: Harcourt, Brace, 1934), 231; caption for commentary on plate XI. The film *The City* appeared in 1938 as part of a series of New Deal documentaries produced by Pare Lorenz, often with scripting prepared by Mumford.

21. See, for example, various issues of the *Saturday Evening Post* from 1939 and 1940.

22. On this philanthropic effort, see Richard P. Hallion, *Legacy of Flight: The Guggenheim Contribution to American Aviation* (Seattle: University of Washington Press, 1977). The story of von Kármán's recruitment and his subsequent influence is recounted in Paul Hanle, *Bringing Aerodynamics to America* (Cambridge, Mass.: MIT Press, 1982), and in Michael Gorn, *The Universal Man: Theodore von Kármán's Life in Aeronautics* (Washington, D.C.: Smithsonian Institution Press, 1992), which illuminates the pervasive impact of von Kármán's advisory roles.

23. My remarks on the significance of the Battle of Britain have been prompted by my reading of surveys such as Samuel Eliot Morison, *Strategy and Compromise: A Reappraisal of the Crucial Decisions Confronting the Allies in the Hazardous Years, 1949–1945* (Boston: Little, Brown, 1958), and Kent Roberts Greenfield, *American Strategy in World War II: A Reconsideration* (Baltimore, Md.: Johns Hopkins University Press, 1963). The battle itself has generated a considerable number of books. Richard Hough and Denis Richards, *The Battle of Britain: The Greatest Air Battle of World War II* (New York: W. W. Norton, 1989), combines detailed coverage and analysis. Richard Overy, *The Battle of Britain: The Myth and the Reality* (New York: W. W. Norton, 2000), represents an instructive synthesis. As Overy remarked (xiv), had Britain not survived the battle, "The consequences . . . would have been a calamity not just for the British people but for the world as a whole."

24. Clark G. Reynolds, *The Carrier War* (Alexandria, Va.: Time-Life Books, 1982).

25. Roger E. Bilstein, *The American Aerospace Industry: From Workshop to Global Enterprise* (New York: Twayne, 1996), 66–67; Wesley Frank Craven and James L. Cate, eds., *The Army Air Forces in World War II*, vol. 6, *Men and Planes* (Washington, D.C.: Office of Air Force History, 1983), 191–301, 313.

26. Bilstein, *American Aerospace Industry*, 68–70.

27. John B. Rae, *Climb to Greatness: The American Aircraft Industry, 1920–1960* (Cambridge, Mass.: MIT Press, 1968), 16, 77–98; Bilstein, "American Aviation Technology," 216–17.

28. Hansen, *Engineer in Charge*, covers both engine and aerodynamic developments, especially chapter 8. On swept wings, see Richard P. Hallion, "Lippisch, Gluhareff and Jones: The Emergence of the Delta Planform and the Origins of the Swept Wing in the United States," *Aerospace Historian* 26 (Mar. 1979): 1–10; for engines, see Virginia

Dawson, *Engines and Innovation: Lewis Laboratory and American Propulsion Technology*, NASA Special Publication 4306 (Washington, D.C.: U.S. Government Printing Office, 1991), 65–72, 78–101, 141–44.

29. Wesley Frank Craven and James L. Cate, eds., *The United States Air Force in World War II*, 7 vols. (Chicago: University of Chicago Press, 1948–55), 6:202–4; Jacob Vander Meulen, *Building the B-29* (Washington, D.C.: Smithsonian Institution Press, 1995). The quote is from Thomas Collison, *The Superfortress Is Born* (New York: Duell, Sloan and Pearce, 1945), 3.

30. Office of History, Military Airlift Command, *Anything, Anywhere, Anytime: An Illustrated History of the Military Airlift Command, 1941–1991* (Scott Air Force Base, Ill.: Military Airlift Command Office of History, 1991), 23–54. See also Robert Serling, *When the Airlines Went to War* (New York: Kensington Books, 1997). For a fascinating contemporary account packed with colorful photography, see Ivan Dmitri, *Flight to Everywhere* (New York: Whittlesey House, 1944).

31. Michael S. Sherry, *The Rise of American Air Power: The Creation of Armageddon* (New Haven, Conn.: Yale University Press, 1987), 351–56.

32. Postwar changes are covered in an official history by Alfred Goldberg, ed., *A History of the United States Air Force, 1907–1957* (New York: D. Van Nostrand, 1957). Discussion of the Ground Observer Corps is based on the author's personal recollections as an adolescent GOC volunteer and on a GOC pamphlet, "The Aircraft Flash: Official GOC Magazine," Sept. 1953, in the author's collection.

33. Several books on the history of popular culture took note of public fears stemming from the possibility of nuclear attack. See, for example, John and Gordon Javna, *60s! A Catalog of Memories and Artifacts* (New York: St. Martin's Press, 1988), 69; Jane and Michael Stern, *Sixties People* (New York: Alfred A. Knopf, 1990), 217–19.

34. Bilstein, *Flight in America*, 167–69.

35. Goldberg, *History of the United States Air Force*, 235–41; Roger G. Miller, *To Save a City: The Berlin Airlift, 1948–1949* (College Station: Texas A&M University Press, 2000).

36. Two books in the Time-Life series The Epic of Flight provide a cogent discussion and superb illustrations covering this watershed era in aviation history. Richard P. Hallion, *Designers and Test Pilots* (Alexandria, Va.: Time-Life Books, 1983), 78–171, discusses European sources of aerodynamic advances and their legacies to postwar aviation technology. Bryce Walker, *Fighting Jets* (Alexandria, Va.: Time-Life Books, 1983), provides details on principal military trends. The elaboration of aerospace electronics and military systems is dissected in Ann Markusen, ed., *The Rise of the Gunbelt: The Military Remapping of Industrial America* (New York: Oxford University Press, 1991). Many periodicals were intensely interested in the postwar whiz-bang aviation technology, including *National Geographic*, which not only devoted a major portion of its September 1965 issue to the U.S. Air Force, but also reprinted the section as a separate publication in the same year. The Department of the Air Force then sent free copies to "responsible leader(s) in the civilian community." A copy is in the author's collection.

37. Bill Gunston, *Bombers of the West* (New York: Charles Scribner's Sons, 1973), 190–92, includes valuable insights on Cold War military programs, as well as the quote.

38. Goldberg, *History of the United States Air Force*, 235–41.

39. Postwar trends and internationalization in the industry are summarized in Bil-

stein, *American Aerospace Industry*, 95. See also Keith Hayward, *The World Aerospace Industry: Collaboration and Competition* (London: Duckworth, 1994), and Donald M. Pattillo, *Pushing the Envelope: The American Aircraft Industry* (Ann Arbor: University of Michigan Press, 1998), which follows the fortunes of major manufacturers. For a superb analysis of a major program, see Glenn Bugos, *Engineering the F-4 Phantom II: Parts into Systems* (Annapolis, Md.: Naval Institute Press, 1996).

40. On the origins of ICAO, see Henry Ladd Smith, *Airways Abroad: The Story of Commercial Aviation in the World*, Smithsonian History of Aviation Series (Washington, D.C.: Smithsonian Institution Press, 1991), 163–204; on its subsequent role, see Eugene Sochor, *The Politics of International Aviation* (Iowa City: University of Iowa Press, 1991).

41. Carl Solberg, *Conquest of the Skies: A History of Commercial Aviation in America* (Boston: Little, Brown, 1979), 345–46, is still a first-rate survey and an excellent source on salient postwar trends. R. E. G. Davies, *Airlines of the United States since 1914*, rev. ed. (Washington, D.C.: Smithsonian Institution Press, 1982), includes remarkable detail that constitutes a veritable encyclopedia of airline evolution and statistical miscellany. The quote is from Max Lerner, *American As a Civilization* (New York: Simon and Schuster, 1957), 97.

42. Robert Serling, *The Jet Age* (Alexandria, Va.: Time-Life Books, 1982), integrates outstanding illustrations with intelligent commentary on the evolution and dynamics of this phenomenon. The quote is from Daniel Boorstin, *America: The Democratic Experience* (New York: Vintage, 1974), 517. For engrossing commentary on the cultural ramifications of air transportation, see Kenneth Hudson and Julian Pettifer, *Diamonds in the Sky: A Social History of Air Travel* (London: Bodley Head, 1979).

43. Bilstein, *Flight Patterns*, 59–96, covers business flying and utility aviation of the 1920s. See also Frank Joseph Rowe and Craig Miner, *Borne on the South Wind: A Century of Kansas Aviation* (Wichita, Kans.: Wichita Eagle and Beacon Publishing, 1994). Because the general aviation manufacturing industry has so many firms in Kansas, this well-written survey is valuable.

44. On cars, see Thomas Hine, *Populuxe: From Tailfins and TV Dinners to Barbie Dolls and Fallout Shelters* (New York: MJF Books, 1999), 83–106. The Plymouth ad is in the author's collection. On the Barbie business, see Paris and Susan Manos, *The World of Barbie Dolls* (Paducah, Ky.: Collector Books, 1990). A jet airliner play set and Barbie pilot are in the author's collection. In 1986 an advertising crew working for Macy's Department Stores journeyed to the Pima Air Museum in Tucson, Arizona, to shoot photos for a high-style, twenty-two-page advertising supplement that featured the retailer's spring line of women's fashions, using models dramatically posed with aircraft in the background. A copy is in the author's collection.

45. Bilstein, *Flight in America*, discusses many elements of postwar general aviation progress. William F. Trimble, "The Collapse of a Dream: Lightplane Ownership and General Aviation in the United States after World War II," in William F. Trimble, ed., *From Airships to Airbus: The History of Civil and Commercial Aviation*, vol. 2, *Pioneers and Operations* (Washington, D.C.: Smithsonian Institution Press, 1995), 129–45, analyzes postwar pricing difficulties. The General Aviation and Manufacturers Association (GAMA) distributes informative material on this colorful segment of aviation. See the GAMA Web site at <www.generalaviation.org>.

46. Bilstein, *Flight in America*, 225, 330, 341–43. The list of downside issues pertaining to aviation includes aerial pollution and airport noise. My commentary on aviation headline stories for 2001 comes from miscellaneous issues of *Aviation Week and Space Technology*. On September 11, 2001, as I finished this essay, overseas terrorists carried out horrendous attacks in New York City and Washington, D.C. Airliners, which had been targets of hijackers and bombers, suddenly had become instruments of fear and destruction.

CHAPTER 2

THE WRIGHT BROTHERS, GOVERNMENT SUPPORT FOR AERONAUTICAL RESEARCH, AND THE EVOLUTION OF FLIGHT

In Stephen E. Ambrose's best-selling biography of Meriwether Lewis, *Undaunted Courage*, the author makes the point that at the beginning of the nineteenth century everything moved at the speed of a horse. "No human being," he observed, "no manufactured item, no bushel of wheat, no side of beef (or any beef on the hoof, for that matter), no letter, no information, no idea, no order, or instruction of any kind moved faster. Nothing ever had moved faster, and, as far as [Thomas] Jefferson's contemporaries were able to tell, nothing ever would. And except on a racetrack, no horse moved very fast."[1]

It is an insightful comment, at once obvious and obscure. But the nineteenth century portended enormous changes in transportation. Those living in it saw the movement from horsepower to steam-driven trains, then the rise of the internal combustion engine and the automobile, and finally at century's end the dawning of a new age of flight. Thanks to the invention of the airplane by the two brothers from Dayton, for the first time in human history the people of the twentieth century would enjoy the thrill of flying. In the process their lives would fundamentally change, and in ways that no one could have foreseen, as the airplane became a sine qua non of a modern, mobile, magnificent, and sometimes monstrous way of life.

Historians of five hundred years hence may well characterize successful human flight, and all that followed in both air and space, as the most significant single technology of the twentieth century.[2] Has it fundamentally reshaped our world, at once awesome and awful in its affect on the human condition? Has it made easy, even luxurious, movement about the globe? At the dawn of the twentieth century, which I would remind you also had mechanized means of transportation, everyone had to allow multiples of days and sometime weeks for travel. Jules Verne's character Phineas Fogg of *Around the World in Eighty Days* was a creature of railroad and steamship timetables that took him throughout the globe with some ease, but certainly on a much longer schedule.[3] At the dawn of the twenty-first century, when one plans a transcontinental or even transatlantic trip, one may allow only one day for travel. We rightfully scoff at eighty days being required to circle the Earth; after all, anyone can do it in a few days by airplane and in a few minutes by spacecraft.

Anyone and anything can be virtually anywhere in the world within one day's travel. This is both a blessing and a curse. Lifesaving personnel and equipment and supplies can reach a region of the world just as easily as bombs and destructive viruses. High-priority passengers and cargo can be almost effortlessly transported wherever they are required, but as the events of September 11, 2001, demonstrated, the destructive power of an airplane in the wrong hands is enormous. The capability of human flight has wrought enormous changes to the human condition that we have never had to deal with before.

So how did we get to this point? I would suggest that the importance of aerospace technology was recognized even before the point that the Wright Brothers first flew. The Wrights were an exceptional pair of brothers who undertook cutting-edge cooperative research—building very well on what had already been accomplished—without any government involvement whatsoever.[4] No doubt they deserve sustained celebration for their contributions as lone inventors. As Americans we love that kind of ingenuity. It's the same kind of ingenuity we have seen throughout our history, as enterprising individuals from Eli Whitney to David Packard to Steve Jobs accomplished great technological breakthroughs.[5] Of course, at the same time the Wrights were undertaking their work, Samuel Langley had significant dollars from the military for his aerodrome and failed to achieve any success whatsoever. Consequently, the Wrights' story is a variation on the Horatio Alger phenomenon in American life.[6]

But it seems that in something as complex as aeronautics and space, for marked progress to take place significant investment beyond the capability of most individuals or even corporations is necessary. Various European governments invested heavily in the technology, and the result was that their airplanes far outstripped American capabilities by the time of World War I. It required significant U.S. government investment to advance aeronautics in this country. I want to trace that investment and to show the government role in fostering the technology of flight. I also want to note that since the 1930s a consistently positive balance of trade has been possible in the aerospace industry, and this is largely because of government investment. I will comment on the relationship between government investment and research and development (R&D) advances in aeronautics.

Lone Inventors versus Standing Armies of Researchers

If there is a folklore in the public mind about the history of engineering, it is the story of genius and its role in innovation. Americans love the idea of the lone inventor, especially if that inventor strives against odds to develop some revolutionary piece of technology in a basement or garage. There have been enough instances of this in U.S. history to feed this folklore and allow it to persist. The "Renaissance man" who with broad background can build a technological system from the ground up permeates this ideal. And it is an especially compelling vision for a nation of overachievers such as the United States.

This sort of inventor does exist in unique personalities and situations. Individualism and versatility characterizes this sort of engineering. Its quintessential expression in American history came in the work of Thomas A. Edison, whose many accomplishments in technology have been recognized as seminal to modern life. These same virtuoso expressions of engineering mastery have also been recognized in the work of U.S. rocket pioneer Robert H. Goddard, who spent most of his career as a lone researcher designing and testing rockets on a piece of isolated land near Roswell, New Mexico.[7]

This tale of the lone inventor, working in solitude, coming up with a hugely significant invention without either assistance or hindrance from others is especially appropriate when speaking of the Wright Brothers and the process of invention that led to the airplane.[8] At the same time, the Renaissance man has never been very common in the history of science and technology. Lone wolves of the kind that make up the folklore, reinforced by the reality of a few bona fide geniuses, are rare indeed, especially in the case of twentieth-century aerospace engineering. It is in part the logical consequence of the increasing depth of information required in the individual disciplines. No one person can master the multifarious skills necessary in the research, design, development, and building of a piece of aerospace hardware.

Accordingly, the vast majority of breakthroughs in aerospace technology have been the result not of the Wrights or of other independent inventors but of well-heeled R&D organizations usually funded by government largesse. Great leaps forward in technological capability almost always require significant long-term investment in R&D, R&D that does not have explicit short-term return to the "bottom line" and may not yield even long-term economic return. Without that large-scale investment in aerospace technology, however, the United States will become a second-class aerospace power. I would suggest that today we are on the road to becoming one. That is the result of two related misperceptions. First, since the end of the Cold War, with the belief that the United States stood alone as the world's only superpower, there has been an erosion in the amount of R&D investment that the federal government made in the technology of flight. That investment was no longer viewed as necessary for national defense. Second, many public officials believe—mistaken though they are—that aerospace technology is mature and that private industry should be able to sustain aerospace advances without significant government investment.

The Importance of U.S. Government R&D Investment

Although the Wrights invented the airplane without external assistance, we should not forget that other researchers were well funded by governments. For instance, in the United States the Smithsonian Institution secretary Samuel P. Langley received $100,000 from the War Department for his experiments at virtually the same time that the Wrights were inventing the airplane.[9] He undertook tests over the Potomac River just a few days before the Wrights' first successful flight at Kitty Hawk, tests that ended in discouraging failure. These discouraging results raised doubts over the propriety of spending government money on Langley's experiments.[10] Langley, it seems, was not only a scientist but also a government agent supplied with public money for the purpose of building a machine that the federal government might use as a military weapon.[11]

In the aftermath of the Langley fiasco there was no public clamor to devote more government money in the uncertain quest for elusive innovation. Nor was any government agency eager to begin a new project that might call to mind the previous waste. Langley's research organization was soon closed down and stood for years as "a silent monument to the political hazards of aeronautical research."[12] The ghost of Langley reminded people in Washington of the uncertain outcome of research aimed at innovation and the political vulnerability of government patronage for such research. For years after Langley's death in 1906, whenever aeronautics was mentioned in Congress, "some gray-haired Senator would whisper 'Remember Langley,' and that ended the talk about all things aerial at the Capitol."[13] As one influential report stated, even though the U.S. led the world into the air age, by 1915 it was "the only first class nation in the world that does not have an Advisory Committee or Board on Aeronautics, and one or more aeronautical laboratories devoted to the solution of problems which the manufacturer and practical aviator meet with in connection with the advancement of aerial flight."[14]

To a very real extent, the result of Langley's lack of success—I hesitate

Aeronautical researcher Samuel Pierpont Langley (1834–1906), at right, with chief mechanic and pilot Charles M. Manly. (NASA photo no. L-1990-04340)

to call it failure, for aerospace R&D is fundamentally a process of trial and error, build-test-retest, that ultimately leads to an advancing in the state of the art—served to stunt the development of the airplane in America. Although the United States invented the airplane, by the time of World War I it was obvious that the knowledge required to fly efficiently, to do much of anything with airplanes, had moved offshore and resided in Europe. This was true for two reasons.

First, European governments, as well as European industrial firms, tended to be more supportive of what might be called applied research. As early as 1909, the internationally known British physicist John William Strutt, Lord Rayleigh, was appointed head of the Advisory Committee for Aeronautics; in Germany, Ludwig Prandtl and others were beginning the sort of investigations that soon made the University of Göttingen a center of theoretical aerodynamics. Other programs were soon underway in France and elsewhere on the continent. As Smithsonian Institution secretary Charles D. Walcott wrote to Congress in 1915:

> As soon as Americans demonstrated the feasibility of flight by heavier-than-air machines, France took the matter up promptly, and utilized all the available agencies, including the army, navy, and similar establishments, both public and private. Large sums were devoted to the research work by wealthy individuals, and rapid advance was made in the art.
>
> Germany quickly followed, and a fund of one million seven hundred thousand dollars was raised by subscription, and experimentation directed by a group of technically trained and experienced men.

Walcott added that England and Russia followed suit, leading the way into the air age. He noted that when World War I began in 1914, about 1,400 military aircraft existed, of which only 23 belonged to the United States.[15]

Second, fueled by military necessity, the nations of Europe invested heavily in aeronautical technology and built flying machines of great complexity and significant capability, capability far outstripping anything that the United States could accomplish in the mid-1910s.[16] As a result of World War I the small, fast, maneuverable, and heavily armed fighter emerged as a major component of the battlefield. Although powered flight had been possible since 1903, as late as 1914 there was little understanding of what might be possible in warfare by extending battle into the third dimension with the use of the airplane. European combatants on both sides transformed airplanes into "warplanes," evolving these vehicles through five essential generations during the Great War. Each stage represented a major technological breakthrough and was dominated by one side of the belligerents. It also forced the development of fighter tactics to make aerial combat more effective. In turn, each stage was made obsolete by its successor, and while vestiges of aircraft types and tactics might remain throughout the rest of the war, they became less significant as they were surpassed by later developments.[17]

Similar progress in the United States was slow in coming. Aware of European activity, Secretary Walcott of the Smithsonian obtained funds to dispatch two Americans on a fact-finding tour overseas. Albert F. Zahm taught physics and experimented in aeronautics at Catholic University in Washington, D.C., while Jerome C. Hunsaker, a graduate of the Massachusetts Institute of Technology, was developing a curriculum in aeronautical engineering at the institute. Their report, submitted to Congress early in 1915, emphasized the galling disparity between European progress and American inertia. The visit also established European contacts that later proved valuable to the National Advisory Committee for Aeronautics.[18]

That American flying lagged far behind European aviation was particularly galling to many aviation advocates in the United States, the home of the Wright Brothers. The European success not only was documented in a growing record of achievements but was also underscored by a lack of organized research in the United States. The best the United States could do for aviation in the war was build the Liberty Engine.[19] Indeed, as late as 1914 the United States stood fourteenth in total funds allocated by nations to military aviation, far behind even Bulgaria and Greece.

Well into the twentieth century there was in the United States little appreciation for scientific and technical research and even less inclination to allocate government funding for such an uncertain activity. But because of this truly poor response at the time of World War I, the United States decided to create the National Advisory Committee for Aeronautics (NACA) in 1915. Sentiment for some sort of center of aeronautical research had been building for several years. At the inaugural meeting of the American Aeronautical Society, in 1911, some of its members discussed a national laboratory with federal patronage. But the American Aeronautical Society's dreams were frustrated by bureaucratic in-fighting and questions about the appropriateness of government investment in technological R&D. Something drastic had to be done, and the stalemate ended with the passage of enabling legislation for the NACA on March 3, 1915, as a rider to the Naval Appropriations Act.[20]

The NACA had emerged by 1920 as a small, loosely organized, and elitist nonbureaucracy that provided aeronautical research services on an equal basis to all. An exceptionally small headquarters staff in Washington—so

The National Advisory Committee for Aeronautics (NACA) was created in 1915 to help the United States return to preeminence in aeronautics. Its laboratories at Langley Field, Virginia, provided new knowledge supporting the continuous improvement in the performance, efficiency, and safety of American aircraft. At this meeting in 1938, Dr. Joseph S. Ames, president emeritus of Johns Hopkins University, was reelected chairman, and Dr. Vannevar Bush, president-elect of the Carnegie Institution of Washington, was elected vice chairman. *Left to right*: Hon. C. M. Hester, administrator, Civil Aeronautics Authority; Captain S. M. Kraus, U.S.N.; Brig. Gen. A. W. Robins, chief, Materiel Division, Army Air Corps; Dr. L. J. Biggs, director, National Bureau of Standards; Dr. E. P. Warner; Dr. Orville Wright; Ames; Dr. C. J. Abbot, secretary, Smithsonian Institution; J. F. Victory, secretary; John J. Ide, NACA European representative; Rear Adm. A. B. Cook, U.S.N., chief, Bureau Aeronautics Authority; Bush; Dr. J. C. Hunsaker; Dr. G. W. Lewis, director of Aeronautical Research. Absent: Col. Charles A. Lindbergh and Maj. Gen. H. "Hap" Arnold, chief, U.S. Army Air Corps. (NASA photo no. 1938NACA)

small in fact that it could be housed in a corner of the Navy Building—oversaw the political situation and secured funding for research activities. A committee of appointees who served without pay, making it one of the most untraditional organizations in Washington, governed it.[21] Moreover, its small Langley Memorial Aeronautical Laboratory, with only 100 employees by 1925, collocated with the Army Air Corps near Hampton, Virginia, and conducted pure research, mostly related to aerodynamics, receiving advice and support from the headquarters director of research, George W. Lewis.[22] Those who remember the agency between the two world wars speak of it in idyllic terms. They were able to develop their own research programs along lines that seemed to them the most productive, handle all test details in-house, and carry out experiments as they believed appropriate. They issued "Technical Notes" partway through many investigations containing interim results and "Technical Reports" with major research conclusions at the end of the effort. No one and no political issue, the old NACA hands recollected, infringed upon their work. Thus they believed that partly for this reason the organization was the premier aeronautical research institution in the world during the 1920s and 1930s.[23]

At the same time, the U.S. military began to perceive, albeit reluctantly, the significance of aircraft in the conduct of modern warfare. This led to a rapid expansion of federal spending for aeronautics. When the United States entered World War I in April 1917, the process accelerated and the government made significant investments in the aviation industry and expanded procurement of military aircraft from 350 on order to an ambitious program to develop and produce 22,000 modern military aircraft by July 1918. Though U.S. manufacturers did not achieve this goal—they delivered 11,950 planes to the government during the war—the massive military appropriations gave the nascent aviation industry a huge boost.[24]

Even with its efforts during World War I the United States refused to make a significant sustained investment in aeronautical technology thereafter. Many people found reason to question these government expenditures virtually as soon as the war was won. The comments of Glenn L. Martin come to mind in considering this situation: "Only a failure of the United States government to place orders with our successful airplane designers and builders will cause our aircraft industrial strength to slip back into the position it occupied three years ago. A vital point is being overlooked by the American people. It is immediately evident that the industrial strength of the United States must be at the war strength all the time. . . . The government must stimulate and aid in the application of aircraft industrially, and also aid in foreign trade, furnishing sufficient outlet for industrial aviation and guaranteeing a continuity of production at the required rate." Martin probably had it pretty much right when he said this in 1920. He complained that the government required a strong aerospace industry as a guarantee of national defense and should put money into it as a matter of industrial policy.[25] Only with war clouds gathering for what became World War II did the federal government become a major patron of science. Again, only because of perceived needs in national defense.

To a very real extent the United States had to be jarred out of lethargy in supporting aeronautical R&D. In 1936 John J. Ide, the NACA's European representative since 1921, fired off an alarming report on the state of aeronautical science on that continent. Ide, the sometime technology expert, sometime intelligence analyst, and sometime expatriate socialite, reported on greatly increased aeronautical research activities in Great Britain, France, Italy, and especially Germany. He observed that new and quite modern wind tunnels were being erected to aid in the development of higher-performing aircraft and suggested that the NACA review its own equipment to determine whether it met contemporary demands.[26] Charles A. Lindbergh, a NACA executive committee member living in seclusion in England, confirmed Ide's report in a May 1936 letter to committee chairman Joseph S. Ames.[27] Because of this, Lewis inserted a deft warning to the government in the NACA's 1936 annual report. Commenting on the arms race in Europe, Ames suggested that "increased recognition abroad of the value and of the vital necessity of aeronautical research has led to recent tremendous expansion in research programs and to multiplication of research facilities by other progressive nations. Thus has the foundation been laid for a serious challenge to America's present leadership in the technical development of aircraft."[28]

Because of these developments, in September–October 1936 George W. Lewis traveled to Europe via the *Hindenberg* to learn about aeronautical development firsthand. He toured with Dr. Adolph Baeumker, the German government's R&D head, several aeronautical facilities in Nazi Germany and was both impressed and disquieted by their activities. He learned that Luftwaffe chief and Adolf Hitler's stalwart Hermann Goering was "intensely interested in research and development" and greatly expanded it. He decentralized it at three major stations: one for research on new aircraft, one for fundamental research without application to specific aircraft designs, and one for the development of new propulsion systems. Lewis remarked: "It is apparent in Germany, especially in aviation, that everyone is working under high pressure. The greatest effort is being made to provide an adequate air fleet. Every manufacturer is turning out as many airplanes as possible, and the research and development organizations are working on problems that have an immediate bearing on this production program." To ensure American competitiveness in aviation, Lewis advised, the nation should immediately start the expansion of R&D capabilities.[29]

These epistles of warning brought moderate action. In 1936 the Congress funded construction of another wind tunnel at Langley and the lengthening of a tank used for seaplane research. It provided the impetus for additional funding through a special "Deficiency Appropriation Act" to fund the construction of new facilities, all because of war sentiment in Europe.[30]

Once the United States got into a shooting war against the Axis, however, the treasury opened and funds flowed to advance the frontiers of flight. War was good business, proving a great boon both to advancement of the technology of flight and to the industry that developed it. World War II was the only time the aircraft industry really enjoyed exceptional success. As an example of what all the industry enjoyed, Douglas Aircraft made the decision in 1942 to expand its manufacturing greatly, opening plants in El Segundo, Santa Monica, and Long Beach, California; Tulsa and Oklahoma City, Oklahoma; and Chicago. By 1944, Douglas employed a total of 160,000 personnel, compared to less than 2,000 when it started building the DC-3 aircraft in the 1930s. By 1945, Douglas Aircraft Company had built 29,385 aircraft in four years and had become one of the largest aircraft firms in the nation.[31]

This success did not last long. Virtually every study of the aerospace industry in World War II speaks of its devastation with demobilization. From late 1943 on the Joint Chiefs of Staff was sure of eventual victory and began to trim defense contracts for aircraft and other war materiel. Between 1942 and 1945 an average of 138 airplanes rolled off the assembly lines each day. The result of this activity was that 1944 became the peak year of production, with 95,272 aircraft delivered, and another 48,912 delivered in the last year of the war. This was nearly half of the total of 316,495 aircraft produced during World War II. But this investment declined significantly, and in 1946 production was just over 36,000 aircraft, the vast majority of which were commercial aircraft purchased by air carriers after years of waiting for newer models. In 1947 the production declined by more than half of its 1946 level. Employment in the industry demonstrated similar declining trends, losing nearly 75 percent of its workforce during the last two years of the war and the first year of peace.[32]

Because of these trends, the industry became much more dominated by a few mega-firms while less lucrative companies went out of business or merged with the giants of the industry. Indeed, the mobilization and demobilization of World War II *made* the aerospace manufacturing concerns that commanded the industry in the Cold War era. It also ensured the demise, merger, or consolidation of several weaker firms. The top fifteen manufacturers of aircraft, by numbers of aircraft delivered, included firms that are familiar to most of us. The top ten of these eclipsed all others in the postwar era, and some of the smaller ones shown in table 1—notably Ford and Taylor—got out of the aerospace business altogether. And these were exceptionally healthy firms throughout the wartime period. Defense industry procurement patterns, at least for the World War II period, placed a heavy reliance on contracting with large, established firms that had a

TABLE I. *Top Fifteen Aeronautical Manufacturers during World War II*

Manufacturer	AAF Accepted	Navy Accepted	Percent of Total
North American	41,839	0	17
Consolidated	27,634	3,296	13
Douglas	25,569	5,411	13
Curtiss	19,703	6,934	11
Lockheed	17,148	1,929	8
Boeing	17,231	291	7
Republic	15,663	0	6
Bell	12,941	1	5
Martin	7,711	1,272	4
Beech	7,430	0	3
Ford	6,792	0	3
Fairchild	6,080	300	3
Piper	5,611	330	3
Cessna	5,359	0	2
Taylor	1,940	0	1

long history in the field. In essence, those with longstanding ties to the government and with exceptional resources already available got much richer. Those without those attributed may have survived but not in as lucrative a manner. All that would change with demobilization, however, as every firm suffered enormous reductions in government contracts as soon as the war ended.[33]

Post–World War II demobilization devastated the aircraft industry and only the beginnings of the Cold War seemed to help it recover from a post-war depression. As historian Charles D. Bright appropriately concluded: "By the latter part of 1947 the industry had run out of money, ideas, courage, and hope."[34] Industry leaders began lobbying the government for relief, and relief was not long in coming because rising tensions between the United States and the Soviet Union brought renewed military procurement in the latter 1940s. Despite the very real demobilization in 1945–47, the United States' Cold War rivalry with the Soviet Union precipitated an arms race both desperate and tragic during the period between 1948 and 1989.[35] With it came an enormous expansion of military aerospace activities. Couple this defense increase—which also had an increasingly important component of rocket development and space systems beginning in the 1950s—with the rise of commercial air travel, especially after the advent of the jet age in

The Boeing 777, the most recent of the American-built wide-body airliners, is poised to compete with the Airbus A-340 built in Europe. (NASA photo no. 1995-694)

the 1950s and early 1960s, and the aerospace industry enjoyed a relatively stable existence for the next forty years.

At the same time, the industry could be viewed as a tragic creature. As Charles Bright commented: "The aerospace industry since Would War II, then, is the story of an increasingly capital-intensive business whose manufacturing function has steadily declined, and its product cost has risen so high that it has almost priced itself out of its market. It is an industry which has . . . not been able to diversify adequately to shelter itself from its captive status in relation to its dominant customer, the government." The industry's reliance upon a dominant client, the U.S. government and its Cold War aerospace needs, has ensured an industry that is over capacity and inefficient.[36]

In the last decade the situation has gotten worse. Market share in all major aerospace sectors has declined. In commercial space launch, which the United States dominated until the advent of the Ariane launcher built by the European consortium Aérospatiale in the early 1980s, the market collapsed for the United States in the aftermath of the *Challenger* and has not recovered. In passenger aircraft, Airbus Industrie's analysis suggests that to satisfy an expected average annual growth rate in passengers and cargo of

5.2 percent during the next ten years, the number of passenger aircraft in service will increase from some 10,350 in 1999 to 14,820 in 2009 and 19,170 in 2019. Satisfying that requirement is Airbus's objective for the indefinite future. And they are showing remarkable staying power there. At the Paris air show this last summer, they nailed down 110 orders for new aircraft to Boeing's less than 40.[37]

Commercial aviation is quickly evolving, both technologically and economically. Pacing world economic growth, air travel is making profound changes to provide better service at lower cost. The world fleet is currently three times as large as it was twenty years ago, and today's fuel-efficient jetliners offer airlines greater choice in range, passenger capacity, and operating economics. More than half of all flights are made on routes bounded by few, if any, regulatory constraints. Flag carriers are privatizing, "open skies" agreements are replacing bilateral air service agreements, and global alliances are on the rise. As a result, airlines now have unprecedented flexibility to pursue strategies that meet the needs of the next century's global community. They will not purchase American aircraft because we want them to do so. Indeed, with the growth of overseas carriers, there is an ideological reason to refrain from buying American if for no other reason than to thumb their noses at the last remaining superpower. Accordingly, U.S. technology must be clearly superior.

Possible Reasons for Stagnation

Between the 1960s and the 1990s, the share of the market enjoyed by American aerospace manufacturers has fallen sharply as foreign corporations—either private or state-run—gained greater portions of the market. In 1986, for example, U.S. high-technology imports exceeded exports for the first time. The aerospace industry was one of the only remaining fields with a trade surplus, 90 percent of which was attributable to the sale of aircraft and aircraft parts. Compared to an overall U.S. trade deficit in manufactured goods of $136 billion in 1986, the aerospace industry had a surplus of $11.8 billion. But the U.S. lead in aerospace was shrinking rapidly. In 1980 the U.S. market share of large civil transport sales was 90 percent. By 1992 that percentage had dropped to 70 percent and was in danger of falling even further. The lead in the commuter aircraft market had already been lost. During the 1990s the U.S. lost its lead in the space launch market as well.

The reasons for this loss of market share are complex. From my perspective, the aerospace community is in the doldrums for the following five key reasons.

First, there are the inherent difficulties of the aerospace marketplace.

As aerospace technology became more complex and expensive, it also became more difficult for individual companies to shoulder the entire financial burden of researching and developing new technology and products themselves. Aerospace technology has always been a marginal economic enterprise in all its myriad permutations. Aerospace manufacturers literally bet the company on a new design because of the enormous cost associated with developing an aircraft or rocket. Malcolm Stamper, former president of Boeing Aircraft Corporation, remarked that "Locating the break-even point is like finding a will-o'-the-wisp."[38] Knowledgeable individuals have concluded that it is not until 20 to 35 production aircraft have actually been manufactured that production costs become predictable. For rockets and other space technology, which does not have large production runs, the economics of manufacturing are even more problematic.[39]

Second, American aerospace executives were too often complacent in maintaining their competitive technological edge. Aerospace corporations, like a lot of other organizations, have a decided "not invented here" syndrome. Ideas emanating from beyond the recognized corporate structure too often get short shrift. I can cite numerous examples ranging from Northrop Aircraft Corporation's hesitancy to embrace retractable landing gear in the 1920s to Boeing's rejection of the so-called "glass cockpit" technology in the 1980s. While the "glass cockpit" offered cutting-edge avionics displays, this American-made technology found its first use at Airbus Industrie in Europe. Airbus made it a centerpiece of its newest generation of transports, in the process helping itself compete more effectively in the marketplace. Losing market share as a result, Boeing raced to adopt the new technology into its own designs.[40]

Third, there has been a lack—indeed a celebration of that lack—of coherent industrial policy in the United States. Because of the nature of our republic and citizenry, Americans have been loath to adopt anything approaching a centralized, rational, long-term industrial policy because of its inherently undemocratic and remarkably technocratic and elitist characteristics. Such a policy would recognize that the health of the American aerospace industry—and perhaps other industries—was important both for national security and economic competitiveness. Accordingly, it is something of a truism to suggest that anything that has passed for aerospace policy in this nation, a subunit of that largely nonexistent industrial policy, has been both ad hoc and expeditious.[41]

Fourth, there has been the success of industrial policy by other nations aimed at securing greater market share for non-U.S. aerospace companies. Their governments, especially in command economies such as the communist bloc, often directly subsidized their national manufacturers. There is no question but that one of the major reasons for the European community to invest in aerospace technology has been to wrest economic market share from the United States. The Europeans have developed an industrial policy aimed at this goal, and they have been quite successful.[42] The Japanese, for another instance, have long pursued policies, and directly subsidized key industries, to help move the fruits of basic research into the marketplace for the purpose of gaining economic advantage vis-à-vis the United States.[43]

Finally, a major problem of the aircraft business was its cyclic nature, leading to boom and bust periods. Complicated by the enormous infrastructure necessary to support the design and manufacture of aircraft, these firms were exceptionally limited in their markets and their capabilities. President Ronald Reagan's science adviser noted in 1982 that "aircraft are now the dominant common carrier for inter-city travel, and the safety and control of that travel are a federal responsibility."[44] He recommended pressing hard for government support of basic research that could then be transferred to American private firms.

A Modest Proposal?

So how do we get out of the current doldrums? There are many things one could suggest, ranging from globalization to a new European-like attitude toward cooperation. Even so, I shall concentrate on the ensuring of the technical superiority of American aerospace technology. There is a direct correlation between governmental R&D investment and excellence in technology. Since the 1960s the percentage of investment by the United States in aerospace technology has stabilized at about 1 percent of the federal budget. The aerospace corporations and some universities invest in R&D as well, but that is a decidedly small amount and at least in the case of the private sector limited almost entirely to short-term research. So let us do one thing that will yield a positive result. The American nation should decide to double its investment in aerospace R&D during the first decade of the twenty-first century. This is fully within the bounds of our capability, and it will help assure American economic, military, and cultural competitiveness in the new century. Not to do so would be to turn our backs, as we did in the 1900s on the legacy of the Wrights and their enormously significant invention.

Let me conclude with a comment made famous by Tom Hanks in the baseball film *A League of Their Own*—and no, it is not "There's no crying in baseball"—but it is one that is equally insightful. He told the Geena Davis character that what they were doing was hard, and of course it had to be because otherwise "everyone would do it." Like baseball, flying is hard and flying with the latest technology is harder still. If it were not, everyone would do it. The United States is a nation with the high quality of economic, political, social, and knowledge base necessary to bring forward the next generation of aerospace technology. All it takes is will. All we have to do is make the decision to do so and follow that with the investment necessary to further the frontiers of flight. No one knows where that might lead, but I believe it will lead to a hypersonic plane, jumbo jets unlike any we have ever seen before, and trips beyond this planet. And we will do so in a safe and environmentally friendly manner.

NOTES

1. Stephen E. Ambrose, *Undaunted Courage: Meriwether Lewis, Thomas Jefferson, and the Opening of the American West* (New York: Simon and Schuster, 1996), 52.

2. I do not want to suggest that this is absolutely the case; the jury is still out. It is possible to also make an argument for anesthesiology, blood transfusion, antibiotics, radio, electronics, or any number of other advances that have had a significant direct impact on individual lives. To understand the impact of flight on the century, one must not only consider war and commerce arguments but also deeply analyze the manner in which the ability to fly has fundamentally altered everyday lives. From that perspective, the airplane transformed humanity from ground-hugging mortals to creatures who will one day roam the universe. In that way it had a revolutionary impact.

3. Jules Verne, *Le Tour du monde en quatre-vingts jours* (*Around the World in Eighty Days*), English translation by George Makepeace Towle (London: Porter and Coates, 1873).

4. That is certainly the position of the preeminent historian to write about the Wright Brothers. See Tom D. Crouch, *The Bishop's Boys: A Life of Wilbur and Orville Wright* (New York: W. W. Norton, 1989).

5. Thomas P. Hughes, *American Genesis: A Century of Invention and Technological Enthusiasm, 1870–1970* (New York: Viking Press, 1989), 47–52, 92–95, 214–16.

6. Norriss S. Hetherington, "The Langley and Wright Aero Accidents: Two Responses to Early Aeronautical Innovation and Government Patronage," in Roger D. Launius, ed., *Innovation and the Development of Flight* (College Station: Texas A&M University Press, 1999), 18–51.

7. See Milton Lehman, *This High Man* (New York: Farrar, Straus, 1963); Neil Baldwin, *Edison: Inventing the Century* (Chicago: University of Chicago Press, 2001); Paul Israel, *Edison: A Life of Invention* (New York: John Wiley & Sons, 1998).

8. This heroic aspect of the Wrights is stated in two very fine books on the brothers: Crouch, *Bishop's Boys*, and Peter L. Jakab, *Visions of a Flying Machine: The Wright Brothers and the Process of Invention*, Smithsonian History of Aviation Book Series (Washington, D.C.: Smithsonian Institution Press, 1990).

9. Archibald D. Turnbull and Clifford L. Lord, *History of United States Naval Aviation* (New Haven, Conn.: Yale University Press, 1949), 1–3; Norriss S. Hetherington, "Langley's Aerodrome," *W.W. I Aero* 131 (Feb. 1991): 3–16; Samuel P. Langley, *Langley Memoir*

on Mechanical Flight, part 2, *1897 to 1903*, edited by Charles M. Manly (Washington, D.C.: Smithsonian Institution, 1911).

10. "Langley Wants More Money," *New York Times*, Sept. 20, 1903, 10.

11. Letter to the editor, *New York Times*, Aug. 30, 1903, 9.

12. Alex Roland, *Model Research: The National Advisory Committee for Aeronautics 1915–1958*, NASA Special Publication 4103 (Washington, D.C.: U.S. Government Printing Office, 1985), 1:2.

13. "Aeroplanes to Be Put to a Government Test," *New York Times*, June 28, 1908, part 5, 8.

14. Charles D. Walcott, secretary of the Smithsonian Institution, to Sen. Benjamin R. Tillman, chairman of the Committee on Naval Affairs, "Memorandum on a National Advisory Committee for Aeronautics," Feb. 1, 1915, reprinted in Roland, *Model Research*, 2:593–97; the quote is from 594–95.

15. Ibid., 595.

16. John H. Morrow Jr., *German Air Power in World War I* (Lincoln: University of Nebraska Press, 1982), 3–13; John H. Morrow Jr., *The Great War in the Air: Military Aviation from 1909 to 1921*, Smithsonian History of Aviation Book Series (Washington, D.C.: Smithsonian Institution Press, 1993).

17. Richard P. Hallion analyzes this transformation in his masterful *Rise of the Fighter Aircraft, 1914–1918* (Baltimore, Md.: Nautical and Aviation Press, 1984).

18. Walcott to Tillman. See also William F. Trimble, *Jerome C. Hunsaker and the Rise of American Aeronautics* (Washington, D.C.: Smithsonian Institution Press, 2002).

19. Rated at 410 horsepower, this engine weighed only two pounds per horsepower, far surpassing in both power and efficiency similar types of engines mass-produced in England, France, Italy, and Germany at that time. During the war, 20,478 Liberty 12s were produced by Packard, Lincoln, Ford, General Motors (Cadillac and Buick), Nordyke, and Marmon. They were used primarily in U.S.-built DH-4s, the only American-made airplane to get into combat over the Western Front. After the war, the Air Corps used the engine for more than a decade in numerous types of airplanes. Some, which were sold to civilians as war surplus, were illegally used in speedboats for "rum running" during the Prohibition era of the 1920s; others were even used in Russian and British tanks during World War II.

20. The story of the NACA's creation is told in Roland, *Model Research*, 1:1–25; Roger E. Bilstein and Frank Anderson, *Orders of Magnitude: A History of the NACA and NASA 1915–1990*, NASA Special Publication 4406 (Washington, D.C.: U.S. Government Printing Office, 1989), chapter 1.

21. The NACA was a unique agency of the federal government throughout the interwar period, but it must be acknowledged that much of Franklin D. Roosevelt's New Deal government was also loosely structured. His constant use of nongovernment employees for many tasks is well documented. See Robert Dallek, *Franklin D. Roosevelt and American Foreign Policy, 1932–1945* (New York: Oxford University Press, 1979).

22. See also George W. Gray, *Frontiers of Flight: The Story of NACA Research* (New York: Alfred A. Knopf, 1948); Arthur L. Levine, "United States Aeronautical Research Policy, 1915–1958: A Study of the Major Policy Decisions of the National Advisory Committee for Aeronautics" (Ph.D. diss., Columbia University, 1963); Ira H. Abbott, "A Re-

view and Commentary of a Thesis by Arthur L. Levine entitled United States Aeronautical Research Policy, 1915–1958: A Study of the Major Policy Decisions of the National Advisory Committee for Aeronautics," Apr. 1964, HHN-36, NASA Historical Reference Collection, NASA History Office, Washington, D.C.

23. William Phillips, "Recollections of Langley in the Forties," n.d., oral presentation, History Collection, Langley Research Center, Langley, Va.; oral histories with Paul E. Purser, Walter S. Diehl, and W. Kemble Johnson by Michael D. Keller, all in NASA Historical Reference Collection, NASA History Office, Washington, D.C.

24. Roger D. Launius, "Aerospace," in Donald C. Bacon, Roger H. Davidson, and Morton Keller, eds., *The Encyclopedia of the United States Congress* (New York: Simon and Schuster, 1995), 1:11–13.

25. Quoted in Ellen M. Pawlikowski, "Surviving the Peace: Lessons Learned from the Aircraft Industry in the 1920s and 1930s," 1, master's thesis, Industrial College of the Armed Forces, National Defense University, Fort McNair, Washington, D.C.

26. "John Jay Ide, 69, Air Pioneer, Dies," *New York Times*, Jan. 13, 1962; NACA Executive Committee Minutes, Mar. 3, 1936, 8–9, Record Group 255, National Archives and Records Administration, Washington, D.C.; Gray, *Frontiers of Flight*, 22–23.

27. Charles A. Lindbergh to Dr. Joseph S. Ames, May 20, 1936; John F. Victory to Charles A. Lindbergh, June 18, 1936, both in NASA Historical Reference Collection, NASA History Office, Washington, D.C.

28. *Twenty-Second Annual Report of the National Advisory Committee for Aeronautics, 1936* (Washington, D.C.: U.S. Government Printing Office, 1937), 3.

29. George W. Lewis, "Report on Trip to Germany and Russia, September–October, 1936," NASA Historical Reference Collection, NASA History Office, Washington, D.C.

30. "Some Important Facts Regarding Expansion of NACA Research Facilities and War-time Status of NACA," Jan. 17, 1946, NASA Historical Reference Collection, NASA History Office, Washington, D.C.; A. Hunter Dupree, *Science in the Federal Government: A History of Policies and Activities to 1940* (Cambridge, Mass.: Harvard University Press, 1957), 363.

31. Frank Cunningham, *Sky Master: The Story of Donald Douglas and the Douglas Aircraft Company* (Philadelphia: Dorrance, 1943), 57–65, 98–99; Denis Mulligan, *Aircraft Manufacture in Chicago* (Chicago: Aircraft Manufacturers Association, 1939), 32–33; Jeffrey A. Fadiman, "Dreamer of the Drawing Board: Donald Wills Douglas (1892–1981)," in Ted C. Hinckley, ed., *Business Entrepreneurs in the West* (Manhattan, Kans.: Sunflower University Press, 1986), 83–93.

32. See Irving B. Holley Jr., *Buying Aircraft: Materiel Procurement for the Army Air Forces* (Washington, D.C.: Office of the Chief of Military History, 1964), 550–61.

33. Roger D. Launius, "End of a Forty Year War: Demobilization in the West Coast Aerospace Industry after the Cold War," *Journal of the West* 36 (July 1997): 85–96.

34. Charles D. Bright, *The Jet Makers: The Aerospace Industry from 1945 to 1972* (Lawrence: Regents Press of Kansas, 1978), 13.

35. On the Cold War, generally three books stand out as excellent syntheses: Walter LeFeber, *America, Russia and the Cold War, 1945–1975* (New York: John Wiley & Sons, 1976); Stephen E. Ambrose, *Rise to Globalism: American Foreign Policy, 1938–1980* (New

York: Penguin, 1980); and Adam B. Ulam, *The Rivals: America and Russia since World War II* (New York: Viking Press, 1971).

36. Bright, *Jet Makers*, 22.

37. *Global Market Forecast, 2000–2019* (Blagnac, France: Airbus Industrie, 2000), 4–5.

38. Quoted in John Newhouse, *The Sporty Game* (New York: Alfred A. Knopf, 1982), 4.

39. Senate Committee on Armed Services, *Hearings, Weapons Systems Acquisition Process*, 92d Cong., 1st sess., 1972, 152.

40. See Walter G. Vincenti, "The Retractable Airplane Landing Gear and the Northrop 'Anomaly': Variation-Selection and the Shaping of Technology," *Technology and Culture* 35 (Jan. 1994): 1–33; Lane E. Wallace, *Airborne Trailblazer: Two Decades with NASA Langley's Boeing 737 Flying Laboratory*, NASA Special Publication 4216 (Washington, D.C.: U.S. Government Printing Office, 1994), 26–39.

41. The problem of aerospace policy is related to the larger problem of industrial policy. For discussions of this issue, see Malcolm L. Goggin, ed., *Governing Science and Technology in a Democracy* (Knoxville: University of Tennessee Press, 1986); Manfred Stanley, *The Technological Conscience: Survival and Dignity in an Age of Expertise* (New York: Free Press, 1978); Sylvia Doughty Fries, "Expertise against Politics: Technology As Ideology on Capitol Hill, 1966–1972," *Science, Technology, and Human Values* 8 (Spring 1983): 6–15; David McKay, *Domestic Policy and Ideology: Presidents and the American State, 1964–1987* (New York: Cambridge University Press, 1989).

42. See, for example, the Convention for the Establishment of a European Space Agency (CSE.CD(73)19. rev. 7: Paris, May 30, 1975). Article VII (I) (b) states: "The industrial policy which the Agency is to elaborate and apply by virtue of Article II (d) shall be designed in particular to: . . . b) improve the world-wide competitiveness of European industry by maintaining and developing space technology and by encouraging the rationalisation and development of an industrial structure appropriate to market requirements, making use in the first place of the existing industrial potential of all Member States."

43. Norman E. Bowie, *University-Business Partnerships: An Assessment* (Lanham, Md.: Rowman & Littlefield, 1994), 19.

44. Quoted in ibid., 24.

CHAPTER 3
INNOVATION IN FLIGHT FROM
THE PERSPECTIVE OF EUROPE

The task of reflecting on innovation in flight from a European perspective is quite broad. Accordingly, I shall look only at certain aspects of it. My main question is, Was there anything distinctly "European" in the development of aviation? What is—and this is closely related to that question—the relationship between some innovations in aviation originating in Europe and their later application in the United States?[1] As examples, I shall briefly deal with the origins of aerodynamics, of jet propulsion, and of the European Airbus. There are other relevant themes which would have merited treatment, like the development of all-metal construction or the practical foundations of the space age. But these topics have been treated abundantly in the literature[2] so that I will not go into them here. I will argue that not too much should be made of distinctly European features in the history of aviation. Generally engineers in the United States have managed to catch up on some—short-lived—European leads and, on their part, have established a lead after World War II. This lead has, however, been challenged since the 1970s. I will confine myself to aeronautics in this essay, leaving astronautics out, which is a huge topic in itself.

Europe and the Origins of Aviation

When one looks at the beginnings of heavier-than-air flying machines, it soon becomes clear that one can hardly speak of divergent national styles of research, development, and design of aircraft. Although there were different centers of development in Europe and North America, the "center of gravity"[3] shifted rapidly, and differences in aeronautical know-how were rather small.

The engineering science of aerodynamics came into existence in the late nineteenth century. There was progress of research in various Western industrialized countries, but also in Russia, and it never took long before researchers managed to catch up with those in the lead. Government funds, for example in university research institutes or other government-funded organizations, as well as private money—for example the Guggenheim Fund to boost aeronautical research—were essential in creating the personal and material conditions for advancement in this field.

In the last decade of the nineteenth century the German Otto Lilienthal carried out the first systematic series of experimental measurements of the aerodynamic properties of cambered airfoils.[4] His more than 2,000 successful glider tests caught the attention of the Russian researcher Nikolai E. Joukowski, head of the department of mechanics at the University of Moscow. Joukowski visited Lilienthal in Berlin, brought a Lilienthal glider with him, and then examined the dynamics and aerodynamics of flight mathematically. His special interest was in the calculation of lift. In 1906 he demonstrated the calculation of lift on an airfoil with mathematical exactness. Another researcher, like Joukowski motivated by the flying successes of Lilienthal, but independent of Joukowski's work, was Wilhelm Kutta. In 1902 Kutta handed in his Ph.D. thesis at the University of Munich on the subject of aerodynamic lift. With their "Kutta-Joukowski Theorem," both researchers made possible a mathematical and scientific understanding of the generation of lift.[5]

But in 1903 the center of gravity in experimental aerodynamics was, as John D. Anderson Jr. points out in his admirable *History of Aerodynamics*, in the United States. Langley's work at the Smithsonian Institution between 1887 and 1903, wind tunnel tests carried out by the Wright Brothers during 1902–3 and Albert Zahm's work at the Catholic University in Washington, D.C., were of special importance. Yet this state of affairs did not continue for long. After 1903 the center of gravity rapidly shifted from the United States to Britain and Europe.[6]

Whereas Joukowski and Kutta had concentrated on lift, Ludwig Prandtl obtained a major breakthrough in the understanding and prediction of aerodynamic drag.[7] Prandtl had studied applied mechanics at the Technical University of Munich, wrote his doctoral dissertation on the strength of materials, made his first experiments with air currents at the M.A.N. works in Nürnberg, and, in 1904, moved to Göttingen University as a professor of applied mechanics.[8] In that year he also gave a pathbreaking paper at the International Congress of Mathematicians in Heidelberg, in which he stated "that for fluids of small viscosity, such as air or water, the viscosity will substantially affect the flow only in a thin layer adjacent to the surface. Outside this layer, viscosity can be neglected and the flow can be described to a higher degree of accuracy by the mechanics of nonviscous fluids. . . . The small thickness of the boundary layer permits essential simplifications in the equations of motion of a viscous fluid, so that the problem of frictional drag becomes accessible to mathematical analysis."[9]

Prandtl had many highly gifted students who continued where he had left off, closely collaborated with him, and brought the teachings of the "Göttingen School" to all parts of the world. Theodore von Kármán was one of them.

In 1906 von Kármán, who was to become one of Prandtl's most prominent disciples, came to Göttingen after graduating from Budapest University. In 1908 he obtained his Ph.D. with Prandtl on research into what later became known as the "Von Kármán Vortex Sheet."[10] In his thesis and later publications he provided a mathematical explanation for the stable, asymmetrical arrangement of vortices shed by an object subject to viscous flow. He also connected the momentum carried by the vortex system with the drag and demonstrated how the creation of such a vortex system can represent the mechanism of wake drag.[11] In 1914 the German aviation engineer and industrialist Hugo Junkers suggested to von Kármán, then professor of mechanics and aerodynamics at the Aachen Institute of Technology and aeronautical adviser to both Junkers and Count Zeppelin, to mathematically compute the aerodynamic properties of airfoils of any arbitrary shape. This led to the development of the "Kármán-Treffz airfoil," which was of great practical relevance to airfoil design.[12]

The Wright Brothers made important contributions to control systems and propeller design, but also to airfoil theory. In their work they used a wind tunnel outlined in an 1871 publication by the Englishman Francis Wenham.[13] In the late nineteenth and the early years of the twentieth century Gustave Eiffel of Eiffel Tower fame built a large wind tunnel of advanced design on the Champs de Mars in Paris.[14] Eiffel was also the first to carry out wind tunnel tests using models of complete airplanes. Moreover, he showed conclusively the correspondence between those tests and the actual performance of an airplane in flight. At Göttingen University Ludwig Prandtl and his collaborators built a closed-circuit return-flow tunnel in 1908. The closed-circuit design required less power, provided a more uniform airflow, and also permitted pressurization and humidity control.[15]

During World War I and immediately afterward researchers in aerodynamics achieved significant advances in the calculation of airfoil and wing aerodynamics. Combining the results of his investigation of lift with Kutta and Joukowski's circulation theory, Prandtl provided a comprehensive theoretical treatment of the aerodynamic properties of a finite wing, i.e., a real wing with wing tips. With this "lifting line theory" he made possible the calculation of lift and induced drag.[16] From now on a wing could be scientifically rather than empirically designed and its aerodynamic lift and resulting drag precisely calculated.[17]

A British Handley Page Heyford heavy bomber being "attacked" by a Hawker-Fury fighter during air exercises in 1937. (NASA photo)

Prandtl and his research group at Göttingen managed to bridge the gap between academic research and airplane design. But the transfer of this knowledge into airplane design was slow: It took twenty years for Prandtl's boundary layer theory to become appreciated by aircraft designers.[18]

During World War I and later, cooperation between European aeronautical laboratories extended to laboratories in the United States and Japan with the purpose of comparing and standardizing wind tunnel data.[19] In these "international trials" the National Physical Laboratory in Britain, Eiffel's laboratories in Paris, and the MIT labs played the main role. But it soon became clear that the Aerodynamic Institute in Göttingen should not be left out. The Royal Aircraft Establishment[20] in Britain sent two researchers, Hermann Glauert[21] and Robert McKinnon Wood, to visit Prandtl's laboratory in Göttingen. Glauert, whose family was of German origin, had followed Prandtl's work closely. The wind tunnel tests in Göttingen were particularly important for the international trials, because researchers in Britain had perceived a discrepancy between French and British testing of the same wing. Glauert and McKinnon Wood hoped that by applying Prandtl's correction for the aerodynamic interference due to wind tunnel walls this discrepancy could be resolved. The French had already applied this "Prandtl correction," which was also standard practice in many foreign laboratories.

Leading members of the British Design Panel engaged in aeronautical research put up strong resistance to applying the Prandtl correction. Leonard Bairstow of the Engineering Department of the National Physical Laboratory[22] as well as others argued that any discrepancies were not attributable to scale effects but to various errors, including errors in full-scale tests themselves. Obviously, Bairstow and his colleagues feared that the application of the Prandtl correction would undermine the significance of aerodynamic research in which they had been engaged for a long time. Finally, however, the British Design Panel had to admit that the application of the Prandtl correction caused a marked improvement in agreement between full-scale and wind tunnel tests and the controversy over scale effects was finally settled.[23]

Streamlined Airplanes and Air Races

Two other factors in aviation stimulated its development: streamlining airplanes and air races. Gliding, the means by which the first airplanes had been developed, had fallen into disregard until 1920, when it was revived in Germany. The main reason was that the Treaty of Versailles prohibited private flying in Germany, but it was also important that Oscar Ursinus, editor of the popular flying magazine *Flugsport*, advocated relying on light, aerodynamically efficient aircraft rather than on heavy airplanes with strong engines. In the United States it was the Lockheed Vega which in 1927 demonstrated the value of streamlining for commercial aircraft. The need of gliders to fly without power and to fly as fast as possible made designers think hard about eliminating unnecessary drag.[24]

Although Prandtl's work was of great importance, the researchers with whom he collaborated at the Aerodynamic Experimental Institute had a similar impact on aerodynamic research. They included Theodore von Kármán,[25] whom the Guggenheim Fund brought to the United States when he accepted a Caltech offer in 1929, and Max Munk, who joined the National Advisory Committee for Aeronautics (NACA) in 1921. Munk was the driving force behind designing of NACA's first variable-density tunnel, which went into operation in 1923.[26] It yielded basic data for some pathbreaking NACA reports on wing improvement and was a model for similar wind tunnels in the United States and abroad. Von Kármán published a comprehensive treatment of skin friction drag in 1930. His flow theory, together with the research results of B. M. Jones, professor of aeronautical engineering at Cambridge, England, led to a substantial redesign of aircraft. After World

War II, in 1947, von Kármán introduced a comprehensive supersonic theory equivalent to Prandtl's subsonic work.[27]

Ernst Prandtl and his Göttingen School also exerted a large influence on swept-back wing design. In the early 1930s swept-back design was already familiar to aircraft designers like A. M. Lippisch or the Horten Brothers in Germany and to G. R. T. Hill in Britain, who designed all-wing and tailless airplanes. Their main reason for using sweepback was to improve longitudinal stability of aircraft. Soon after World War II the designs of engineers like Lippisch and Hill, but also those of the German aerodynamicists and aircraft designers Adolf Busemann and Kurt Tank, were adopted in the United States. They were first most noticeable in the B-47 and the F-86, which were designed on the basis of German wind tunnel results.[28]

What conclusion can be drawn from these brief remarks on the beginnings of aerodynamic research for the issue of international leadership in this field? Although different centers of gravity existed in aerodynamics, there was no clear dominance of any of them lasting for a long time. Prandtl and his collaborators in Göttingen were in the lead for some time, performing pathbreaking research and setting up fundamental theories in aerodynamics like the boundary layer theory. Although interested in application, Prandtl, very much in the German scientific tradition, was fascinated by setting up theories. But scientists and engineers working with him brought his approaches and findings to other parts of the world, particularly to the United States. Here, superb institutional conditions for research and development as well as the personal factor—excellent researchers like Munk and von Kármán—gave the United States a prime position in aerodynamics. Apart from those two there were many other aeronautical engineers in America who had received their education and training in Europe, many Russians and also Germans like Ed Schmued, who, working for North American, designed the P-51 in World War II.[29] Because of this variety of ethnic backgrounds and mobility of aeronautical engineers, it does not make much sense to try to make clear distinctions between U.S. and European research practice.

Europe and the Origins of the Turbojet Revolution

The story of the origins of jet propulsion has also significant aerodynamic aspects. German and British researchers played an important role in its development. In the late 1920s, A. A. Griffith and Frank Whittle in Britain, independently of each other, made the first practical proposals for using gas turbines in aircraft. In 1926 Griffith developed a new aerodynamic theory of turbine-blade design and in 1929 suggested an axial compressor for a gas turbine driving a propeller, the so-called turboprop engine.[30]

Frank Whittle had entered the Royal Air Force (RAF) as a craftsman, repairing and maintaining aircraft. He was selected to become an officer cadet, went to the Officer's School of Engineering at RAF Henlow, and, at the age of twenty-seven, was sent to Cambridge University to study engineering. The young RAF officer Frank Whittle came to the conclusion that piston engines and propellers could not be used for high speed and high altitude flight. Applying gas turbines seemed to be a solution.[31]

Apart from Britain, the center of gravity in the development of jet propulsion was Germany. Three researchers have to be mentioned in particular: Hans Pabst von Ohain, Herbert Wagner, and Helmut Schelp. Von Ohain went to Göttingen University in the early 1930s, where he studied general physics and also received his doctorate in it. He became a teaching and research assistant to Robert W. Pohl, head of the Science Institute and a scientist noted for his ability to set up clear basic concepts. Although von Ohain took courses in aerodynamics at Göttingen, he did not belong to the Aerodynamic Experimental Institute. His interest in jet propulsion was derived from his knowledge in aerodynamics. He worked along lines similar to those of Frank Whittle in England. In 1934 von Ohain took out a patent on a centrifugal engine. Herbert Wagner was head of airframe development at the Junkers Aircraft Company and from 1930 to 1938 professor of aeronautics at the Technical University of Berlin. Helmut Schelp had some American connections. He had earned his master's degree at the Stevens Institute of Technology at Hoboken, New Jersey, in 1926, returned to Germany, and later joined the German Ministry of Aviation.[32]

The first Ohain jet engine ran in 1937 at about the same time as the first Whittle engine. The first jet plane, the German Heinkel He-178, flew on August 27, 1939, and was followed by the British Gloster E-28/39. Although a centrifugal type powered the first gas turbine aircraft flight by the He-178, the axial-flow jet, more efficient and capable of greater thrust, was used in the Messerschmitt Me-262 fighters that entered service in the autumn of 1944.[33]

Only at the beginning of 1942 did Whittle's firm Power Jet adopt an expensive nickel-chromium alloy called Nimonic 80 for producing turbine blades that essentially solved the daunting problem of turbine blade failure. In Germany the relative backwardness in aero engine design[34] and especially the lack of strategic materials were bottlenecks. Still, the only German turbojet to achieve mass production, the Junkers Jumo-004 powering the Me-262, achieved high performance with minimal use of strategic materials. The

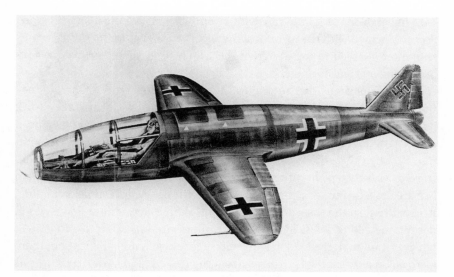

A postwar artist's conception of the He-176, the first rocket-powered aircraft, in flight. (Photo courtesy of the Deutsches Museum)

Jumo-004B required little chromium and no nickel at all but only aluminum alloy and some steel. Although, under those circumstances, its performance was remarkable, there were serious setbacks: an extremely short service life and a short range.[35]

But in the late 1930s advances in aero jet development in Germany surpassed those in Britain. One of the reasons that von Ohain progressed more rapidly than Whittle was that he had chosen hydrogen as an experimental fuel. But there was another, even more important, reason: compared with the scarce means Whittle received, von Ohain had the ample resources of the Heinkel aircraft company at his disposal. Besides, Ernst Heinkel, enthusiastic about progress in aviation, supported von Ohain as best he could.[36] As a project of a private aviation company, the first von Ohain engine did not have to undergo rigorous government scrutiny and certification tests like those Whittle experienced in Britain.[37]

Independently of early jet research in Britain and Germany, some work had begun in the United States in 1940, but the successful attempts in Britain led the U.S. Army Air Forces to import a Whittle centrifugal jet in the spring of 1941. Copies of this engine, produced by General Electric, were used in the twin-engine Bell P-59A.[38]

As mentioned before, researchers and aircraft manufacturers in the United States were already familiar with jet propulsion for aircraft. On the basis of documents captured in Germany in 1945, they studied aerodynamic data for high-speed jet aircraft with swept wings. The information gained there coincided with the ideas of some aerodynamicists in the United States and, inter alia, led to the design of the widely acclaimed F-86 with swept-back wing and tail surfaces.[39]

What conclusions do the origins of jet propulsion yield for my theme? British and German researchers were in the lead; they performed their work largely independently of each other. The first aerojet engine was produced in Germany. This had to do with personal factors—the successful cooperation between the gifted engineering scientist Hans Pabst von Ohain, who started with theory, and the vigorous aviation engineer and businessman Ernst Heinkel—but, at a later stage, also with heavy government funding. In order to turn the tides of war after 1942–43, the German Air Force was in desperate need of a fast fighter. After a short interlude after World War II, when Britain took the lead in jet propulsion research and application, there were again aeronautical engineers from the United States who, on the basis of know-how from Germany and Britain, shifted the center of jet propulsion research and application to the United States. This happened despite the fact that manufacturers in the United States were reluctant to embark on this new field, because piston aero engines were still profitable. But the fact that this was possible within a relatively short time again shows that gaps in aeronautical know-how were only short-lived.

Airbus and the Rise of European Commercial Airplanes

"The Airbus Program is a most spectacular European success story. In 1967 the governments of Germany, France, and Britain agreed on developing the Airbus. The first aircraft of the type A300 B took off in 1972. Since its foundation in 1970 Airbus Industrie has become the second largest wide-body commercial aircraft manufacturer worldwide and is now competing with Boeing for the No. 1 place. From 1970 until now Airbus has received orders for over 4,100 aircraft from more than 170 customers."[40]

The above is not my prose but comes from Internet information provided by Airbus Industrie. Has Airbus really been such a success story? What political, economic, and technological factors influenced its development? I will restrict myself to a few remarks on the origins and early years of the first Airbus model, the A300.

In 1964 several British and French firms started talking about the possibility of a wide-body aircraft designed for short trips. In December 1965 some German firms cooperating already on the C-160 Transall, the successor of the military transport aircraft Noratlas, founded a German Airbus Working Group (Arbeitsgemeinschaft Airbus). In 1970, when Airbus Industrie

The Airbus family of airliners has been successful in capturing a large share of the commercial airliner market, and the percentage seems to be growing. (Photo courtesy of Airbus Industrie)

was created, the American share of the world commercial aircraft market amounted to 94 percent. The main European airlines—Air France, British European Airways, and Deutsche Lufthansa—predicted that their medium-haul traffic would soon increase to such an extent that a 200- to 250-seat aircraft would be required on their trunk routes. Because there had been cooperation between the French Sud Aviation and BAC (British Aircraft Corporation) already on the Concorde, cooperation between Britain and France in the commercial aircraft sector was nothing new. Close cooperation was, indeed, necessary, because Europe's aircraft manufacturers lacked the financial strength and perhaps also the technical know-how to design and build a major airliner nationally.[41]

What then, were the motives of the major countries involved in this project? France has to be mentioned first, because it provided powerful de facto leadership. The French president Charles de Gaulle did not mince his words. He was, as he made clear, determined to break the "American colonization of the skies."[42] The French government also insisted on Sud Aviation being made the pilot contractor for the airframe. The two state-owned companies in France had already begun to merge into Aérospatiale, a firm that had all the makings of managing such a large project.[43]

Germany's interest in the Airbus had several economic, political, and technological factors.[44] By the late 1960s the German aircraft industry had acquired substantial technical know-how again in state-of-the-art aircraft design and production, and the controversial Starfighter project had contributed to this. The Starfighter project had already been a global effort with seven nations and twenty-one major aircraft companies involved.[45] The German aviation industry wanted to apply the expertise it had gained, especially in view of the fact that economic prospects in commercial aviation seemed to be bright. Besides, close collaboration with France would enhance the cause of European unification with its economic and political prospects.[46]

The case with Britain was not so clear-cut. The British aero industry was not very optimistic about the success of a program in which it did not have the main say. Moreover, in case of success, the Airbus would be a direct competitor of the British BAC 3-11. On the other hand, there was the attractive proposition of the Airbus carrying Rolls-Royce engines, but those were still in the development process. Also, Rolls-Royce encountered severe design problems. The company therefore concentrated its efforts on designing the RB-211 for the Lockheed L-011 Tristar. But this did not prove successful either. In the end Rolls-Royce had lost so much money that it went bankrupt and was nationalized in 1972.[47]

Although Britain officially left the Airbus project in 1969—it rejoined in 1979—another British company made its interest felt. This time the subject was not the Airbus engine, but its wings. In 1970 Hawker Siddeley had urged the British government to become a member of the Airbus consortium. French and German consortium members agreed that Hawker had first-class expertise in designing and building aircraft wings. So the firm was given the job, although as a subcontractor.[48]

Italy is an interesting case. Although heavily involved in the MRCA Tornado project, it had a negative attitude toward the Airbus, preferring to cooperate with McDonnell Douglas on the DC-9 and the DC-10 and later with Boeing on the 767. Quite early Aeritalia pursued the strategy of not trying to sell aircraft to American airlines but to produce parts that American aircraft manufacturers would use on their aircraft they sold in North America and abroad.[49]

Toward the U.S. companies the Airbus consortium's attitude was pragmatic. When the European firms lacked design expertise, they did not hesitate to buy it in the United States. About 40 percent of the equipment and spare parts for the first Airbus were bought there. The major U.S. companies saw no problem with that. In the beginning, manufacturers like Boeing and McDonnell Douglas do not seem to have taken European competition seriously: In the late 1960s Douglas licensed Airbus to produce its DC-10 engine nacelle, which connected the General Electric (GE) Airbus engines to the wings.[50] Also GE and the French SNECMA company cooperated on the development of the new turbofan jet engine CFM-56.[51]

From the early 1970s onwards the Airbus program had concentrated on only one objective, the production of a competitive civil airframe. Vis-à-vis the United States the choice of the GE CF-6 engine proved to be an important element in A300's success. There was also a good match between airframe and engine development, and the engine's American origins contributed to its attractiveness for the U.S. market.[52]

In producing the Airbus, the consortium appointed contracts geographically: Deutsche Airbus managed all subcontracts with German firms and all the research into the center and rear fuselage; Aérospatiale was in charge of the forward fuselage and the cockpit; Hawker Siddeley was responsible for the wings; and CASA, the Spanish company, concentrated on the tail.[53]

It is not surprising that several coordination problems between the producers occurred, but the companies managed to overcome them. The consortium structure had several advantages for managing this large international civil aircraft project: its flexibility, its capacity to give financial reassurance to potential purchasers, and its capacity to centralize control of key functions, for example, marketing.[54]

Still, in May 1974, one year after Federal Aviation Administration (FAA) certification, Airbus Industrie had sold only twenty A300s, and those mainly to Air France. In July 1975 it made its first non-European sales to Air India and South African Airways. But the A300's fuel-efficient design, the absence of a direct U.S. competitor, and an aggressive sales campaign did bring some success. Airbus was also lucky because the fuel price increase of the mid-1970s made a large twin-engine airliner more attractive, and at that time the A300 was the only one available.[55] In the late 1970s and early 1980s Boeing, with its 767, adopted some features of the Airbus.[56]

Whether the Airbus will continue to be a success story remains to be seen. So far, the performance of its various models has, indeed, been respectable, although there were severe setbacks, for example with the Airbus's 320 fly-by-wire system.[57] But, as a whole, the case of the Airbus has shown that cooperation between major European countries on a large industrial project can work. Governmental support for Airbus Industrie was to a large extent motivated by the objective of enhancing national aircraft industries. Business and military strategic interests require a large aircraft industry with a range of capacities and projects in order to keep research, design, and production teams fully employed and not to lose ground on state-of-the-art aviation technology know-how. The encouragement and support of aircraft industries have therefore been prime tools of technology policy, and specific economic goals were pursued by airliner exports. Among foreign policy achievements, European collaboration stands out, but national governments also had to respond to strong and well-organized industrial pressure-group activities and to national pride, especially in France.[58] The heavy involvement of national governments has come under severe criticism from competitors, especially in the United States.

Conclusion

In conclusion it can be said that from the mid-1920s onward the United States quickly caught up with any lead aeronautics as practiced in Europe might have had. Many aeronautical engineers trained in Europe contributed to this accomplishment; and also, the NACA and the rapid implementation of technical innovations, which partly had their origin in Europe, played important roles. In the late 1920s and early 1930s U.S. commercial aviation had established a lead that the European aircraft producers watched with some trepidation. At that time the efforts of European countries to equal the performance and economy of American airplanes were hampered by the predominance of military influence on the European aircraft industry. In the United States manufacturers benefited from the greater share of their sales in civil aircraft; this encouraged them to concentrate on the needs of commercial aviation.[59]

An important factor making this lead possible was clearly the competitive market situation in the United States. This differed significantly from the situation in Europe, with its comparatively small, dispersed markets and national government regulation. But, in contrast to Miller and Sawers in their sound book *The Technical Development of Modern Aviation*, I would not emphasize the market aspect too much. After all, given the smallness of European industries, government involvement and support was probably essential for competing with the larger aviation industry in the United States.

Government involvement was also vital for Europe to compete with the United States after World War II. In the 1950s and 1960s European aircraft manufacturers absorbed the U.S. model of managing large technological projects. They therefore met the American challenge by imitating the United States but also by cooperating in consortia. There were setbacks, partly due to some unpleasant national rivalries between countries. But the outcome— for example, the recent models of the Airbus family and the 3XX project— is respectable. Future developments, and especially the competition with Boeing, will be interesting to watch.

What we see today in the civil aviation sector—partly also in the military sector—are globally produced aircraft. Today British Airways and American Airlines, together with some other airlines, are united in what they call the One World Alliance. Lufthansa and United Airlines are together in the Star Alliance, a term which seems to be only partly justified, because the Star Alliance's aircraft are not designed to fly to the moon. In the aviation sector things have, indeed, gone into the direction of a One World Alliance. However, the scenario of perhaps having only one giant left one day—as an aircraft manufacturer or as an airline—or, to let fantasy soar even further, as a giant airline manufacturing its own aircraft—is probably not such a good prospect.

NOTES

1. Of course, this is not the first time these questions have been explored. See, among other works, Ronald E. Sawers and David Miller, eds., *The Technical Development of Modern Aviation* (Boston: AMS Press, 1969); Paul A. Hanle, *Bringing Aerodynamics to America* (Cambridge, Mass.: MIT Press, 1982); John D. Anderson Jr., *A History of Aerodynamics and Its Impact on Flying Machines* (Cambridge, Eng.: Cambridge University Press, 1997).

2. See, for example, Ronald Miller and David Sawers, *The Technical Development of Modern Aviation* (London: Routledge & Kegan Paul, 1968).

3. This term is taken from Anderson, *History of Aerodynamics*, 294–95.

4. Otto Lilienthal, *Der Vogelflug als Grundlage der Fliegekunst* (Berlin: R. Gaertners, 1899). On Lilienthal, see especially Werner Schwipps, *Lilienthal: Die Biographie des ersten Fliegers* (Gräfelfing, Ger.: Aviatic Verlag, 1979).

5. John D. Anderson Jr., "The Evolution of Aerodynamics in the Twentieth-Century: Engineering or Science?" in Peter Galison and Alex Roland, eds., *Atmospheric Flight in the Twentieth Century* (Dordrecht, Neth.: Kluwer Academic, 2000), 243–44.

6. Anderson, *History of Aerodynamics*, 294–95.

7. Anderson, "Evolution of Aerodynamics," 246. On Prandtl, see Julius C. Rotta, *Die Aerodynamische Versuchsanstalt in Göttingen, ein Werk Ernst Ludwig Prandtls. Ihre Geschichte von den Anfängen bis 1925* (Göttingen: Vandenhoeck & Ruprecht, 1990).

8. Ernst Heinrich Hirschel, Horst Prem, and Gero Madelung, eds., *Luftfahrtforschung in Deutschland* (Bonn: Bernard & Graefe, 1985), 38–39; Edward W. Constant II, *The Origins of the Turbojet Revolution* (Baltimore, Md.: Johns Hopkins University Press, 1980), 104.

9. Ludwig Prandtl, "Über Flüssigkeitsbewegung bei sehr kleiner Reibung," *Proceedings of the Third International Mathematical Congress* (Leipzig: International Mathematical Congress, 1905); Theodore von Kármán, *Aerodynamics* (Ithaca, N.Y.: Cornell University Press, 1954), 88; Constant, *Origins*, 105; Hugh L. Dryden, "Fifty Years of Boundary Layer Theory and Experiments," *Science* 121 (1955): 375–80; Helmuth Trischler, *Luft- und Raumfahrtforschung in Deutschland 1900–1970, Politische Geschichte einer Wissenschaft* (Frankfurt-am-Main, Ger.: Campus, 1992), 53.

10. Hirschel et al., eds., *Luftfahrtforschung*, 33.

11. Von Kármán, *Aerodynamics*, 69; Constant, *Origins*, 108.

12. Werner Heinzerling, "Windkanäle," in Ludwig Bölkow, ed., *Ein Jahrhundert Flugzeuge: Geschichte und Technik des Fliegens* (Düsseldorf, Ger.: VDI Verlag, 1990), 310.

13. On Wenham, see J. Laurence Pritchard, "Francis Herbert Wenham, Honorary Member, 1824–1908: An Appreciation of the First Lecturer to the Aeronautical Society," *Journal of the Royal Aeronautical Society* 62 (1958): 571–96; Roger E. Bilstein, "American Aviation Technology: An International Heritage," in Galison and Roland, *Atmospheric Flight*, 207.

14. Gustave Eiffel, *Recherches expérimentales sur la résistance de l'air exécutées à la tour* (Paris: Maretheux, 1907); Joseph Black, "Gustave Eiffel—Pioneer of Experimental Aerodynamics," *Aeronautical Review* 94 (1990): 431–44; Anderson, "Evolution of Aerodynamics," 248–50.

15. Bilstein, "American Aviation Technology," 260.

16. Anderson, "Evolution of Aerodynamics," 251.

17. Constant, *Origins*, 105.

18. Anderson, *History of Aerodynamics*, 432.

19. For this and the following, see Takehiko Hashimoto, "The Wind Tunnel and the Emergence of Aeronautical Research in Britain," in Galison and Roland, *Atmospheric Flight*, 223–39.

20. See Percy B. Walker, *Early Aviation at Farnsborough: The History of the Royal Aircraft Establishment*, 2 vols. (London: Macdonald, 1971).

21. Glauert's main publication is *The Elements of Airfoil and Airscrew Theory* (Cambridge, Eng.: Cambridge University Press, 1926).

22. See Leonard Bairstow, *Applied Aerodynamics* (London: Longmans, Green, 1920).

23. Hashimoto, "Wind Tunnel," 235.

24. Miller and Sawers, *Technical Development*, 58–59.

25. On von Kármán, see Michael H. Gorn, *The Universal Man: Theodore von Kármán's Life in Aeronautics* (Washington, D.C.: Smithsonian Institution Press, 1992).

26. On the transfer process of aerodynamic knowledge, see Paul A. Hanle, *Bringing Aerodynamics to America* (Cambridge, Mass.: MIT Press, 1982).

27. Roger E. Bilstein, *Flight in America: From the Wrights to the Astronauts*, 3d ed. (Baltimore, Md.: Johns Hopkins University Press, 2001), 70–73; Constant, *Origins*, 105–8.

28. Miller and Sawers, *Technical Development*, 169, 172.

29. Roger E. Bilstein, *The American Aerospace Industry: From Workshop to Global Enterprise* (New York: Twayne, 1996), 69.

30. Peter W. Brooks, "Aircraft and Their Operation," in Trevor I. Williams, *A History of Technology*, vol. 7 (Oxford, U.K.: Clarendon Press, 1978), 819.

31. David Edgerton, *England and the Aeroplane: An Essay on a Militant and Technological Nation* (Basingstoke, U.K.: Macmillan, 1991), 56.

32. Constant, *Origins*, 201–4.

33. Brooks, "Aircraft," 821.

34. Hans-Joachim Braun, "Aero-engine Production in the Third Reich," *History of Technology* 14 (1992): 1–15.

35. Brooks, "Aircraft," 821.

36. On Heinkel, see Hans D. Köhler, *Ernst Heinkel—Pionier der Schnellflugzeuge: Eine Biographie* (Koblenz, Ger.: Bernard & Graefe, 1988).

37. Constant, *Origins*, 201.

38. Bilstein, *Flight in America*, 179.

39. Ibid., 179–80. On Britain and the United States, see Virginia P. Dawson, "The American Turbojet Industry and British Competition: The Mediating Role of Government Research," in William M. Leary, ed., *From Airships to Airbus: The History of Civil and Commercial Aviation*, vol. 1, *Infrastructure and Environment* (Washington, D.C.: Smithsonian Institution Press, 1995), 127–50.

40. Homepage Airbus Industrie.

41. Desmond Hickie, "Airbus Industrie: A Case Study in European High Technology," in Ulrich Hilpert, ed., *State Policies and Techno-Industrial Innovation* (London: Routledge, 1991), 190, 210; Roger Béteille, "Introduction: Airbus; or, the Reconstruction of European Civil Aeronautics," in Leary, *From Airships to Airbus*, 1:5.

42. Glenn A. Bugos, "The Airbus Matrix: The Reorganization of the Postwar European Aircraft Industry," in Francis W. Heller and John R. Gillingham, eds., *The United States and the Integration of Europe: Legacies of the Postwar Era* (New York: St. Martin's Press, 1996), 385.

43. Keith Hayward, "Airbus: Twenty Years of European Collaboration," *International Affairs* 64 (1988): 21.

44. On the part Germany played in the origin of the airbus, see Ulrich Kirchner, *Geschichte des bundesdeutschen Verkehrsflugzeugbaus. Der lange Weg zum Airbus* (Frankfurt-am-Main, Ger.: Campus, 1998).

45. Bilstein, *American Aerospace Industry*, 100.

46. Christopher Magnus Andres, *Die bundesdeutsche Luft- und Raumfahrtindustrie, 1945–1970: Ein Industriebereich im Spannungsfeld von Politik, Wirtschaft und Militär* (Frankfurt-am-Main, Ger.: Peter Lang, 1996), 354–59.

47. Béteille, "Airbus," 7.

48. Bugos, "Airbus Matrix," 389.

49. Bilstein, *American Aerospace Industry*, 190; Bugos, "Airbus Matrix," 396.

50. Bugos, "Airbus Matrix," 393–95.

51. Bilstein, *American Aerospace Industry*, 192.

52. Hayward, "Airbus," 19.

53. Bugos, "Airbus Matrix," 389.

54. Hickie, "Airbus Industrie," 211.

55. Hayward, "Airbus," 19; Bugos, "Airbus Matrix," 389.

56. Bilstein, *American Aerospace Industry*, 180–81.

57. Henry Petroski, *Invention by Design: How Engineers Get from Thought to Thing* (Cambridge, Mass.: Harvard University Press, 1996), 132–33.

58. Hickie, "Airbus Industrie," 210.

59. Miller and Sawers, *Technical Development*, 20.

Part 2

**CIVIL AERONAUTICS
AND GOVERNMENT
POLICY**

CHAPTER 4
HERBERT HOOVER AND COMMERCIAL
AVIATION POLICY, 1921–1933

The interwar decades are remembered as a golden age in the history of aviation, a dramatic, even glamorous era that laid the foundation of a vital new industry. Its luminaries were the pilots, business leaders, and industrialists whose often spectacular achievements gave form to commercial flying. But the growth of aviation in the United States inevitably created significant questions about the relationship between this fledgling enterprise and the nation's legal and political structures. One of the major figures in defining the relationship between commercial aviation and the federal government was the mining-engineer-turned-public-servant Herbert Hoover. Involved with aviation issues from the time he entered Warren Harding's cabinet as secretary of commerce in 1921 until he relinquished the presidency in 1933, Hoover strove both to nurture commercial aviation during the Great Prosperity and later to rescue it during the Great Depression. In the process he played a crucial role in shaping both the statutory and bureaucratic mechanisms that guided federal policy. Although often neglected even by his champions, Hoover's part in the development of commercial aviation is one of his most enduring legacies.[1]

Hoover entered the Harding cabinet with very firm ideas about nurturing economic progress through positive but restrained government action. As Ellis Hawley has demonstrated, Hoover was a pioneer in developing an American version of the corporatist philosophy that influenced most industrial societies in the years after World War I. This outlook valued a society organized into functionally independent economic units, such as labor, agriculture, and management, which were voluntarily decentralized and self-regulating. Motivated by a sense of community and social responsibility and led by an enlightened elite, these large units would work harmoniously together creating efficient, technologically driven solutions to society's problems. Political and economic elites were closely interlinked, and the state served as partner and facilitator rather than regulator.[2]

Drawing on this general concept, Hoover envisioned an "associative state" in which the government, particularly the Department of Commerce, fostered the growth of cooperative institutions designed to provide a voluntary network of self-regulation. In this way, Hoover intended to avoid the growth of monopoly capitalism on the one hand and the appearance of a strong, interventionist state bureaucracy on the other. Either of those, he believed, would destroy the national heritage he called American individualism. Hoover assigned the federal government a definite role in this system. Besides encouraging the growth of these associations, it should prod them to put aside selfish impulses, resist efforts to enhance economic power with political power, and act to preserve opportunity and initiative. The government's authority to coerce these units, however, should be carefully limited. It should work toward its ends by avoiding direct regulation and concentrating on promotional conferences, expert inquiries, and publicity designed to stimulate private solutions to problems.[3]

Through the boom years of the 1920s, Hoover saw aviation as a struggling business whose development was impeded by feeble market conditions, technical limitations, and a failure to develop responsible leadership. Guided by his concept of the associative state, he believed that commercial aviation should pursue scientific rationalization under the stewardship of industrial statesmen working through voluntary associations. The federal government should support these industry-generated solutions by offering publicity and pertinent economic and scientific data. Moreover, taking transportation as a model, Hoover also believed the government should provide and maintain the infrastructure of air transportation just as it provided navigational aids for the nation's waterways.[4]

The field of commercial aviation posed a special challenge for Hoover and his theories. Like other transportation industries, flying had a certain public dimension that set it apart from most private businesses. Aircraft manufacturers relied heavily on the government to be the major consumer of their wares, while the planes themselves crisscrossed the nation's airspace in the same way commercial ships traveled its navigable waterways. Also, as an underdeveloped industry, commercial aviation faced problems that were not susceptible to macroeconomic solutions but rather demanded mechanisms uniquely geared to the needs of a fledgling enterprise. Limited capital, straitened markets, and a dearth of consumer confidence all worked to stunt the development of a viable industry.[5]

During his years as secretary of commerce, Hoover promoted aviation through two basic strategies. One approach was the generation and the dissemination of information for professional constituencies and for the general public. Through the Department of Commerce, Hoover worked to provide the industry with economic and technical data designed to serve as the basis for better management decisions and better service. Simultaneously, he sought to educate the public, especially the business community, about the potential value of commercial aviation. Many Americans in the 1920s saw aviation as a suitable arena for military pilots and civilian daredevils but not for serious financial investment. To counter this view, Hoover's Department of Commerce launched a publicity campaign designed to persuade the nation that air travel boosted efficiency and enhanced profits.[6]

Beyond his public-relations campaign to make the nation "air-minded," however, Hoover realized that skepticism about flying stemmed not only from ignorance about its possibilities but also from legitimate concerns about its safety. The inadequate training of pilots and the improper maintenance of planes contributed to an alarming accident rate, which in turn encouraged the idea that flying was too dangerous for most people. In devising his approach to this problem, Hoover embraced a larger role for the federal government than he was typically willing to accept because he realized that the industry itself could not credibly certify the safety of its own service. Only a disinterested party such as the government could assume that responsibility, so Hoover played a substantial role in committing the Department of Commerce to the regulation of civilian flying. Obviously, such a role could not be undertaken without some legal sanction, and Hoover became an influential advocate for a federal air law—ultimately the Air Commerce Act of 1926—that could serve as a charter for the industry and a guide for its relationship with the government.[7]

Hoover's efforts to build the infrastructure of the industry around government–industry cooperation were highly successful, but his plans for economic regulation met a different fate. His early strategies, first as commerce secretary and then as president, were designed to secure the place of the new industry in an expanding economy, but the collapse of 1929 changed the thrust of Hoover's efforts from development to rescue. With Postmaster General Walter F. Brown as chief policymaker, the Hoover administration used the airmail contracts to restrict entry to the industry, to encourage the consolidation of small firms, and to limit competition. Brown came to function as a one-person public utility board, anticipating in some ways the future Civil Aeronautics Board. While highly successful in shaping the industry, Brown's approach became a political lightning rod. The New Deal overturned the Hoover-Brown system in the mid-1930s amid a storm of controversy, but it was later resurrected in modified form by the Civil Aeronautics Act of 1938, and its principles influenced federal aviation policy for decades to come.[8]

Nurturing a New Industry

In one important respect, Hoover's plan for aviation represented a special application of his corporatist philosophy in that he assigned an unusually large role to government. Although Hoover believed that solutions to aviation problems should be generated by the industry, he agreed with industry leaders that aviation would not progress without federal assistance. Influenced by historical precedent in the transportation field, Hoover justified this departure from his usual attitude by comparing aviation with shipping. The public owned the airways just as it owned the navigable waterways; and just as the government provided basic services to shippers on water, so it should provide precisely the same services to shippers by air. This meant supplying navigational aids such as airways equipped with emergency landing fields properly lighted, marked, and mapped. Hoover considered lighting especially important because it opened the way to night flying, which in turn meant the faster delivery of goods and passengers and a stronger competitive position for air transport.[9]

Hoover also believed the federal government should uphold the safety of flying by licensing pilots and certifying the airworthiness of aircraft. Once again, his reasoning flowed from his shipping analogy: the government performed the same services for that industry. Only the government imprimatur, Hoover felt, could counteract the tremendous amount of bad publicity the industry had received and restore a measure of public confidence. By reducing accidents and building public respect, government inspection would help to reduce insurance rates and attract more capital for investment.[10]

Still pursuing his comparison with shipping, Hoover staunchly opposed federal involvement in establishing airports. Each locality was responsible for its own docks, so each city, he felt, must provide for it own airport, although the facility would have to conform to federal standards. In this area Hoover was closer to a more conventional application of his philosophy on business and government. He approved a federal campaign to publicize the economic advantages of airports and stressed that they should be close to centers of population so that the time gained by flying would not be lost again, but he refused to endorse the use of federal money in their construction. This, in his view, was a proper area for local initiative.[11]

Hoover also had definite ideas about the financing involved in his proposals. He was strongly opposed to direct subsidy on the European model. Instead he advocated using the airmail as a form of indirect subsidy to stimulate growth through private initiative. Hoover argued that the government should stop flying the airmail and instead contract various routes to private companies at generous rates. He envisioned a system wherein mail planes would carry passengers and freight as well, with airmail contracts ensuring that the enterprise would be profitable. These airmail contracts would provide a firm base for the creation of self-sufficient commercial aviation. Hoover admitted that navigational aids, safety inspection, and airmail contracts would be expensive, but he believed that these steps would save the government money in the long run. Although his plan would cost roughly one to four million dollars, depending on the volume of traffic, he predicted a tenfold savings in military expenditures if commercial aviation developed. Encouraging commercial aviation, he said, was "a most constructive drive for immediate economy in government."[12]

One of Hoover's most important contributions was his role in drafting and implementing the landmark Air Commerce Act of 1926. Although the legislation itself was the work of many hands, Hoover and the Department of Commerce worked closely with Senator Hiram Bingham in preparing the initial draft of the bill that Bingham and Representative James Parker presented to Congress. Influenced by Hoover, the proposed bill moved from the creation of a bureau of civil aeronautics and toward the idea that services to aviation could best be provided by extending the functions of existing department agencies and simply creating an assistant secretary of commerce to coordinate them. For example, the Bureau of Lighthouses could assume responsibility for lighting airways, or the Bureau of Standards could make research facilities available. Such an approach, Hoover argued, would be cheaper and more efficient than a new government bureau. In a letter to Senator Wesley Jones, Hoover described the new bill as a "much simplified method of setting up a civilian aviation agency."[13]

Passed on May 20, 1926, the Air Commerce Act represented an important milestone in the growth of commercial aviation in the United States. The National Advisory Committee on Aeronautics hailed it as the "legislative cornerstone of the industry," and a later historian styled it the "Bill of Rights of the aviation industry." Its impact was enormous, but as pathbreaking as it was, the act laid out only broad regulatory principles and provided a meager $50,000 for implementation. However, Hoover and the new assistant secretary of commerce for aeronautics, aviation lobbyist and attorney William MacCracken, set about implementing the law within the framework of Hoover's corporatist principles. Drawing on the Hooverian concept of the associative state, MacCracken established numerous precedents for aviation regulation and left an enduring imprint on commercial aviation. Applauding Hoover's policy of "conferring with industry about all matters in which they have a mutual interest," MacCracken rejected the "erroneous ideas that the interests of the Government and of business are diametrically opposed, and that they should deal with each other at arms length . . . rather than in a spirit of mutual confidence and helpfulness."[14]

Hoover took great pride in the department's achievements under Mac-Cracken's leadership. During the period 1922 to 1926, the nation added only 369 miles of regular air service operated by private enterprise and 3,000 miles of airmail lines run by the Post Office that did not carry passengers or express. After the passage of the act, Hoover later recalled: "We went at it with great zest." He summarized progress by 1929 with an outpouring of his beloved statistics: 25,000 miles of government-improved airways of which 14,000 were lighted; 1,000 airports built and 1,200 in progress; 6,400 licensed planes making 25,000,000 miles in regular flights annually; a manufacturing output of 7,500 planes a year. "I know of no satisfaction," Hoover concluded, "equal to the growth under one's own hand of a great economic and human agency."[15]

Although clearly modified by historical precedent, Hoover's blueprint for aviation drew heavily on the corporatist approach characteristic of much of his thinking. Far from blindly copying the past, however, Hoover left his unique stamp on government regulation of flight by applying the ideology of the associative state. Working with trade organizations that expressed the consensus of the aviation industry, Hoover strove to regulate commercial flying through cooperative rather than adversarial proceedings. Under his guidance, the Department of Commerce dispensed expert analysis of pertinent questions and waged publicity campaigns touting the future of flight. Most important, Hoover played a major role in industry efforts to secure and implement a federal air law. The Air Commerce Act, in the words of Ellis Hawley, was a "discretionary measure under which regulatory law could emerge from 'experience' and from 'sincere cooperation' between the public and private sectors." Such "sincere cooperation" lay at the heart of Hoover's associative state. Encouraged by their success in this arena, Hoover and his colleagues now applied the principles of the associative state to the marketplace.[16]

The Associative State in the Marketplace

Hoover's market development strategies for aviation largely focused on the airmail. A strong critic of direct government subsidies to the industry, Hoover was an early champion of the Kelly Air Mail Act of 1925, which permitted the Post Office to contract with private carriers to haul the airmail. Although enormously significant for the development of commercial aviation, the Kelly Act had serious flaws as an instrument of market regu-

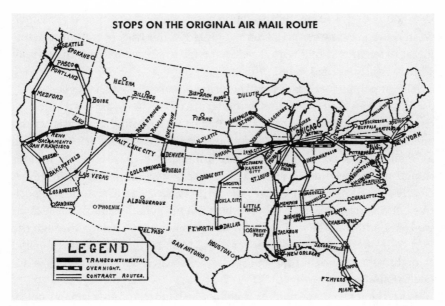

STOPS ON THE ORIGINAL AIR MAIL ROUTE

This map shows the stops on the original airmail route. The transcontinental route appears as a solid black line. The contract routes are in white, and the overnight sections are dashed in black and white and run at this time only between Chicago and New York. (NASA photo)

lation. While the government intended the airmail payments as a temporary prop for the industry until private support developed, the law actually discouraged the contractors from seeking private sector customers. For example, carrying passengers meant higher costs in the form of larger aircraft and more support personnel. Moreover, while a 200-pound man would pay about $300 for a coast-to-coast ticket, the Post Office Department would pay about $1,800 to transport 200 pounds of mail that distance. Far from seeking private sector customers, lines occasionally bumped ticketed passengers to make room for mail.[17]

The payment structure of the Kelly Act created financial problems for the government and for many carriers. Although mail volume increased significantly, postal revenues dropped below the cost of the service, and by 1929 the government was losing $7 million a year on its airmail operation. Moreover, the system dealt inequitably with the carriers. Companies flying short routes linking large urban centers and using inexpensive equipment prospered, while those flying more difficult routes in modern equipment received no additional compensation. Thus the rate per pound of mail was $3.00 between New York and Boston while the rate was $0.86 on the longer New York–Chicago route. As a result, the airmail network tended to be a hodge-podge of small firms flying short routes, an arrangement which wasted the airplane's only advantage over ground transportation. Operating at a deficit and failing to develop the air transport market, the airmail system clearly needed major reform.[18]

The impending expiration of the original contracts with the carriers—established for four years under the terms of the Kelly Act—provided an opportunity to overhaul the system. Reflecting his usual emphasis on the advice of experts, President Hoover advocated "a comprehensive report on air mail rates as to the amount we are losing and as to whether these rates should not be readjusted upward before they become too crystallized." A closely related matter, in Hoover's mind, was the need to determine "positive national routes" and to "build those routes in as substantial fashion as we can, both from the point of view of equipment and support through the Post Office for air and passenger services." "All of it requires some time and thought," the president told reporters. "We are in a new field, and it is difficult to find one's way about."[19]

With the focus on airmail as the chief instrument of government policy in the commercial aviation marketplace, Hoover increasingly relied on Walter Brown to develop the administration's aviation policy. Brown's appointment as postmaster general capped a long and successful political career built largely around his considerable administrative ability. Although not so glamorous as other cabinet portfolios, the position demanded an incumbent who had two rather different kinds of skills, namely the political instincts to keep the party machinery well oiled and the managerial talent to supervise 360,000 employees and a $700,000,000 budget. Walter Brown was particularly well suited to handle what Herbert Hoover called "the tangled skein of politics and business inherent in the Post Office." A Toledo businessman and attorney, Brown had been active in the Ohio Republican party, rising steadily through its hierarchy to become state chairman. Committed to the progressive movement, he was a close adviser to Theodore Roosevelt and followed TR into the Progressive party in 1912. Returning to the Republican fold after the Bull Moose debacle, Brown planned to support Roosevelt for president again in 1920 but, after the Rough Rider's sudden death, he switched to fellow Ohioan Warren G. Harding and worked as Harding's floor manager at the tumultuous Republican convention in Chicago.[20]

Brown's interest in government reorganization brought him to the attention of Herbert Hoover, who made him an assistant secretary of commerce in late 1927. The pairing of Hoover and Brown marked the beginning of a fruitful collaboration. In personal terms, both were distant, rather austere men with great intelligence and a zeal for administration. Each had supported

Roosevelt in 1912 and remained committed to applying progressive principles to the problems of the 1920s. Hoover was particularly impressed with Brown's grasp of the bureaucracy, declaring he possessed "a greater knowledge of the Federal machinery than any other man in the United States." Moreover, as a widely known party professional, Brown linked Hoover with the Republican leaders who would be so important to his bid for the White House in 1928. After a brief tenure, Brown left the Department of Commerce to manage Hoover's successful campaign, and the president-elect ultimately named him postmaster general.[21]

Although Brown knew virtually nothing about flying when he took office, he quickly developed definite ideas about the future of commercial aviation. Like Hoover, he concluded that strong commercial aviation was central to the national defense. In his first major public address on the future of commercial aviation, Brown told his audience that if commercial flying prospered, "we need have no anxiety about being able to defend ourselves in the air if the occasion should ever arise." A prosperous commercial aviation "will keep our country abreast of the flying" and guarantee equipment and personnel would be available in case of a national emergency. He made the link between civilian and military flying explicit in a 1931 speech to the New York City Bond Club. "The design of the large plane that flies several times a day between Washington and New York is that of an army bomber," Brown told the financiers. "It would be a matter of a very few weeks for the factory that turns out that plane to be turning out the finest bombers that have yet been produced." A thriving aviation industry would thus serve as a defense auxiliary and reduce the cost of a military air branch.[22]

Brown also decided that America's rickety aviation industry could not hope to provide such a defense auxiliary. In the future, the United States would need "air lines which will carry not only the fast mail but express and passengers as well, to establish a comprehensive air transport system." For Brown, this meant an emphasis on large, well-financed, well-equipped companies flying the latest equipment over long routes linking major cities, routes that "have been traveled by ox team, pony express, railroad, automobile, and airplane . . . since white men have inhabited North America."[23]

Unfortunately, the shape of the industry in 1929 was nothing like that. As Hoover wrote years later in his memoirs, "We were threatened with a permanent muddle such as had resulted from our chaotic railway development with all its separation into short and long lines, duplication, and waste." The United States airway system was serviced by forty-four companies flying fifty-three established routes, forty-three of which covered less than five hundred miles each. Only two routes were longer than a thousand miles.

An airmail pilot on route 5 between Seattle and Salt Lake City flies over the wasteland of the Great Salt Lake in 1927. (USAF photo)

Besides being too short to be profitable, the routes were poorly arranged to meet the transportation needs of the nation. For example, American Airways, an umbrella group for Aviation Corporation transport subsidiaries, offered flights from New York to El Paso, but only in the most roundabout fashion. The Texas-bound traveler first flew to Albany, then Cleveland, Fort Worth, and finally El Paso, using the planes of three different companies in the process. Passengers continuing to Los Angeles finished their trip on the Southern Pacific Railroad. According to an industry joke, American Airways went from "nowhere to nowhere."[24]

Brown also concluded that the government's method of awarding airmail contracts actually exacerbated some of the problems facing the industry. Would-be contractors filed compensation-per-pound bids with the Post Office Department, and authorities then gave the route to the lowest bidder. Distance was a factor only if the route was more than a thousand miles, and the department completely ignored such factors as weather, terrain, or type of aircraft. Even more pressing for Brown was the need to induce airmail contractors to create nonpostal revenues to relieve the Post Office Department of a subsidy burden that Brown feared would "become intolerable." The weight-based method of payment made mail more attractive to companies than either passengers or private freight.[25]

In late 1929, Hoover and Brown unveiled their strategy for managing the situation. Under the existing system, Hoover concluded, "the air transport system was rapidly developing complete chaos." To solve the problem, Brown declared, "The method of determining the compensation of air mail contractors must be revolutionized" in order to permit "the dispatch of air mail on regularly scheduled passenger flights." To do this, he proposed to junk the weight-based formula in favor of paying a fixed rate per mile for space in the aircraft. Furthermore, the department would pay extra for night flying, for flying mountainous or fog-enshrouded routes, and for carrying passengers. Such a rate system would encourage operators to buy larger planes and establish around-the-clock schedules to lure private-sector customers as well as mail from the government. According to the postmaster general's design, the airmail contracts would serve to nurture the industry until the American people "realize the safety and advantage of travel by air on regular scheduled routes and . . . give to the air passenger carriers the support necessary to put the aeronautical industry permanently on a sound financial basis."[26]

In his approach to aviation problems, Brown largely shared the corporatist views expressed by Herbert Hoover. Most presidents, Brown observed, have been men with legal training, which meant they tended to stress laws and regulations in framing public policy. Such an attitude by politicians "probably accounts in some measure for the hostility of business and industry to any species of governmental interference or control." Now, however, the American people expected their government to implement "the principles and practices of management which have proved so successful in the commercial, industrial, and scientific world," by which Brown meant "the methods of research and cooperation which have contributed so largely to our national progress." Such practices were the essence of good management in the modern industrial state. By adopting the "scientist's method," public officials could relinquish haphazard approaches to national problems and move toward efficient, rational solutions.[27]

Brown's aviation policy was rooted in a certain set of assumptions about the American political economy. Prizing efficiency more than competition, Brown sought to replace competition with a close partnership between industry representatives and government bureaucrats. Eschewing an adversary tone and stressing consultation with the industry, he envisioned a regulatory mechanism rooted in administrative law enforced by officials whose careers shuttled smoothly between the public and private sectors. Moreover, the goals of Brown's aviation policy reflect his corporatist concern for advancing the public interest by rationalizing a chaotic industry. To a large extent, as Robert van der Linden notes, Brown's ideas reflect the New Nationalism of Theodore Roosevelt that also accepted limits on competition if corporations acted in the public interest.[28]

The major obstacle to Brown's corporatist approach was the inability of the acrimonious aviation community to achieve the kind of industry-wide consensus the postmaster general wanted to see. The situation was complicated by two prime sources of tension. First, the major aviation holding companies included a wide variety of aeronautical interests, most notably manufacturing concerns, transport operations, and various accessory companies. Consequently, many important industry leaders had administrative and financial commitments that reached beyond air transport. For example, North American Aviation and United, conglomerates that controlled a number of manufacturing interests, often took a different view of government policy than did the Aviation Corporation, which mostly controlled transport companies. A second source of tension was the rivalry among the transport companies themselves. Scrambling for scarce resources in a tight market, the firms bitterly distrusted each other. In the words of aviation student Henry Ladd Smith, they were "suspicious, jealous of each other, and only too willing to stab a rival."[29]

Hoover and Brown believed this plan would work only if they could curtail competition in the industry and guarantee the contractor that the investment in passenger-related planes and equipment would be protected. They also feared bids from companies with too little flying experience or from firms not interested in becoming passenger carriers. Such companies did not fit their vision of a flourishing, responsible industry, and they wanted to make sure these companies could not obtain a government contract. Therefore, Brown wanted to end competitive bidding for contracts and negotiate terms with firms of "good character and financial responsibility." Hoover agreed with Brown's assessment, complaining that "a great many distortions have grown up" in the airmail system. Describing competitive bidding in the airmail business as being "of doubtful value and more or less a myth," Brown feared businessmen with no experience in aviation would establish paper companies and underbid established operators who understood the real costs of the industry. He did not want to be compelled to award a contract to stock-manipulating promoters "who will pick up a flying personnel and such equipment as can get by the Department of Commerce" while simultaneously "doing the most unbusinesslike thing of throwing away . . . the experience of these men in the pioneering period of the aviation industry."[30]

Such a plan required revision of the existing airmail laws, and so in November 1929, in the aftermath of the Great Crash, Brown sought the "co-operation" of the operators in drafting a bill. He assembled a small committee of aviation experts from the private sector and put them to work preparing new legislation based on his ideas. The three were William MacCracken, Paul Henderson, and Mabel Walker Willebrandt, all people with prior experience in government who now worked in the industry. MacCracken, the first assistant secretary of commerce for aeronautics, represented several aviation clients, most notably Western Air Express, Willebrandt represented Aviation Corporation, and Henderson was an executive with National Air Transport. Second Assistant Postmaster General Washington Irving Glover and Brown himself joined the deliberations. Neither the small operators nor the nonmail operators played any role in the drafting process. These sessions were illustrative of Brown's attitude toward government regulation of the industry. Like Hoover, Brown preferred to deal with "industrial statesmen," business leaders who supposedly were aware of the public dimension of their enterprises, in preparing a regulatory system based on cooperative rather than adversarial proceedings.[31]

Within a few weeks the group had drafted a bill drastically altering the relationship between the government and the industry largely in accord with Brown's perspective. "The basic ideas of it were mine," Brown summarized, "somewhat modified and embroidered as we went along" by the other participants. It scrapped the weight formula and replaced it with a space per mile formula, which paid a generous $1.25 per mile for the total capacity of a plane. It gave the postmaster general the authority to award contracts without competitive bidding, and it retained the route certificate provisions of the 1928 law, although the new bill would permit the postmaster general to negotiate rate increases as well as decreases. The bill addressed the problem of the short route by authorizing the postmaster general to consolidate routes if such a step served the public interest. On January 14, 1930, Brown announced the substance of the proposed legislation in a speech before the Cleveland Chamber of Commerce.[32]

Brown's proposal received a hostile reception on Capitol Hill, largely because of the opposition of Clyde Kelly. In hearings before the House Committee on the Post Office and Post Roads, the "father of the airmail" strongly opposed Brown's plan to award contracts through negotiation. Convinced "such arbitrary power granted to any man would lead to abuses which would endanger the entire airmail service," he insisted that Congress could not compromise the principle of competitive bidding. Kelly's point of view won strong support from the testimony of Comptroller General John McCarl, who declared that Congress had already embraced competitive bidding in previous airmail legislation and that he saw no reason to abandon the concept. Brown's backers countered that if the government was going to assist an industry, then the executive officer responsible should have adequate authority to carry out his mandate. After a bitter discussion, the committee reported the bill out, but Kelly's influence with the House leadership kept it from reaching the floor.[33]

Stymied, the frustrated Brown realized that Kelly had effectively blocked him. Swallowing his pride, he met with Kelly, Laurence Watres of the House Post Office Committee, and a few others in a six-hour session that resulted in H.B. 11704, a revised bill that Watres introduced in mid-April. The new measure eliminated the provisions that would have given Brown the power to forgo competitive bidding. Kelly then withdrew his opposition, and the bill quickly passed both houses of Congress. Hoover signed the measure on April 29, just a few days before the old airmail contracts would have started to expire.[34]

Herbert Hoover hailed the Watres Act as "a revolution in the relations of the government with commercial aviation." Its purpose, Hoover explained to reporters on May 2, 1930, "was to enable the Postmaster General . . . to encourage passenger traffic and to bridge over from solely a postal aviation to passenger carrying airplanes." Such government-assisted efforts at industrial rationalization and development meshed neatly with the president's corporatist regulatory strategies, especially in light of Brown's insistence on involving industry leaders in the planning process. The president noted approvingly that the "heads of various aviation companies have been called in to work out the details" of the new government policy toward commercial aviation.[35]

Despite the president's public optimism, the government–industry relationship faced a major test in the wake of the Watres Act. While the act was originally conceived as an instrument of rationalization and development, the onset of the Great Depression made the issue one of industrial salvage rather than industrial development. The changing agenda placed greater stress on the Hooverian regulatory mechanisms that had worked effectively in safety-related issues but had proven much less effective in regulating the marketplace. The precarious financial condition of most aviation companies sharpened differences over market issues and made consensus virtually impossible. Deeply factionalized, industry leaders sought corporate survival rather than Brown's goal of a comprehensive, integrated national air system. The long argument over the Watres Act itself worsened these divisions. Its final version limited the postmaster general's power to mold the industry through negotiation and instead stressed more adversarial methods of regulation such as competitive bidding. These conditions posed a serious challenge to the corporatist model of industrial regulation with its emphasis on consensus building and government–industry collaboration.

As Hoover had noted, the postmaster general began by convening a meeting of aviation leaders to discuss the implementation of the Watres Act. Brown opened the conference on May 19, 1930, by outlining his plans. Gesturing toward a map of the United States, he proposed twelve airmail routes. Two of these would be transcontinental, supplementing the transcontinental air-rail route operated by Transcontinental Air Transport that ran from Los Angeles to New York. Although the meeting was restricted to the industry elite, Brown said he would limit the role each firm could play in the total airmail picture. For example, United could not operate either of the other two transcontinental routes. The other routes on the map would serve as cross lines or feeders to the major transcontinental routes.[36]

The plan's success depended on the ability of the operators to apportion the routes—especially the transcontinental routes—among themselves and to decide how to take care of the passenger lines. While the Watres Act required competitive bids before placing mail on passenger flights, the postmaster general could issue route certificates without competitive bidding to contractors who already held contracts. If the conferees could agree among themselves on who should fly various routes, Brown could issue route certificates to the big contractors with the understanding that they in turn would sublet contracts to deserving passenger lines. Thus Brown would be able to put mail on passenger lines without competitive bidding, thereby subsidizing the companies he valued while excluding the rest.[37]

The conference failed to meet the goals Brown had set for it, and on June 4 the participants gave up and adjourned. "The committee has made a study of twelve routes, and has agreed upon recommendations as to seven of these," conference chair William MacCracken reported to Brown, "while as to the remaining five there are still some matters in controversy." The operators, MacCracken wrote, "have all agreed to submit the issues to the Postmaster General in the hope that a satisfactory solution may be reached." Inconclusive as it was, this two-week session with aviation executives became known as the "spoils conference." Brown's critics charged that the big operators met to divide the spoils of federal largesse by conspiring to thwart competitive budding. His backers contended that Brown was simply inviting the industry, in MacCracken's words, "to co-operate with the Postmaster General in trying to make some sense out of the Watres Act so that aviation service in the country could continue." Perhaps more accurately, the Watres Act authorized the postmaster general to use the airmail contracts as an instrument of industrial rationalization, and Brown was inviting the airmail contractors to join him in the process. In that sense, the charges of collusion directed against the spoils conference are difficult to gauge. On the one hand, groups involved in the conferences received 14,700 miles of new airmail routes while nonparticipants received no contracts. On the other hand, the operators were so deeply distrustful of one another that collusion among them was virtually impossible. As *Fortune* magazine noted, "every air line operator on the map despised every other" and "nearly all shared an abiding professional hatred of . . . Walter Brown." Moreover, Brown later testified that he did not expect the operators to reach agreement but instead anticipated they would ask him to adjudicate their differences, opening the way for him to offer his own solution.[38]

In retrospect, the charges of collusion seem irrelevant to the real purpose of the meeting, namely formulating a coordinated response to economic trauma created by the Depression. In that sense, the spoils conference not only typified the recovery mechanisms devised by the Hoover administration but also anticipated the industry-wide conferences arranged by the New Deal under the National Industrial Recovery Act. However, the outcome of the meetings clearly reflected the failure of an important element of the administration's corporatist philosophy, specifically the need for the industry to formulate answers to its own problems. The acrimony and ultimate deadlock of the spoils conference severely undercut the basic working mechanisms of corporatism.[39]

With the contractors unable to agree among themselves on a division of the routes, Brown turned to other strategies for rationalizing and apportioning routes. The Watres Act authorized the postmaster general to extend routes in the public interest, a power Brown planned to interpret broadly. Such extensions would permit Brown to referee route disputes by generously extending the lines of favored companies. His plan was to use his extension power to award the transcontinental routes without opening them for bids. However, Comptroller General John McCarl, a strong critic of the Watres bill as originally written, distrusted Brown and was known to interpret the statute more narrowly. As the sole authority for government disbursements, McCarl could refuse payments on contracts he considered improper or illegal. In early June, the postmaster general tested McCarl's attitude by extending the Chicago–Minneapolis route held by Northwest Airways all the way to Winnipeg, Manitoba, an action that roughly doubled the original route. Within a week, McCarl rejected the change. He claimed Brown had exceeded his authority and further asserted that extensions must be fairly modest and not major additions to the original route.[40]

Undeterred, Brown now chose to manipulate the bidding process. In the late summer of 1930, he reconvened representatives of selected companies to consider how to restrict access to the contracts. Requirements that bidders post a bond of $250,000 and begin operation within thirty days excluded small companies and those not already flying. Neither provision, however, would automatically preclude bids from passenger lines. Consequently, at MacCracken's suggestion, Brown wrote into the bid specifications a requirement that all bidders have at least six months experience in night flying over a route 250 miles in length, a requirement only the mail operators could meet. No passenger lines flew at night.[41]

A storm of controversy centers on the night flying requirement. Brown insisted that the requirement was legal and appropriate. He pointed out that the postmaster general was authorized to determine the safety standards of railway mail cars, and the original Air Mail Act had directed him to "make such rules, regulations, and orders" as were necessary to implement the law. Brown also cited, a bit disingenuously, paragraph 7, section 1330, of the *Postal Laws and Regulations* in support of his actions. However, Brown himself had written that section the day the contracts were issued. Writing in 1972, noted aviation scholar Ronald E. G. Davies found "much to be said in Brown's defence." Airmail obviously had to move at night or its advantage over ground transportation vanished. Experience flying a route at night "was demonstrably a reasonable proof" of a bidder's ability to fulfill that responsibility.[42]

The events of 1930 represent the high tide of Hooverian associationalism in commercial aviation policy. As market conditions in the aviation industry grew more precarious, Walter Brown led the administration toward a strategy for regulating the marketplace modeled on the strategy Hoover had used so successfully to regulate safety matters when he was secretary of commerce. Because Congress had granted Brown only limited authority to use the Watres Act as an instrument of industrial rationalization, the postmaster general tried to supplement his legally delegated power with bureaucratic power rooted in informal agreements between public officials and industry leaders. Brown's strategy was enormously effective, and he fashioned the enduring outlines of the nation's commercial aviation industry. However, the Brown model was also susceptible to political attack, especially as the Great Depression eroded the credibility of the Hoover administration in economic matters.

Although Walter Brown struggled to maintain the corporatist model for industrial rationalization, events forced substantial modifications in the administration's policies. Beset by a depression economy, air transport operators proved less able than ever to generate the kind of cooperative solution to common problems that was crucial to the functioning of the model. Brown still sought to operate through consultation with industrial statesmen and trade associations, but such mechanisms were less and less effective because he could not establish a consensus for his vision of the industry. The mail contractors remained highly suspicious of each other and did not embrace Brown's vision of the future. Small firms without mail contracts added to the industry-wide acrimony, with their angry and increasingly effective complaints about the administration of airmail policies. Finally, Congress took a larger role in airmail policy by cutting appropriations, specifying routes to be funded, and investigating Brown's actions. In the face of this discord, the corporatist model faded, and the postmaster general increasingly operated as a "one man public service commission," establishing routes, setting rates, and restricting entry to the industry. His actions are clearly an important step toward the regulatory policy established by the Civil Aeronautics Act of 1938.[43]

By the end of Herbert Hoover's term in March 1933, the corporatist model that had guided much of the administration's aviation policy had seriously eroded, but to a remarkable extent Brown had achieved his goals through an expansive and energetic interpretation of the Watres Act. The nation was bound together by eight north–south lines and three transcontinental routes. Perhaps most striking, the once-frail passenger-carrying lines prospered in the midst of the nation's worst depression. Between 1930 and 1932, the latter being Brown's last full year in office, the number of passenger-miles flown rose from 84,000,000 to 127,000,000. Scheduled aircraft miles during the same period went from 16,200,000 to 34,500,000. Simultaneously, Brown cut the rates the government paid to the haulers through periodic renegotiation of the contracts. The average rate per mile had been $1.09 when he took office, but the Post Office Department was paying only $.54 per mile in March 1933. Clearly Brown's efforts mark one of the most important achievements of the Hoover administration.[44]

Whatever its merits as an exercise in economic development, the Hoover-Brown aviation policy proved to be a political failure. Brown's efforts to eliminate the smaller carriers proved unsuccessful, and his enemies within the industry grew stronger politically as the Hoover administration slid toward defeat. In early 1934, Brown's actions in awarding airmail contracts became the focal point of a major congressional investigation chaired by Senator Hugo Black. Deeply suspicious of large concentrations of economic power in private hands, Black stressed competition as the surest regulator of economic activity, and despite his usually fervent support for President

Postmaster General Walter Folger Brown is sworn in before testifying at a Senate committee hearing investigating airmail contracts in February 1934. (NASA photo)

Franklin D. Roosevelt, he had strongly opposed a major Roosevelt program, the National Recovery Administration, because he believed it sanctioned collusion and price-fixing. His sensational investigation of Hoover-Brown policies attracted extensive news coverage and made the young Alabama senator a national figure.[45]

In his investigation of the airmail, Black was certainly concerned with exposing corruption, but his real aim was to challenge the assumptions that underlay Republican aviation policy and to generate a Democratic alternative. Like Brown, Black believed a strong commercial aviation was important to America's future, and he favored "governmental assistance to a young and growing and necessary industry." Under Brown, however, "these huge Governmental expenditures have in great part found their way into the pockets of profiteers, stock manipulators, and powerful financial groups, who never flew a plane, who never invented an engine, who never improved an airplane part." Repudiating Brown's emphasis on large, well-financed corporations, Black argued that aviation would benefit from competition among small firms. Furthermore, the senator insisted that any firms holding airmail contracts should sever their connections with holding companies or any other business engaged in transportation or manufacture. Only if the companies were free of the "vast power of holding companies to sell their own products, at their own price, to their own subsidiaries" could the public be assured the contractors were operating at the lowest possible cost and not feathering the nests of their business partners. Finally, Black demanded an open system of competitive bidding to guarantee that all interested firms would have a chance at a contract. Permitting market forces to operate in the industry would prod its development more efficiently than would the government–business partnership of the Hoover-Brown years.[46]

Black moved quickly to establish these themes early in the hearings. The list of first-day witnesses included Paul Henderson, a former second assistant postmaster general in the Calvin Coolidge administration and current vice president of United Aircraft and Transport. Widely recognized as a leading spokesman for aviation interests, Henderson provided Black with a morning of very useful testimony. Denying he had anything to do with the Watres Act, he declared the draft of legislation was "gotten up in the Postmaster General's office and submitted to those of us who were interested in it." To Henderson, the spoils conference, aptly named because "it spoiled a number of things," represented obvious collusion. The postmaster general's plan to extend the line of existing airmail contractors and then arrange the subletting of those extended routes to financially starved passenger lines which had no contracts struck Henderson as "so contrary to the spirit of the law . . . that I personally took the thing as a joke." When the conference deadlocked and failed in its efforts to divide the airmail pie, the postmaster general stepped in to achieve his goals by restricting competition and extending existing routes.[47]

Henderson's viewpoint drew support from two other opening-day witnesses, Erle Halliburton and Thomas McKee, both disgruntled small operators undercut by Walter Brown's policies. A highly successful Oklahoma entrepreneur, Halliburton had gotten into the aviation business by organizing Southwest Air Fast Express or Safeway, a skillfully managed line that offered to fly the mail at bargain-basement rates. To forestall that possibility, American bought out Safeway, giving Halliburton a very substantial profit. Halliburton had made such a nuisance of himself that at one point W. Irving Glover, the second assistant postmaster general stormed, "I will ruin you if it is the last act of my life; you have tried to buck this thing all the way through and you are not going to do it." McKee, an attorney for the defunct Wedell-Williams Air Service Corporation, added to Halliburton's intimation that public policy was being made by conspiracy. The spoils conference, according to McKee, who had been thrown out of the meeting by MacCracken "in his very cordial and pleasant way," was made up of "fourteen or fifteen men talking and discussing in an atmosphere of blue smoke." Taken together, the testimonies of Halliburton and McKee established precisely the sort of sinister image Black hoped to plant in the public mind.[48]

Reacting to the hearings, the Roosevelt administration concluded that the airmail contracts were fraudulent. In a startling move, Postmaster General James Farley announced that, effective February 19, all domestic airmail contracts would be canceled. The Army Air Corps would fly the mail until new contracts could be negotiated. The system Walter Brown had established was swept aside, and the Democrats now set about constructing a new model. The announcement represents the zenith of the Black investigation; its findings had widely discredited the architects of the existing airmail structure and prompted the administration to embrace a new policy, which largely reflected the assumptions that had undergirded the investigation.[49]

The same day Roosevelt's executive order took effect, the embattled Walter Brown testified before Senator Black's committee. Opening with a lengthy statement, Brown vigorously defended his airmail policies. "The major purpose," he began, "was not to transport the mails at the lowest possible cost to the Government, but to foster the maritime and aeronautical industries. . . . The ultimate goal . . . is to create an economically independent aeronautical industry by enabling air-transport operators to recoup in the form of mail pay their out-of-pocket losses while they are building up adequate passenger and express revenues from the public and are developing transport airplanes capable under competitive conditions of earning their costs of operation." Far from representing collusion, the spoils conferences of May and June 1930 resulted in a "tentative suggestion" which was "impracticable and unsound." Indeed Brown declared that he had suspected such a result and had convened the meeting so the competing firms would thereby be forced to acknowledge their conflicting claims. The way would then be open for Brown to step in and structure the various routes.[50]

Brown also defended his reliance on the large firms at the expense of the small operators. In his opinion, "there was no justification in sound business practices" for supporting firms operating short lines. From the history of railroad development, Brown said, "We had discovered that all short lines were absorbed by the larger lines, . . . and I could see nothing to be gained by covering the map over with . . . little, disjointed services." Very properly then, the Post Office Department "exerted every proper influence to consolidate the short, detached, and failing lines into well-financed and well-managed systems." All such consolidations "authorized by me were in the public interest," Brown declared, and "resulted in improved public service and ultimately in lower flying costs."[51]

Questioning the witness sharply, Black examined Brown extensively about the background of the Watres Act, a piece of legislation that the latter admitted he "probably had as much to do with it as anybody." Prodded by the chairman, Brown read from his own testimony at a 1930 House hearing in which he had described competitive bidding on airmail contracts as being "of doubtful value and more or less of a myth" because there is "only little real substantial bidding by men of experience able to carry on in an industry of this kind in the present state of the art." Black emphasized that although the bill as originally drawn had permitted negotiated contracts, Congress had not passed the bill until it had been rewritten specifically to include competitive bidding.[52]

Black next raised the question of the spoils conference, asking "was not the object and purpose of the proceeding . . . for the operators to attempt to decide among themselves as to which ones should have which lines?" He then read into the record a letter written by Lewis H. Brittin of Northwest Airways in which Brittin stated that Brown had "called the operators together, handed them this map and instructed them to settle among themselves the distribution of these routes." As further evidence, Black referred to testimony by Hainer Hinshaw of American Airways to the effect that the operators had no choice but to accept the division of routes, which the postmaster general intended to make, regardless of competitive bidding. If such evidence was true, Black demanded, would the contracts awarded under such circumstances be illegal? Brown refused to answer, saying the question was hypothetical: "I am not going to answer a question as foolish as that."[53]

Black also assailed Brown's reliance on the major aviation holding companies in constructing his air network. He repeatedly demonstrated how Brown had disregarded equities in awarding mail contracts to large operators at the expense of smaller ones. Using C. E. Woolman's Delta Air Service as an example, Black pointed out that Woolman was the "first man and the only man" to fly between Birmingham and Dallas. However, the mail contract went to the giant Aviation Corporation with the result that "the man who had pioneered the route, the only man who had ever flown over it on any schedule of any kind or type and who was flying on it at that time was compelled to sell out because he could not operate in competition with mail contractors." Brown countered that "a short line could not pay the expense of maintaining a ground force and the supervisory force that would be necessary." Woolman needed a line that "started some place and got some place"; in other words, he needed to fly on to Los Angeles, something Woolman was unwilling to do.[54]

Increasingly the hearings became a bitter personal clash between prosecutor and witness. Black repeatedly interrupted answers he considered irrelevant by telling the clerk to read the question to Brown. Finally the exasperated Brown snapped, "Let me finish." "Answer these questions," the chairman responded. "If you don't care to listen to me, I won't answer any more questions." "I don't want any more speeches." "You want to make the speeches." This personal antipathy reached something of an anticlimax late in the hearings when Brown, describing a conversation with Postmaster General James Farley, referred to a personal remark Farley allegedly made about Black. Brown refused to repeat the remark without Farley's approval, so the committee summoned him to a Saturday afternoon session to release Brown publicly from his pledge of discretion. Brown then quoted Farley as saying of Black, "He is just a publicity hound but don't tell anybody I said so, because I have to get along with him." The hearing room exploded in laughter although Farley diplomatically denied making the remark. The hostility between Brown and Black was not so easily expunged, however; nearly a decade later Brown was still referring to "that Ku Kluxer Black" in his private correspondence.[55]

The committee concluded its deliberations on April 25, 1934, with Hugo Black's report to the Senate, but the impact of its work lasted for nearly a decade after the final gavel. Most immediately its findings shaped a new federal air law, the Black-McKellar Act of 1934. Reflecting Senator Black's concerns about the industry, the measure limited the rates to be paid to airmail contractors, outlawed interlocking interests between contractors and other aviation interests, prohibited mergers between competing companies, and restricted the salaries of company officials. Another highly controversial provision blacklisted the participants in the 1930 spoils conference, saying no company that employed any of those people could hold an airmail contract. "In a sense," historian Ellis Hawley has written, "the act represented a qualified victory for the champions of small business and enforced competition." The committee also received a form of approval in the U.S. Court of Claims. Several carriers, including United, sued the government over the cancellation of their airmail contracts. The others settled out of court, but United pursued the case until 1942, when a three-judge panel voted 2–1 that the participants in the spoils conference were indeed guilty of collusion and the Roosevelt administration had acted properly in suspending their airmail contracts.[56]

Although successful in an institutional sense, the aviation policy generated by the Black Committee failed in the economic arena, and the Brown model later achieved a reaffirmation of sorts. The Black-McKellar Act dealt only with airmail while ignoring the broader problem of commercial aviation in general. Under its provisions, the carriers lost money and many provided inferior service. The blacklisting of those in attendance at the 1930 spoils conference raised civil liberties questions in some circles. Moreover, one of the first actions taken by Postmaster General Farley in implementing the new law was to assemble a meeting of the contractors, a session some found similar to the much-criticized spoils conference of the Brown years. Indeed many argued that the 1930 conferences resembled the industry-wide meetings held under the auspices of the National Industrial Recovery Act to draw up the Blue Eagle Codes. In a sense, then, the Black-McKellar Act with its emphasis on competition represented a sharp departure from New Deal policy in most other areas of the economy.[57]

Finally, in 1938 Congress repudiated the Black-McKellar approach and passed the Civil Aeronautics Act, which restructured the industry according to the assumptions of the once discredited Hoover-Brown policy. The new law emphasized government-corporate linkages, limited competition, and restricted entry to the industry. As aviation historian Henry Ladd Smith put it, the Civil Aeronautics Act "tossed overboard the old conception of competition, thereby coming back to the policy advocated by Postmaster General Brown," an interpretation reiterated by a more recent scholar, Ellis Hawley, who described the act as restoring the "legitimacy" of the Brown approach to the industry.[58]

Herbert Hoover clearly stated his opinion of the Black Committee hearings. He described the investigation as an effort by the New Deal to demonstrate "by any method, honest or dishonest, that corruption or malfeasance marked my administration." Writing to a friend, he declared that "my blood boils at much evident misrepresentation," and he fumed that the "so-called 'conspiracy' meeting of the airmail contractors was fully reported in the press at that time." Nevertheless, Hoover ultimately found a note of vindication in the hearings. Despite the committee's best effort, Hoover concluded, "They were unable to find an atom of corruption."[59]

Although often overlooked even by his champions, Herbert Hoover's work on behalf of commercial aviation is among the most important legacies of his public career. When Hoover became secretary of commerce in 1921, the federal government had no formal connection with aviation at all. Twelve years later, however, the Department of Commerce was primarily responsible for assuring the safety of commercial flying, and the Post Office Department was regulating the economic side of the industry, an industry growing rapidly in the midst of the Great Depression. Drawing on their shared heritage of progressivism and the concept of the associative state, Hoover and Walter Brown played a major role in shaping these policies. "Far from restraining and injuring aviation," Robert van der Linden has written, "Hoover's policies actively promoted the new industry and through rational regulation and judiciously applied subsidies and incentives, which resulted in a national transportation infrastructure within a remarkably short span of only four years."[60]

NOTES

I would like to acknowledge the support of the Smithsonian Institution, the National Endowment for the Humanities, the Herbert Hoover Presidential Library Association, and Western Kentucky University in the preparation of this essay. Portions of this article have previously appeared in *Business History Review*, in *The Historian*, and in William M. Leary, ed., *Aviation's Golden Age: Portraits from the 1920s and 1930s* (Iowa City: University of Iowa Press, 1989).

1. David D. Lee, "Herbert Hoover and the Rise of Commercial Aviation, 1921–1926," *Business History Review* 58 (1984): 78–102; F. Robert van der Linden, *Airlines and Air Mail: The Post Office and the Birth of Commercial Aviation Industry* (Lexington: University Press of Kentucky, 2002); Ellis Hawley, "Three Facets of Hooverian Associationalism: Lumber, Aviation, and Movies, 1921–1930," in Thomas McCraw, ed., *Regulation in Perspective* (New York: Cambridge University Press, 1981), 95–123; Henry Ladd Smith, *Airways: The History of Commercial Aviation in the United States* (New York: Alfred A. Knopf, 1942), 99; Herbert Hoover, *The Memoirs of Herbert Hoover: The Cabinet and the Presidency, 1920–1933* (New York: Macmillan, 1952), 132–33, 243–45; R. E. G. Davies, *Airlines of the United States since 1914* (London: Putnam, 1972), 110–11.

2. Joan Hoff Wilson, *Herbert Hoover: Forgotten Progressive* (Boston: Little, Brown, 1975), 38–39, 73; Ellis Hawley, *The Great War and the Search for a Modern Order: A History of the American People and Their Institutions, 1917–1933* (New York: Waveland Press, 1997), 3–15; Ellis Hawley, "Herbert Hoover, the Commerce Secretariat, and the Vision of an 'Associative State,' 1921–1928," *Journal of American History* 61 (June 1974): 116–40.

3. Hawley, "Herbert Hoover," 116–40; Ellis Hawley, "The Great War and the Search for a Modern Order," in Martin Fausold and George Mazuzan, eds., *The Hoover Presidency: A Reappraisal* (Albany: State University of New York Press, 1974), 101–19; William Appleman Williams, *Some Presidents: Wilson to Nixon* (New York: New York Review, 1972), 33–49; William Appleman Williams, *The Contours of American History* (Chicago: World, 1959), 425–38; James Weinstein, *The Corporate Ideal in the Liberal State* (Boston: Little, Brown, 1968); Herbert Hoover, *American Individualism* (Garden City, N.Y.: Doubleday, 1922), 32–62.

4. Lee, "Herbert Hoover," 79–80, 92–95; Hawley, "Three Facets," 108–15, 120–23; Hawley, "Herbert Hoover," 116–40.

5. Hawley, "Three Facets," 95–123; Smith, *Airways*, 94–102; Donald R. Whitnah, *Safer Skyways: Federal Control of Aviation, 1926–1966* (Ames: Iowa State University Press,

1966), 15–23; Nick A. Komons, *Bonfires to Beacons: Federal Civil Aviation Policy under the Air Commerce Act, 1926–1938* (Washington, D.C.: U.S. Government Printing Office, 1978), 7–25.

6. Lee, "Herbert Hoover," 84, 94–95, 97; Hawley, "Three Facets," 111.

7. Lee, "Herbert Hoover," 92, 95, 100–102; Smith, *Airways*, 99; Komons, *Bonfires to Beacons*, 25, 28, 45–56, 80–82.

8. Komons, *Bonfires*, 197–216; Smith, *Airways*, 306; Hawley, "Three Facets," 114; van der Linden, *Airlines and Air Mail*, vii–xii.

9. Lee, "Herbert Hoover," 78–81, 88–89; Hawley, "Three Facets," 108–15; Herbert Hoover, "Statement on Commercial Aviation," in Aviation 1925 file; Herbert Hoover, "Speech before San Francisco Chamber of Commerce," in Aviation 1926 file, both in Box 40, Commerce Papers, Herbert Hoover Presidential Library, West Branch, Iowa.

10. Hoover, "Commercial Aviation"; Hoover, "Speech before San Francisco Chamber of Commerce."

11. Hoover, "Commercial Aviation"; Hoover, "Speech before San Francisco Chamber of Commerce."

12. Hoover, "Commercial Aviation"; quote is from Hoover, "Speech before San Francisco Chamber of Commerce."

13. Hoover to Hiram Bingham, Sept. 23, 1925; James Wadsworth Jr. to Hoover, Dec. 5, 1925; Clarence Young to Hoover, Dec. 7, 1925; Hoover to Wesley Jones, Dec. 9, 1925, all in Commerce Department: Bureau of Aeronautics; Legislation 1925–1926 file, Box 123, Commerce Papers, Herbert Hoover Presidential Library; *Congressional Record*, 69th Cong., 1st sess., 828–30, 7314–17; *New York Times*, Dec. 9, 1925; Komons, *Bonfires*, 80–81.

14. William MacCracken, "Special Problems in Aeronautical Legislation," 1–2, copy in Speeches file, William MacCracken Papers, Herbert Hoover Presidential Library; Hoover, *Memoirs*, 134; NACA quote is from Smith, *Airways*, 98; Elsbeth S. Freudenthal, *The Aviation Business: From Kitty Hawk to Wall Street* (New York: Vanguard, 1940), 77.

15. Hoover, *Memoirs*, 134.

16. Hawley, "Three Facets," 112.

17. *Aviation*, Apr. 1931, 218; Komons, *Bonfires to Beacons*, 191–94; Smith, *Airways*, 94–96.

18. Komons, *Bonfires to Beacons*, 191–97.

19. The first quote is from Hoover to Walter F. Brown, Aug. 21, 1929, Presidential Papers, Cabinet Offices, Post Office, Correspondence, Box 40; other quotes are from Hoover press conference excerpt, Oct. 15, 1929, Post-Presidential Papers, Subject—Air Mail Cancellation, Box 12, both in Herbert Hoover Presidential Library.

20. *New York Times*, Jan. 27, 1961; Anne Hard, "Uncle Sam's New Mail Man," *New York Herald Tribune*, Apr. 7, 1929, copy in Walter F. Brown file, Box 9, President's Personal File, Presidential Papers, Herbert Hoover Presidential Library; Harvey S. Ford, "Walter Folger Brown," *Northwest Ohio Quarterly* (Summer 1954): 200–209, copy in Walter F. Brown file, Box 9, President's Personal File, Presidential Papers, Herbert Hoover Presidential Library; quote is from Hoover, *Memoirs*, 220; Randolph Downes, *The Rise of Warren Gamaliel Harding, 1865–1920* (Columbus: Ohio State University Press, 1970), 178–80, 346–50; Robert K. Murray, *The Harding Era: Warren G. Harding and His Administration* (Minneapolis: University of Minnesota Press, 1969), 415–16.

21. Hoover quote is from *New York Times*, Jan. 27, 1961; Hard, "Uncle Sam's New Mail Man," in Walter F. Brown file, Box 9, President's Personal File, Presidential Papers, Herbert Hoover Presidential Library; Senate Committee to Investigate Foreign and Domestic, Ocean and Air Mail Contracts, *Investigation of Air Mail and Ocean Mail Contracts, Hearings*, 73d Cong., 2d sess., 1934, 2353 [hereafter Black Committee Hearings]; Hoover, *Memoirs*, 220, 243–45; Murray, *Harding Era*, 415–16.

22. Walter F. Brown, "Commercial Aviation and the Air Mail," address to Cleveland Chamber of Commerce, Jan. 14, 1930, 3, in Post Office Correspondence file, Box 40; Walter F. Brown, untitled address to Bond Club of New York, Apr. 27, 1931, 6–7, in Post Office—Press Releases of the Postmaster General, 1931, file, Box 42, both in Presidential Papers, Herbert Hoover Presidential Library; Black Committee Hearings, 2207, 2863.

23. Walter F. Brown, untitled address over NBC Radio, Sept. 23, 1931, 5, in Post Office—Press Releases of the Postmaster General, 1931, file, Box 42; Walter F. Brown, untitled address to Advertising Club of Washington, Oct. 8, 1929, 9, in Post Office—Press Releases of the Postmaster General, 1929–April 1930, file, Box 42, Presidential Papers, Herbert Hoover Presidential Library.

24. American Airways quote is from Komons, *Bonfires to Beacons*, 197–98; Davies, *Airlines of the United States since 1914*, 114; Hoover, *Memoirs*, 135.

25. Walter F. Brown, untitled address to the Bond Club of New York, Apr. 27, 1931, 1–6, in Post Office—Press Releases of the Post Master General, 1931, file, Box 42, Presidential Papers, Herbert Hoover Presidential Library.

26. Brown, "Commercial Aviation and Air Mail," 7–9, Post Office Correspondence file, Box 40, Presidential Papers, Herbert Hoover Presidential Library; Hoover, *Memoirs*, 243–45.

27. Quotes are from Walter F. Brown, untitled address to the Charitable Irish Society of Boston, Mar. 17, 1930, esp. 5–6, in Post Office Correspondence, March 1930, file, Box 40, Presidential Papers, Herbert Hoover Presidential Library; Hawley, "Herbert Hoover," 116–40; Hawley, "Three Facets," 95–123.

28. Van der Linden, *Airlines and Air Mail*, 63–64.

29. Smith, *Airways*, 241; William (Doc) Bishop to James Woolley, Dec. 7, 1929, in Paul Henderson file, Box 122, records of the Black Committee, Record Group 46, National Archives, Washington, D.C. [hereafter Black Committee records]; Black Committee Hearings, 2349–51, 2371–73, 2451–58.

30. Hoover quote is from press conference excerpt, Oct. 15, 1929, Post-Presidential Papers, Subject—Air Mail Cancellation, Box 12, Herbert Hoover Presidential Library; Brown quote is from U.S. Congress, House, Committee on the Post Office and Post Roads, *Amending the Air Mail Act, Hearings*, 71st Cong., 1st sess., 1930, 1, 8, 23–24; Brown, "Commercial Aviation and Air Mail," 3–9; Hoover, *Memoirs*, 243–45.

31. Black Committee Hearings, 1471–72, 2371–72; Brown, "Commercial Aviation and Air Mail," 7–9; Davies, *Airlines of the United States*, 114; Dorothy M. Brown, *Mabel Walker Willebrandt: A Study of Power, Loyalty, and Law* (Knoxville: University of Tennessee Press, 1984), 198–200.

32. Black Committee Hearings, 1471–72, 2371–72, esp. 2372; Brown, "Commercial Aviation and Air Mail," 7–9; Davies, *Airlines of the United States*, 114.

33. Kelly quote is from U.S. Congress, House, *Congressional Record*, 71st Cong., 2d

sess., 1931, 7373, 7377; Glover notes, Mar. 15, 1930, Earl Wadsworth file, Box 129, Black Committee records; Black Committee Hearings, 2437–39; House Committee on the Post Office and Post Roads, *Amending the Air Mail Act*, 25; Michael Osborn and Joseph Riggs, *"Mr. Mac": William P. MacCracken on Aviation, Law, and Optometry* (Memphis, Tenn.: Southern College of Optometry, 1970), 145–47.

34. Glover notes, Mar. 15, 1930, Earl Wadsworth file, Box 129; Bishop to Woolley, Apr. 6, 1930, Paul Henderson file, Box 122; Florence Kahn to Woolley, Apr. 17, 1930, and undated, Paul Henderson file, Box 122, all in Black Committee records. See also Black Committee Hearings, 2438, and Court of Claims decision, 120, copy in Air Mail file, Box 34, James Farley Papers, Library of Congress, Washington, D.C.

35. Hoover, press conference excerpt, May 2, 1930, Post-Presidential Papers, Subject—Air Mail Cancellation, Box 12, Herbert Hoover Presidential Library.

36. Minutes taken by Earl Wadsworth, Aviation Bid file, Box 112, and Post Office Department press release, May 19, 1930, copy in Brown's Press Statements file, Box 122, Black Committee records; Black Committee Hearings, 1476, 2323–26; Osborn and Riggs, *"Mr. Mac,"* 147–48; William Van Deusen interview with William P. MacCracken, June 3, 1966, transcript in William P. MacCracken Papers, Herbert Hoover Presidential Library.

37. Black Committee Hearings, 1476, 1584.

38. Ibid., 2350–51, 2437–2555; William MacCracken to Walter Brown, June 4, 1930, Aviation Bid file, Box 112, Black Committee records; *Fortune*, May 1934, 142.

39. Hawley, *Great War*, 226–29.

40. *Congressional Record*, 71st Cong., 2d sess., 1931, 7377; "Report of Fulton Lewis, Jr.," Box 118, Black Committee records, 53–55; Paul David, *The Economics of Air Mail Transportation* (Washington, D.C.: Brookings Institution, 1934), 111–14; Smith, *Airways*, 168–69; Komons, *Bonfires to Beacons*, 204–5.

41. Fulton Lewis report, Black Committee records, 53–55; David, *Economics of Air Mail Transportation*, 111–14; Smith, *Airways*, 168–70; Komons, *Bonfires to Beacons*, 204–5.

42. Fulton Lewis report, Black Committee records, 60; Smith, *Airways*, 170; Davies, *Airlines of the United States*, 119.

43. Francis A. Spencer, *Air Mail Payment and the Government* (Washington, D.C.: Brookings Institution, 1941), 43; Van der Linden, *Airlines and Air Mail*.

44. Komons, *Bonfires to Beacons*, 211; Smith, *Airways*, 212.

45. Hugo Black, "Inside a Senate Investigation," *Harper's*, Feb. 1936, 279; John P. Frank, *Mr. Justice Black: The Man and His Opinions* (New York: Alfred A. Knopf, 1949), 63; Virginia Van Der Veer Hamilton, *Hugo Black: The Alabama Years* (Baton Rouge: Louisiana State University Press, 1972), 224–34; M. Nelson McGeary, *The Development of Congressional Investigative Power* (New York: Alfred A. Knopf, 1940), 45.

46. Frank, *Mr. Justice Black*, 65; Hamilton, *Hugo Black*, 220; quotes are from Hugo Black, CBS Radio address, reprinted in *New York Times*, Feb. 17, 1934.

47. Black Committee Hearings, 1457–87, esp. 1472–76.

48. Black Committee Hearings, 1440–57, 1488–1525, esp. 1446, 1447, 1491.

49. Komons, *Bonfires to Beacons*, 254–59; Smith, *Airways*, 249–51; *New York Times*, Feb. 10, 1934.

50. Black Committee Hearings, 2349–51; *New York Times*, Feb. 20, 1934.

51. Black Committee Hearings, 2351, 2599; *New York Times*, Feb. 20, 24, 1934.

52. Quotes are from Black Committee Hearings, 2372–77; James Doran and Eggleton to A. G. Patterson, Feb. 18, 1934, Avigation Bid file, Box 112, Black Committee Papers; *New York Times*, Feb. 20, 1934.

53. Quotes are from Black Committee Hearings, 2437, 2650–51; Doran and Eggleton to Patterson, Feb. 18, 1934, Avigation Bid file, Box 112, Black Committee Papers; *New York Times*, Feb. 24, 1934.

54. Quotes are from Black Committee Hearings, 2450–55; *New York Times*, Feb. 21, 1934.

55. Black Committee Hearings, 2652, 2742; *New York Times*, Feb. 25, 1934; Walter F. Brown to A. E. Benson, June 16, 1942, File 14, Box 1, Collection 12, Walter Brown Papers, Ohio Historical Society. Brown was referring to Black's brief membership in the Ku Klux Klan early in his political career.

56. Paul Godehn and Frank Quindry, "Air Mail Contract Cancellations of 1934 and Resulting Litigation," *Journal of Air Law and Commerce* (Summer 1954): 253–76, esp. 271–74; Komons, *Bonfires to Beacons*, 266–67, 271–72; Ellis Hawley, *The New Deal and the Problem of Monopoly: A Study in Economic Ambivalence* (Princeton, N.J.: Princeton University Press, 1966), 241–43.

57. *Time*, Feb. 26, 1934, 26–28; Komons, *Bonfires to Beacons*, 267, 352–54; Smith, *Airways*, 301–6; Hawley, "Three Facets," 121–23.

58. Smith, *Airways*, 278–79; Hawley, "Three Facets," 114; Hawley, *New Deal and the Problem of Monopoly*, 242.

59. Herbert Hoover to J. C. O'Laughlin, Feb. 16, 1934, O'Laughlin, J. C., file, Box 457B, Post-Presidential Papers, Herbert Hoover Presidential Library; Hoover, *Memoirs*, 245.

60. Van der Linden, *Airlines and Air Mail*, v.

CHAPTER 5
EDWARD V. RICKENBACKER'S REACTION TO CIVIL AVIATION POLICY IN THE 1930S
A HIDDEN DIMENSION

No aeronautical leader reacted more strongly against federal policies of which he disapproved than did Edward V. (Eddie) Rickenbacker, who became general manager of Eastern Air Lines in 1934 and president and chief executive officer after a reorganization took place in 1938. Even before taking charge at Eastern, which under his management became one of the largest and most profitable carriers in the American skies, Rickenbacker had been a stormy petrel of the aviation industry. His views and activities were constantly in the news. A man with strong convictions who never hesitated to speak his mind, Rickenbacker was a passionate defender of free enterprise. He held extremely conservative views about the limits of government power. His animosity toward Franklin D. Roosevelt and the New Deal put him on a collision course with the Civil Aeronautics Board, which emerged by 1940 as the ultimate arbiter of airline development.[1] The career path that led to Rickenbacker becoming a commanding presence in the history of American flight is well known.[2] Nevertheless, there is a pattern in his rise to prominence and the way he reacted to gyrations in federal policy affecting civil and commercial aviation that needs to be better understood. My purpose here will be to show how this pattern arose from his automotive career, and particularly from his connections with General Motors (GM), the colossus of the automobile industry.

Rickenbacker was already famous in the automotive world by the time he went to France in 1917 as a driver on the staff of General John J. Pershing. He was best known as a racer who had competed regularly in the Indianapolis 500 and won seven major championship events.[3] After he switched from being a chauffeur to become an extraordinary combat pilot in the American Air Service, he shifted his horizons and aspired to leadership in the infant aviation industry. But he always had the automobile industry to fall back on, and he did so repeatedly throughout the postwar period.

When Rickenbacker came home in 1919 as a national hero after shooting down twenty-six German aircraft in World War I, a feat that made him America's ace of aces, he hoped to become an aircraft manufacturer. He

Eddie Rickenbacker on the steps of an Eastern Airlines plane. (NASA photo)

considered joining wartime comrade Reed Chambers, high-altitude balloon-ist Rudolph W. "Shorty" Schroeder, and aircraft designer Alfred Verville in plans to establish an enterprise that they hoped would receive a flood of orders for planes from the separate U.S. Air Force that General William "Billy" Mitchell was trying to create.[4] These dreams faded as Mitchell's ef-forts met determined opposition and it became obvious that markets were too flooded with surplus planes left over from the war to profit from making more of them. Instead, Rickenbacker returned to familiar ground by accepting an offer from Detroit millionaire Barney Everitt to back him in bringing out a new automobile, the Rickenbacker, which would be named in his honor. Before going to Detroit to begin manufacturing the car, Rickenbacker spent several months in California learning the ropes by helping set up a dealership network for the Sheridan, an automobile that that recently been added to GM's growing menagerie of motor vehicles.[5] His work drew favorable attention from Pierre S. Du Pont, Alfred P. Sloan, and other leaders in the giant corporation that would figure so importantly in his future.[6] His evolving relationship with GM entered a new phase in 1922 when he married Adelaide Frost Durant, whose previous husband, playboy and race-car driver Cliff Durant, whom she divorced on grounds of mental cruelty, was a son of William C. "Billy" Durant, the founder of GM. Adelaide's former father-in-law continued to dote on her after the divorce and gave her large blocks of shares in General and United Motors as Christmas and birthday presents, helping greatly to strengthen Rickenbacker's finances.[7]

Rickenbacker's heart was in the sky, and he soon tired of being an automobile manufacturer. He spent much of his time flying around the country trying to set speed and distance records and promoting new ideas like duralumin construction, embodied in the Larsen JL-6, an imported version of the Junkers J-13. Partly because he could not focus his attention exclusively on automobiles, the Rickenbacker Motor Company did not thrive and went bankrupt in 1927.[8] Even before it failed, he had been diverted by the idea of producing a light plane that virtually anybody could afford to own for private use, hoping it would also attract military orders. Its most distinctive feature, an ingenious five-cylinder radial engine, showed considerable promise, but the project did not succeed.[9] Rickenbacker then joined a wartime comrade, Reed Chambers, in launching Florida Airways, hoping to develop an airmail and passenger route between Miami and Jacksonville. Investors in the venture included Percy Rockefeller, Henry Ford (whose stake included three single-engine metal planes developed by William B. Stout), and Richard C. Hoyt, a member of the Hayden, Stone financial empire. Its pilots included Ben Eielson, who later became famous for his aerial adventures in Alaska. Nevertheless, it too failed because of adverse circumstances, including the collapse of the Florida real-estate boom, a hurricane in 1926, and a lack of cooperation from the municipal officials in Tampa, who did not deliver on promises to provide an adequate airport. Harold Pitcairn bought the enterprise in receivership and ultimately consolidated it into Eastern Air Transport, linking Miami, Atlanta, and New York.[10] Returning to the automobile industry, Rickenbacker bought the Indianapolis Speedway and became a vice president of GM's Cadillac division to help market a new luxury car, the LaSalle, which GM had created as a counterpart to Rolls-Royce's Bentley.

Still fixated on aviation, Rickenbacker went back to it in 1929 when GM bought a 40 percent interest in the Fokker Aircraft Corporation of America and made him vice president for sales. It was an unenviable job. Fokker's engineering methods were faulty, and the company was building a big new airliner, the F-32, which was too expensive for shrinking markets in the Great Depression. GM tried to remedy the situation by taking full control of Fokker and making an arrangement with Claude Dornier to switch to metal construction, with which Rickenbacker was already familiar. As GM strengthened its grip on Fokker, which ultimately became General Aviation, it also acquired the Allison Engine Company and, at Rickenbacker's urging, the Pioneer Instrument Company, which became better known as Bendix Aviation. Rickenbacker thus became a key player in GM's diversification from automobiles into aircraft.[11]

Rickenbacker was conscious that Charles A. Lindbergh had upstaged him as the nation's most famous aviator by making his nonstop flight to Paris in 1927. In that year Eddie launched a campaign with the help of a congressman from Detroit to receive the Medal of Honor for his valor in a sortie in 1918 in which he had taken on seven German aircraft single-handed. He achieved his goal in 1930, adding to his luster as a national hero. GM, however, had trouble trying to make Fokker a valuable asset, and Rickenbacker became increasingly discouraged with his job. Sales of aircraft were hard hit in the sagging economy despite steep price cuts Rickenbacker was obliged to put into effect. Things went from bad to worse on March 31, 1931, when one of TWA's Fokker F-10A Trimotors crashed in a Kansas thunderstorm. Among those killed was Notre Dame's fabled football coach, Knute Rockne. After an investigation traced the accident to structural failure due to faulty gluing of plywood wing components, the government took the unprecedented step of grounding thirty-five F-10As built at a company plants in West Virginia. Most were later returned to service, but the reputation of Fokker airliners was irreparably damaged. As losses mounted, GM closed its three existing Fokker plants and transferred operations to a former Curtiss-Caproni facility at Dundalk, Maryland.[12] Rickenbacker, whose office was in the General Motors Building on Broadway, thereupon resigned, saying that he liked living in New York City too much to move to Baltimore. He was probably sincere, because he remained a resident of the metropolis for the rest of his life, but he also must have been disenchanted by what had happened in the previous few years. The role played by federal officials in destroying the reputation of the F-10A did nothing to make his experience any more pleasant and foreshadowed later events that only increased his hostility to government regulation.

Rickenbacker now became Vice President for Government Relations of American Airlines, an ill-coordinated enterprise whose routes ran mainly from east to west. He urged its chief stockholders, Averell Harriman and Robert Lehman, to attempt a merger with Eastern Air Lines, whose system went from north to south. In attempting to carry out the idea, they became embroiled in a proxy fight with Errett L. Cord, a master of Wall Street intrigue.[13] Rickenbacker resigned after they lost the battle and went back to being an aviation consultant for GM. Still enamored by the prospects of Eastern Air Transport, which had strengthened its potential to build a lucrative vacation route to Florida by absorbing New York Airways in 1931, he urged GM to acquire North American Aviation, a bankrupt conglomerate that numbered Eastern among its assets. His role in creating one of Eastern's predecessors, Florida Airways, thus came full circle, as did his previous relationship with General Aviation, which by this time had also come under North American's umbrella.

Reestablishing his ties with GM put Rickenbacker in a position to play a vital role in the future of civil and commercial aviation after Franklin D. Roosevelt became president in 1933. Rickenbacker's friends in GM had played important roles in the Democratic party in the 1920s when it was out of power and the Republicans held control of national affairs. As many Democratic leaders moved away from Wilsonian progressivism, the party began returning to the conservative, strict-constructionist ideas of previous standard-bearers like Grover Cleveland. As a result, Big Business became increasingly powerful in the party's inner councils, reflected by its nomination of a prominent corporation lawyer John W. Davis for president in 1924. GM was one of the large corporations that became heavily involved in Democratic politics. Because of their strong opposition to Prohibition, Pierre S. Du Pont and John J. Raskob forged an alliance with Al Smith, the Democratic presidential nominee in 1928. Raskob resigned his position as a vice president of GM to become chairman of the Democratic National Committee. Financial support from Du Pont and Raskob enabled Smith to promote the construction of the Empire State Building, beginning in 1929, after his failed run for the presidency.[14] Despite Smith's estrangement from Franklin D. Roosevelt, Du Pont, Raskob, and other high-ranking members of the GM hierarchy supported FDR in 1932 because he ran on a platform that promised to end Prohibition and condemned the Hoover administration for unduly interfering in economic matters in combating the Great Depression.

Roosevelt himself took a conservative line in some of his campaign rhetoric to ingratiate himself with the party's southern wing and with corporate executives who had rallied to its standard in the 1920s.[15]

From this perspective it is easy to understand why Rickenbacker voted for FDR in 1932. As a prominent member of the GM family, as a heavy drinker who fiercely opposed Prohibition, as a firm believer in untrammeled private enterprise, and as a lifelong critic of centralized government, it would have been impossible for him to do otherwise. He disliked both Hoover's Republican brand of progressivism and his support of Prohibition. It is also understandable from this perspective that he had no reason to like Walter F. Brown, Hoover's postmaster general, whose authoritarian style of airline supervision under the Watres Act complicated Rickenbacker's lobbying efforts for American.[16] In 1994, in an interview with William F. Rickenbacker, one of Eddie's two sons, I asked him why his father voted for Roosevelt in 1932. He responded with a bitter indictment of Hoover, charging that he was the true founder of the New Deal, a squanderer of government money who had trampled on the Constitution by creating the Reconstruction Finance Corporation and other agencies that FDR merely perpetuated.[17] He had learned such lessons from his famous father, who had no way of knowing in 1932 that, from his perspective, FDR was a dangerous man to put in the White House.

Like Du Pont, Raskob, and other people prominently associated with GM, Rickenbacker soon became disenchanted with FDR and the New Deal. The repeal of Prohibition, for example, was gratifying but also disappointing because it was accompanied by the creation of a Federal Alcohol Control Administration that "exercised enormous power . . . including the right to grant and revoke permits to engage in liquor manufacture and the authority to control production." Pierre Du Pont, who had contributed heavily to Roosevelt's campaign, turned against the National Recovery Administration (NRA) after FDR appointed him chairman of the Industrial Advisory Board (IAB) and a member of the National Labor Board (NLB). In mid-1934, disturbed by mounting bureaucracy and the proliferation of federal agencies, Du Pont resigned from the NLB because it did not protect workers from "exploitation," which to him meant pressure on workers to join unions. By August of the same year Du Pont had become convinced that the NRA was "a conspiracy on the part of the New Dealers to change permanently the traditional balance between federal power, state prerogatives, and private business." The result of this disenchantment, which was shared throughout GM, was the formation of the Liberty League, which was heavily financed by Du Pont money.[18] Rickenbacker naturally made common cause with the league and became a bitter critic of FDR. "I had supported him for the Presidency in 1932, believing that his platform, which to me seemed sound and conservative, was what the country needed," he later said. "No sooner had he taken office, however, than he made a complete 180-degree turn and taken off in the other direction toward liberalism and socialism."[19]

Rickenbacker broke decisively from Roosevelt over his handling of scandals that erupted in 1934 concerning the way Brown had dispensed airmail contracts during the Hoover administration. Despite being no admirer of Hoover or Brown, Rickenbacker believed that the airline industry was making great strides in combining passenger and airmail operations as the Watres Act had mandated. "Though the airlines of the United States were just beginning to realize their potential, they were flying two hundred thousand miles in every 24-hour period," he later said, praising the performance of commercial airline pilots and the equipment, including two-way radio, that the industry had adopted.

Rickenbacker was therefore "shocked and astounded" when FDR, acting on the advice of Postmaster General James A. Farley, canceled existing airmail contracts in February 1934 and ordered the army to fly the mail. To cancel contractual agreements was in itself a grave offense against business ethics, but to replace airline pilots with military aviators who were untrained for commercial operations was in Rickenbacker's eyes nothing short of criminal. Speaking to reporters on a foggy day in New York, he expressed concern about "what is going to happen to those young Army pilots on a day like this. . . . Either they are going to pile up ships all the way across the continent, or they are not going to be able to fly the mail on schedule." When three pilots crashed and died on their way to pick up mail for their first day of operations, Rickenbacker did not hesitate to pronounce FDR guilty of "legalized murder." Reporters asked if they could quote him. "You're damned right you can," he retorted.[20]

Rickenbacker made his "legalized murder" remark in California on the eve of a flight that was deliberately designed by North American Aviation and GM to embarrass the Roosevelt administration by demonstrating the high standards of operation attained by privately owned airlines. In 1931, when Fokker aircraft had been discredited by government fiat and United Air Lines was preparing to introduce the revolutionary Boeing 247, TWA, a subsidiary of North American, sought to escape the cloud it was under because of the Rockne crash and increase its competitiveness in one stroke. In a famous letter, Jack Frye, TWA's operational head, asked the Douglas Aircraft Corporation to design an even better airliner than the new Boeing plane. The result was the Douglas DC-1, which was delivered to TWA in December 1933. On February 19, 1934, after Rickenbacker gave a radio address blasting FDR, he and Frye staged a record-setting transcontinental flight in the DC-1 on the final day of regular airmail service, carrying a load of mail from Burbank, California, to Newark, New Jersey. Substantially exceeding the plane's rated speed of 180 miles per hour, they completed the trip in 13 hours, 4 minutes, breaking the previous cross-country speed record by about 58 minutes and bettering the normal transcontinental mail schedule by almost six hours. Averaging 230 miles per hour between Burbank, California, and Albuquerque, New Mexico, they made only two refueling stops, one at Kansas City, Missouri, and the other at Columbus, Ohio. A large throng hailed the crew when they arrived at Newark International Airport, landing just ahead of a storm that was closing in on the East Coast. The plane was jammed with reporters and the venture received much favorable publicity.[21]

Rickenbacker was the hero of the flight despite the fact that Frye did most of the actual flying when the plane was not on automatic pilot. Rickenbacker later candidly admitted that he operated the plane for only a few minutes from the co-pilot's seat and spent the last leg of the flight writing a speech. Criticism by knowledgeable aviation historians about his effrontery in accepting plaudits for breaking the transcontinental speed record are valid but beside the point.[22] Frye, a modest and self-effacing man, was not a national hero whose role would dramatize TWA and North American Aviation as GM and its allies in the Liberty League wished.[23] Rickenbacker, by contrast, was a highly visible symbol of corporate animosity toward the New Deal and inescapably received the lion's share of attention, which was what his fellow Roosevelt haters wanted.[24]

The flight of the DC-1 was merely a prelude to events that made Roosevelt's use of army pilots to fly the mail a political disaster, the first major setback he had suffered after his inauguration. Rickenbacker's predictions came true, as a series of crashes in one of the worst winters on record took the lives of ten pilots and caused serious injuries to others, forcing FDR to back down and return control of the nation's airways to the privately owned carriers that had already operated them. To save face, the administration pushed legislation through Congress mandating that no airlines that had participated in the infamous "spoils conferences" Brown had held in 1930 could receive new route awards and proscribing executives who had participated in the meetings from holding office in firms carrying airmail. The first of these provisions resulted in a charade in which Postmaster General Farley permitted existing carriers to pose as new entities merely by changing their names. Thus Eastern Air Transport became Eastern Air Lines and TWA magically transformed itself by adding "Inc." to its corporate title. The second provision, however, was enforced more rigidly and ended the careers of many capable and creative executives, including Paul Henderson, the father of night airmail delivery, who had actually protested against Brown's policies while attending the ill-fated meetings.[25]

Ironically, disqualifying executives who had attended the spoils conferences opened the way for Rickenbacker, one of the New Deal's most militant critics, to become general manager of Eastern Air Lines. Thomas B. Doe, who had participated in the meetings as president of Eastern Air Transport, was therefore ineligible to remain with the company. Ernest R. Breech, a GM executive who headed North American Aviation, took Doe's title as president but did not wish to be responsible for day-to-day operations and asked Rickenbacker to assume administrative control. Jumping at the offer, Rickenbacker took the position on January 1, 1935.[26]

North American retained control of Eastern despite a provision in the Airmail (Black-McKellar) Act of 1934 that separated air transport from aircraft manufacturing and forbade airmail contractors to hold conflicting interests in other firms.[27] Because North American made only military aircraft, GM successfully claimed that Eastern, a passenger line, could stay under its corporate umbrella. Rickenbacker therefore remained associated with GM, reporting to Breech and staying in frequent touch with Henry Belin Du Pont, who represented his family on North American's board of directors.[28] Under Rickenbacker's leadership, Eastern, which had lost $1.5 million in 1934, began to flourish. Plunging vigorously into his new duties, he sold Eastern's Curtiss Condors and Pitcairn Mailwings and replaced them with modern Lockheed L-10 Electras and Douglas DC-2s. Firing old and jaded executives and station managers, he recruited fresh managerial talent, reorganized the traffic department, instituted a stock-option plan to enhance employee morale, and established pioneering medical, radio, and meteorological departments. He also added Houston to Eastern's route structure in 1936 by acquiring the Wedell-Williams Transport Corporation.[29]

Breech, however, had fallen out with Rickenbacker for reasons too complex to explain here.[30] Wanting to detach Eastern from North American Aviation, he persuaded GM to sell Eastern for $3 million to a syndicate headed by rental-car magnate John Hertz, which had already bought TWA from North American. Rickenbacker was not told about the deal and found out about it from a reporter. Knowing that Eastern could not have been offered to Hertz for such a large price but for the hard work he had done in rebuilding it, Rickenbacker was enraged and decided to fight. Playing his strongest card, resulting from his years of faithful service to GM and the Du Pont family, he went over Breech's head to its general manager and chief executive, Alfred P. Sloan. He also enlisted the aid of GM's president, William S. Knudsen, Donaldson Brown, its ranking financial expert, and other members of the corporate hierarchy whom he had known for many years.

Sloan responded by giving Rickenbacker thirty days to raise $3.5 million in cash to retain control of Eastern. Aided by William Barclay Harding of Smith, Barney, and Co., who was instrumental in securing funding from the Kuhn, Loeb banking empire, Rickenbacker succeeded in raising the money. Anxious about whether his supporters could produce the cash required to cement his victory in time to meet the expiration of his deadline, which expired on a Sunday, Rickenbacker had the temerity to arouse Sloan from bed in the middle of the night to make sure it would be forthcoming. "If I were you, Eddie, I wouldn't worry," Sloan said. The next morning, Frederick Warburg handed Rickenbacker a certified check for $3.5 million. On presenting it to Sloan, Rickenbacker assumed the presidency of Eastern Air Lines. "Congratulations, Eddie, and God bless you, and I wish you every success in the world," Sloan said.[31] It was a final benediction for Rickenbacker's years of service to GM. The transaction was a major turning point in Rickenbacker's life, putting his destiny in the hands of new financial masters. Ending his long association with GM, he now answered to a group of investors led by Laurance S. Rockefeller, who was just beginning his career as a venture capitalist.[32]

As Rickenbacker assumed his new role in Eastern, the airline industry was awaiting passage of the Civil Aeronautics Act, which went into effect in August 1938. Passed in response to fears aroused by the introduction of the Douglas DC-3, the first airliner that could make money without airmail subsidy, the statute gave airlines protection from competition in exchange for submitting to an unprecedented degree of federal control that, after 1940, was exercised by an all-powerful Civil Aeronautics Board. I have recounted Rickenbacker's epic struggles with the CAB in other publications and cannot repeat what I have already said about them.[33] His enmity toward federal regulators cost Eastern dearly and had much to do with its ultimate slide into bankruptcy. What I have tried to demonstrate here is that Rickenbacker's stormy relationship with the CAB was foreshadowed in the 1930s by his association with GM and its political offshoot, the Liberty League. Only by understanding his connections with the automobile industry and its most powerful corporate entity is it possible to see why his career developed the way it did.

NOTES

1. For an extended analysis of Rickenbacker's ultraconservative philosophy and hostile relationship with federal regulators, see my essay, "A Man Born Out of Season: Edward V. Rickenbacker, Eastern Air Lines, and the Civil Aeronautics Board," in W. David Lewis, ed., *Airline Executives and Federal Regulation: Case Studies in American Enterprise from the Airmail Era to the Dawn of the Jet Age* (Columbus: Ohio State University Press, 2000), 242–92. For a valuable assessment of Rickenbacker as an airline leader, with shrewd insights into his personality and character, see Robert J. Serling, *From the Captain to the Colonel: An Informal History of Eastern Air Lines* (New York: Dial Press, 1988).

2. Details on Rickenbacker's life and career in this essay are based largely on Edward V. Rickenbacker, *Rickenbacker: An Autobiography* (Englewood Cliffs, N.J.: Prentice-Hall, 1967). I have also drawn upon Hans C. Adamson, *Eddie Rickenbacker* (New York: Macmillan, 1946), and Finis Farr, *Rickenbacker's Luck: An American Life* (Boston: Houghton Mifflin, 1979). Because one or more of these sources could be mentioned in virtually all the notes that follow, I have chosen not to multiply such references to no good purpose except when making direct quotations. For an extended biographical article about Rickenbacker, see W. David Lewis, "Edward V. Rickenbacker," in William M. Leary, ed., *The Airline Industry* (New York: Facts on File, 1992), 398–415.

3. W. David Lewis, "Eddie Rickenbacker: Racetrack Entrepreneur," *Essays in Economic and Business History* 29 (2000): 85–100.

4. Rickenbacker's experiences in World War I are recounted in his book, *Fighting the Flying Circus* (New York: Frederick A. Stokes, 1919). On Mitchell's unsuccessful campaign for a separate U.S. Air Force after World War I, see particularly Burke Davis, *The Billy Mitchell Affair* (New York: Random House, 1967), which contains several references (263–65) to Rickenbacker's supportive role. For a statement by Rickenbacker himself advocating Mitchell's ideas, see Edward V. Rickenbacker, "Is General Pershing Wrong?" in *U.S. Air Service* 3, no. 4 (Apr. 1920): 14–16.

5. On the history of the Sheridan, see Beverly Rae Kimes and Harry Austin Clark Jr., *Standard Catalog of American Cars, 1805–1942*, 3d ed. (Iola, Wis.: Krause, 1996), 1346.

6. Pierre S. Du Pont withdrew from active management of GM in 1924 but remained chairman of the board until 1928 and retained a powerful role in its corporate affairs throughout his life. John J. Raskob became Du Pont's secretary and stenographer in 1900, advanced steadily in the business ventures in which Du Pont was engaged, and was responsible for persuading Du Pont to acquire a controlling interest in GM. Raskob became its vice president and chief of financial affairs. In 1924 Du Pont entrusted day-to-day control of GM to Alfred P. Sloan Jr., who had been brought into GM by William C. Durant in 1916 when Durant acquired control of the Hyatt Roller Bearing Company, with which Sloan had been an executive. Sloan became one of the greatest executives in American industrial history. On the intertwining careers of Du Pont, Raskob, and Sloan, see particularly Alfred D. Chandler and Stephen Salsbury, *Pierre S. Du Pont and the Making of the Modern Corporation* (New York: Harper and Row, 1971).

7. "Rickenbacker Material Collected by Isabel Leighton," Accession No. 99-047, Record Group 101, Eddie Rickenbacker Papers, Auburn University Archives, Auburn University, Auburn, Ala.; articles of indenture between William C. Durant and Guaranty

Trust Company of New York, Dec. 31, 1918, and California divorce decree of July 10, 1922, provided by Nancy A. Rickenbacker, widow of William F. Rickenbacker; interview with William F. Rickenbacker by W. David Lewis, Oct. 6, 1994. Leighton, a friend of Adelaide Rickenbacker, gathered material about her to provide background for a film about Eddie Rickenbacker, *Captain Eddie*, produced by 20th Century–Fox in 1944. William C. Durant, who had founded GM in 1908, ended his second and final term as its president on November 30, 1920, leaving control in the hands of the Du Ponts and their financial associates, but Durant retained large shareholdings in the company and its divisions, particularly Chevrolet. See Bernard A. Weisberger, *The Dream Maker: William C. Durant, Founder of General Motors* (Boston: Little, Brown, 1979), 264–87.

8. Beverly Rae Kimes, "Hat in the Ring: The Rickenbacker," *Automobile Quarterly* 13 (1975): 418–35; Kimes and Clark, *Standard Catalog*, 1291–92.

9. Numerous clippings about the project are in the second of twenty-six scrapbooks kept by Rickenbacker, now in Box 3, Series 3, Accession No. 97-001, Record Group 101, Auburn University Archives. For examples, see "New Driggs Coupe Takes the Air," *Detroit Free Press*, Aug. 29, 1925, and "The Family AirFlivver: Will Flying Soon Be a National Pastime?" *Los Angeles Sunday Times*, Sept. 26, 1925.

10. Albert LeShane Jr., "Florida Airways," *Journal of the American Aviation Historical Society* 22 (1975): 123–35; Victor Chenea, "Skeleton Outline of the Events Connected with Florida Airways," kindly provided to me by Matthew Rodina; "History of Eastern Air Lines," typewritten ms., Rockefeller Family Archives, Record Group 2, Rockefeller Archives Center, North Tarrytown, N.Y.

11. On the diversification of GM into aviation, see Arthur Pound, *The Turning Wheel: The Story of General Motors through Twenty-five Years, 1908–1933* (Garden City, N.Y.: Doubleday, Doran, 1934), 318–28, and Alfred P. Sloan Jr., *My Years with General Motors* (Garden City, N.Y.: Doubleday, 1964), 362–74.

12. On the Fokker Aircraft Corporation of America, its connections with GM, and Rickenbacker's experience with Fokker during this period, see Marc Dierikx, *Fokker: A Transatlantic Biography*, Smithsonian History of Aviation Book Series (Washington, D.C.: Smithsonian Institution Press, 1997), 108–45.

13. On the history of AVCO, see particularly Henry Ladd Smith, *Airways: The History of Commercial Aviation in the United States* (New York: Alfred A. Knopf, 1942), 147–55. For the best account of the struggle between Cord, Harriman, and Lehman, see Rudy Abramson, *Spanning the Century: The Life of W. Averell Harriman, 1891–1986* (New York: William Morrow & Company, 1992), 201–7.

14. Mitchell Pacelle, *Empire: A Tale of Obsession, Betrayal, and the Battle for an American Icon* (New York: John Wiley & Sons, 2001), 21–25.

15. The interpretation advanced here draws heavily on Douglas B. Craig, *After Wilson: The Struggle for the Democratic Party, 1920–1934* (Chapel Hill: University of North Carolina Press, 1992). A close reading of this book is indispensable to understanding Rickenbacker's fervent opposition to the New Deal.

16. On Brown and his role in administering the Watres Act, see David D. Lee, "Walter Folger Brown," in Leary, *Airline Industry*, 82–89.

17. Interview with William F. Rickenbacker by W. David Lewis, Oct. 6, 1994.

18. Craig, *After Wilson*, 270–95; Robert F. Burk, *The Corporate State and the Broker*

State: The Du Ponts and American National Politics, 1925–1940 (Cambridge, Mass.: Harvard University Press, 1990), 143–277. On the history of the Liberty League, see also Frederick Rudolph, "The American Liberty League, 1934–1940," *American Historical Review* 56 (Oct. 1950): 19–33, and George Wolfskill, *The Revolt of the Conservatives: A History of the American Liberty League, 1934–1940* (Boston: Houghton Mifflin, 1962).

19. Rickenbacker, *Autobiography*, 184.

20. Ibid., 186.

21. In addition to the account of the flight in ibid., 185–88, see Robert J. Serling, *Howard Hughes' Airline: An Informal History of TWA* (New York: St. Martin's/Marek, 1983), 42–43. Among newspaper accounts, see particularly "Private Mail in Final Flourish" and "Rickenbacker Spans U.S. in 13 Hours," *New York Evening Post*, Feb. 19, 1934.

22. See, for example, Robert J. Serling, "Unsung Hero: Frye of TWA," *Airways* 3 (May–June 1996): 24–26. D. W. Tomlinson, assistant to the president of TWA, also piloted the DC-1 during part of the trip. Box 27 of the Edward V. Rickenbacker Papers at the Library of Congress contains many congratulatory letters and telegrams received by Rickenbacker after the flight.

23. On Frye and his career, see Patricia A. Michaelis, "Jack Frye," in Leary, *Airline Industry*, 186–89.

24. To the best of my knowledge, the significance of the record-setting flight in dramatizing opposition by GM and the Du Ponts to the New Deal has not been previously recognized. It is not far-fetched to think of the DC-1 as a cruise missile directed at Franklin D. Roosevelt by the Liberty League.

25. David D. Lee, "Airmail Act of 1934," in Leary, *Airline Industry*, 23–26.

26. In addition to sources already cited, see W. David Lewis, "Eastern Air Lines," in Leary, *Airline Industry*, 161–62.

27. David D. Lee, "Airmail Act of 1934," in Leary, *Airline Industry*, 22–23; R. E. G. Davies, *Airlines of the United States since 1914* (Washington, D.C.: Smithsonian Institution Press, 1988), 170.

28. The Henry B. Du Pont Papers at the Hagley Museum and Library, Wilmington, Del., contain much correspondence between Rickenbacker and H. B. Du Pont.

29. Lewis, "Eastern Air Lines," in Leary, *Airline Industry*, 162.

30. Breech and Rickenbacker quarreled repeatedly after Rickenbacker became general manager of Eastern in 1935. Correspondence and memoranda provided to the author by Nancy A. Rickenbacker indicate that the two men had a dispute about secretarial expenses at Eastern that Breech considered too high. They also had a difference of opinion about a secretary whom Breech wanted to transfer from his office to Eastern and the compensation that Breech wanted Eastern to pay her. Breech also rejected a claim Rickenbacker made for travel expenses incurred on a trip to Newfoundland to help rescue Richard (Dick) Merrill, Eastern's best-known pilot, who had to make an emergency landing in a bog when he ran out of fuel on his return from an attempted record-setting transatlantic flight. Breech thought that the trip was unnecessary and that Rickenbacker was grandstanding by making it.

31. Rickenbacker, *Autobiography*, 190–94; Sloan, *My Years with General Motors*, 427–28.

32. On Laurance Rockefeller's career as a venture capitalist in Eastern Air Lines and other aerospace enterprises, see Peter Collier and David Horowitz, *The Rockefellers: An American Dynasty* (New York: Holt, Rinehart and Winston, 1976), 217–18, 292–303.

33. See particularly my essay "Man Born Out of Season." On the CAB, see Donald R. Whitnah, "Civil Aeronautics Board," in Leary, *Airline Industry*, 105–8.

CHAPTER 6
A PERENNIAL CHALLENGE TO AVIATION SAFETY
BATTLING THE MENACE OF ICE

The aeronautical community has battled the menace of icing since the earliest days of powered flight. From time to time, victory has been declared over this dangerous foe only to find that new challenges lay ahead. The first round of what became a lengthy fight—and one that would continue into the twenty-first century—took place in the 1930s when the appearance of pneumatic deicers seemed to provide the solution to the icing problem.[1]

One of the first group of fliers to encounter icing regularly were the intrepid aviators of the U.S. Air Mail Service. After World War I, the Post Office inaugurated a transcontinental airmail route between New York and San Francisco. As the technology of aviation had not advanced to the point where pilots could control their airplanes or find their way by reference to instruments, the aviators of the Air Mail Service had to keep the ground in sight. To do so, they did their best to avoid clouds. Encounters with icing, therefore, were accidental.[2]

One of these accidental encounters took place on November 17, 1921. Jack Knight, perhaps the most famous of the airmail pilots, departed Omaha, Nebraska, shortly after noon, heading westward with the mail. The ceiling at Omaha was 400 feet. By the time Knight reached the Platte River, twenty miles away, the clouds had dropped to 100 feet. At this point, he found himself in the midst of a sleet storm. Within a few minutes, the wires of the de Haviland DH-4 biplane became coated with ice to a width of one inch and a thickness of a quarter of an inch on the leading edge. "The ship became very loggy," Knight reported, "and I had to open the motor up to maintain altitude."

As he was inclined to do, Knight decided to press on toward North Platte, Nebraska. But the ice got worse, and the wires began to vibrate, then snap. He had to hold up the nose of the biplane in order to maintain altitude. He tried to turn around, but the wing and nose of the airplane immediately dropped "as though about to go into a spin almost out of control." Knight applied full power and managed to raise the nose and regain limited control.

By this time, Knight was ready to land at the nearest emergency field. At 2 P.M. he reached Kearney, Nebraska. Using flat turns, he managed to land safely in the small field. He placed the mail on the train for North Platte. The next morning, it took Knight and a mechanic four hours to remove the ice from the flying wires, propeller, and other parts of the airplane.

In his report of this incident, Knight noted that "the formation of ice on the wires . . . and the subsequent breaking of the flying wires is a new experience for me, and the effect of this combination very nearly placed the ship out of control." As it was, he concluded, "the maneuverability of the ship was practically nothing." He was indeed fortunate to have survived this encounter with freezing precipitation.[3]

The danger from icing for aviators increased in the mid-1920s when the Post Office started an overnight airmail schedule between New York and Chicago. By now, pilots had begun to experiment with flying by reference to instruments. They found that it was possible to climb and descend through cloud layers without losing control of their airplanes. They also quickly learned that when the temperature in the clouds was below 32 degrees Fahrenheit, ice was likely to form on their airplanes.

A typical encounter took place during the early morning hours of December 23, 1926. Pilot Warren Williams was en route from Cleveland to Chicago with 321 pounds of mail. He was flying underneath an overcast until low clouds at Woodville, Ohio, blocked his way. He decided to fly on top, as the cloud layer seemed only 1,000 feet thick. He flew on instruments, monitoring his gyroscopic turn indicator, ball-bank indicator, and airspeed. As the Douglas M-4 biplane began to climb, Williams felt his controls grow "mushy." His turn indicator malfunctioned; his compass began to spin; his altimeter unwound. Williams fought the controls, but without success. As the ground approached, he cut the throttle and jumped. He pulled the rip cord on his recently issued parachute and floated down safely from 300 feet.[4]

William was lucky to have survived his encounter with icing. Fellow pilot John F. Milatzo was not as fortunate. Shortly after midnight on April 22, 1927, while en route from Chicago to New York with the mail, Milatzo crashed into a field near Goshen, Indiana, during a severe snow and sleet storm. He was killed. As one airmail pilot noted at the time about the hazards of the New York–Chicago route, "the greatest of all our problems is ice."[5]

There were a few attempts to deal with the icing problems during the mid-1920s. The Air Mail Service, for example, worked closely with Army Air Corps technicians at McCook Field, the major Air Service research facility near Dayton, Ohio, to find some ways to keep instruments free of ice. In 1925 the army used a small wind tunnel that had been set up in a refrigerated room at McCook to study the formation of ice on pitot tubes. Although the army failed to come with a solution, instrument manufacturers later developed electrically heated pitot-static tubes that worked well.[6]

After the Post Office turned over its routes to private enterprise in 1927, the icing problem seemed to attract more attention. On March 28, 1928, William P. MacCracken Jr., assistant secretary of commerce for aeronautics, called together representatives from the U.S. Army Air Corps, the U.S. Navy's Bureau of Aeronautics, the Weather Bureau, the Bureau of Standards, and the National Advisory Committee for Aeronautics (NACA) for the purpose of "studying the atmospheric conditions which cause ice formation on the wings and other structural parts of aircraft in flight, and of recommending investigations for the development of an instrument or instruments which will indicate to the pilot the formation of ice in flight, which will be of special value in night flying."

The NACA's George W. Lewis, director of aeronautical research, reported to the group that such an investigation was already underway. The NACA had collected reports from pilots, like Jack Knight, who had experienced flight through icing conditions. An analysis of these reports suggested "that ice formation usually occurs when rain from a warmer stratum above falls on an airplane when it is flying in a stratum of freezing temperature." The NACA, he continued, had initiated a test flight program to investigate icing, which would be supplemented by experiments in a special wind tunnel. It was expected that the results from these investigations would indicate before next winter's flying season what measures might be taken, if any, to prevent ice from forming on airplanes.[7]

As Lewis had indicated, the NACA had already launched a flight investigation of icing. The impetus for this program had come from NACA chief pilot Thomas Carroll, who had been attracted to the problem ever since Air Mail Service pilots had begun to report their encounters with icing. "The most difficult aspect of the matter," Carroll wrote to Henry J. E. Reid, engineer in charge at the NACA's Langley Aeronautical Laboratory, on March 17, 1928, "appears to lie in an almost total ignorance of the conditions which must exist to produce the phenomena." He recommended that the NACA's Flight Operations Section conduct a research program into in-flight icing with a high priority. To accomplish this task, he wanted "a supercharged airplane" that would be equipped with automatically recording air temperature thermometers that also would be visible to the pilot. In addition, he sought "an automatically driven motion picture camera placed and focused at short range on certain struts and surfaces on the airplane to be controlled by the pilot or observer to provide additional data to visual observation." He proposed to seek out cloud formations with a range of temperatures that were conducive to icing. Together with gathering information on the phenomenon of icing, Carroll also wanted to experiment with preventive measures, such as "heat control, oiling, etc."[8]

Using a Vought VE-7, Carroll and fellow research pilot William H. McAvoy set out in search of cloud formations in which ice was likely to be encountered. They recognized and described the types of ice that they found. Glaze ice, which formed at temperatures just below freezing, was clear and tended to protrude from the leading edge of airfoils. Rime ice, on the other hand, which occurred at lower temperatures and was opaque, usually took on a streamline shape. Performance penalties, they noted, were caused more by distortions to the shape of the airfoil than by the weight of the ice. Engineers, of course, knew that a cubic foot of water weighed 62½ pounds. When water expanded on freezing, the density decreased to 56 pounds per cubic foot. Even a biplane, with its wires and struts, was unlikely to accrete more than 7 cubic feet of ice, or some 400 pounds, during the most severe icing conditions. While the added weight would cause some performance penalties, it would not constitute a dangerous overload.

Aeronautical engineers also knew that air flowing over the upper surface of a wing produced lift. To ensure that nothing disturbed the even flow of air, the airfoil surface was gently rounded. When ice formed on the surface, especially on the leading edge of the airfoil, turbulence and eddy currents caused by the nonstreamlined shape destroyed lift. At some point, the airfoil would stall, and the airplane would fall out of the sky.

Carroll and McAvoy pointed out that a number of means of preventing or removing ice had been suggested. The most frequently tried method had been the use of oil or grease to reduce adhesion of the ice to vulnerable parts of the airplane. To date, however, the application of oil or grease had failed to produce worthwhile results. In fact, it seemed that the use of these substances might even hasten the formation of ice. Suggestions were also made to pipe the engine exhaust heat through the leading edge of the wing to melt the ice or prevent it from forming. Although this method might be worth further investigation, Carroll and McAvoy concluded, the current state of ice prevention dictated that "safety . . . obviously lies in avoidance."[9]

At the same time that the NACA was exploring the icing problem, William C. Geer, a retired research chemist, was having a good deal of success in devising a method to protect aircraft from it. Geer had graduated from Cornell University in 1905 with a doctorate in chemistry and then joined the B. F. Goodrich Company of Akron, Ohio, in 1907 as their chief chemist. In 1927, two years after he had retired from Goodrich because of ill health, Geer became interested in the airplane icing problem. He knew that there had been sporadic research since 1922 on ice-retardant liquids and wing fabrics, but these early efforts had produced no satisfactory results. Geer decided to try his own experiments. He built a small research laboratory and began to test chemical methods to prevent the formation of ice.[10]

By 1929 Geer's work had showed sufficient promise to attract the attention of the Daniel Guggenheim Fund for the Promotion of Aeronautics. As part of its grant program to enhance aeronautical safety, the Guggenheim fund gave Geer $10,000 to conduct further research. Working with Dr. Merritt Scott of Union College, Geer arranged with the Department of Physics at Cornell University to build a small icing research tunnel. The facility featured a seven-inch-by-seven-inch test section and a three-inch circular throat. With the temperature lowered by ice, it was possible to reach $-4°$ F.[11]

Geer tested various substances in the tunnel, but most of them worked poorly. Soluble materials, especially pine oil, retarded the formation of ice, but rain would wash them off. Waterproofing materials, like waxes, were not effective in cold temperatures. Geer, who was an experienced rubber chemist, next hit upon the idea of applying the pine oil on a rubber base. He put a coating of rubber on the leading edge of a test model in the tunnel and then soaked it with pine oil. Ice would build up on the rubber surface, but when he touched it with a screw driver, it would immediately slide off. A side effect was that the pine oil tended to make the rubber swell. He next tried mixing pine oil with other oils to avoid the swelling problem. He finally came up with a mixture of four parts pine, four parts diethyl phthalate, and one part castor oil. This solved the swelling problem, but he still needed a way to dislodge the ice. His answer, after a good deal of trial-and-error, was to use flat tubes, embedded in the rubber, that could be inflated by air pressure and jar loose the ice. This was the origin of what Geer would label the "ice-removing overshoe."[12]

Working with his former employer, B. F. Goodrich, Geer quickly developed a model of the "oiled expansion device." It received its first flight test on December 20, 1929, thanks to Wesley L. Smith, superintendent of the Eastern Division of National Air Transport (NAT). A veteran Air Mail Service pilot, Smith had been hired by NAT to supervise the New York–Chicago route after the well-funded private company had taken over the airmail from the Post Office in 1927. Interested in technical developments, Smith had been in the forefront of testing instrument flying techniques and radio aids to navigation. He had equipped NAT airplanes with thermometers so that pilots might avoid icing conditions, which he considered "the greatest hazard to aviation." He was happy to cooperate with the Guggenheim Foun-

These ice formations on the propeller and fuselage surfaces of a test unit installed in the Icing Research Tunnel at the Aircraft Engine Research Laboratory of the National Advisory Committee for Aeronautics, Cleveland, Ohio, now known as the John H. Glenn Research Center at Lewis Field, show what can happen to an airplane in flight under certain atmospheric conditions. Ice degrades the performance of an aircraft in flight and can cause loss of control. (NASA photo no. C1945-8833)

dation. "All of us who are cooperating in the effort," Smith noted, "believe that the final solution to the problem will be found at sometime in the near future."[13]

Results of the December flight were promising. Smith reported that inflating the overshoes—or boots, as they more commonly became called—had no noticeable effect on the handling qualities of the airplane. However, Geer noted, "the overshoes on the wires, of which several different types of design were used, caused the wires to vibrate greatly even before the ice formed upon them." Goodrich engineers set to work to reduce the size and weight and to better streamline the rubber attachments to the wires.[14]

By early March 1930 an improved version was ready for flight test. Again NAT's Wesley Smith volunteered for the experiment. A layer of specially vulcanized rubber, thirty-six inches long and fifteen inches wide that exuded the colorless pine oil–based liquid was laced to the leading edge of the wings, together with smaller rubber coverings on supporting struts and the radio mast. Under the rubber were two two-inch flattened rubber tubes through which air could be driven by a pump. Smith took off from Cleveland on March 18 to test the system. Russell S. Colley, a Goodrich engineer, sat on an orange crate in the mail compartment of the biplane and used a bicycle pump to inflate the tubes, alternating air from one tube to the other with a manually operated valve. Two flights were made, one in the morning and one in the afternoon. The main problem was ice forming on the propeller, which Smith dislodged by making violent left and right sideslips. But the boots worked well. "At that moment," the NAT company newsletter proclaimed, "ice became a beverage instead of a hazard to aviation."[15]

Henry J. E. Reid, engineer in charge of the NACA's Langley laboratory, was less optimistic. After a visit to Cleveland in mid-April 1930 to discuss the flight tests with Geer, Reid concluded that "there seems to be a great deal to be done before such a device would be generally applicable to even those planes which would normally be expected to frequently encounter ice storms." Geer agreed. A suitable air pump and automatic air valve had to be developed. While this was being done, there was a need to conduct further tests in a sizable refrigerated wind tunnel to secure data on the most efficient design and the lightest and most durable construction for the boots. Happily, the Goodrich company, encouraged by results of the flight tests, was prepared to spend $20,000 on the icing program, including the construction of a refrigerated wind tunnel. As Lewis reported to Harry Guggenheim upon learning about the Goodrich commitment, "This points to the fact that the device has passed the research stage and is in the development stage, and as such I feel should be carried on by a manufacturer who is interested in marketing the device."[16]

Geer sent a sketch of the proposed tunnel to Lewis and asked the NACA's advice on its design. Although generally in favor of Geer's plans, engineers of the Aerodynamics Division at Langley came out against building the tunnel in a refrigerated room. Because of the NACA's experience with a small icing tunnel, they suggested that it would be more economical to insulate the walls of the tunnel and circulate brine through guide vanes for cooling. They delivered this advice on May 21, when Geer and C. W. Leguillon, manager of Machine and Process Development Departments at Goodrich, traveled to Langley to review plans for the tunnel. Leguillon, who would be responsible for construction of the facility, was receptive to the NACA's suggestions. "As a result of this conference," he wrote to Lewis, "we immediately redesigned our tunnel."[17]

Goodrich decided to build its tunnel on the roof of a large refrigeration plant so that it could be directly connected to a ninety-ton ammonia compressor. Constructed of sheet metal, the insulated tunnel measured forty feet by eighty feet, with a test section of three feet by seven feet by six feet high. Two thirty-six-inch metal propellers, connected in tandem to a drive-shaft that extended outside the tunnel to a fifteen-horsepower motor, supplied a maximum wind speed of eighty to eighty-five miles per hour. The air was cooled by being drawn over 780 feet of 1¼-inch coils of piping through which liquid ammonia circulated. Just before the cooled air passed into the test chamber, water was sprayed into the wind stream. Kept as cool as possible in an ice-filled reservoir tank, the water was atomized into a fine mist by passing it through airbrush nozzles using eighty pounds of compressed air.[18]

The first test of what was the world's largest refrigerated wind tunnel took place on August 22, 1930. A Clark Y airfoil with a chord of three feet, two inches, was placed in the test section and the ammonia was turned on. The tunnel cooled to 26° F in one and a half hours. Next, the fan was started. When the wind speed reached sixty miles per hour, water was sprayed into the airstream. A slush type of ice formed on the airfoil and was removed by the pneumatic boot. After several coatings of ice were successfully removed, the nozzle of the water spray clogged, and the test was terminated.[19]

Over the next three weeks, Goodrich engineers worked to improve the performance of the spray nozzle. By mid-September, Leguillon was able to report that several experiments had been successfully completed. With an airfoil in the test section, the tunnel had attained a speed of 80 miles per hour and temperatures of 20° to 25° F. At slower speeds, a temperature of 1° F had been maintained. Among other things, the tests had suggested that the boots should be inflated three times per minute for the most effective results.[20]

Although Goodrich had hoped to resume flight tests with an improved version of the pneumatic boot before the winter of 1930-31, work in the tunnel took longer that had been anticipated. It was not until early in 1931 that Goodrich was able to install the latest model of deicing equipment on its own test airplane, a Lockheed Vega monoplane named *Miss Silvertown*. Pneumatic boots were snapped onto the wings of the aircraft, fastened to struts with zippers, and laced to tail surfaces. Inflation of the boots would be accomplished by an automatic air compressor that was connected to the aircraft's engine. In March, William S. Brock flew *Miss Silvertown* to determine the aerodynamic effects of the boots. He found that they had no detectable affect on performance. After waiting for the right weather conditions, Charles Meyers took off from Akron and flew into a cloud layer that extended from 2,000 to 8,000 feet. He activated the pneumatic boots, which quickly eliminated the ice that had begun to build up on the airplane. On a second flight, he allowed the ice to accrete and then landed for ground tests. When he activated the boots, the ice cracked and was swept away by the air from the propeller. Meyers's successful flights, the *New York Times* confidently announced, represented "victory" over one of aviation's "most dangerous enemies."[21]

While Goodrich was focusing on the development of pneumatic deicer boots, the NACA was heading in the direction of heat. When Geer and Scott had examined various methods of ice removal in 1929-30, they had rejected the use of engine exhaust gas to melt ice accretion. Their research indicated that if only the leading edge of a wing was heated, melted ice would run back on the airfoil surface and freeze again. "To prevent ice," they concluded, "the outer surface of the entire wing must be maintained at 0° C or above, and it seems doubtful if the arrangements employing inside ducts will absorb sufficient heat to warm the entire wing surfaces without too much cumbersomeness and weight and without offering too much back pressure to the engine." In addition, they registered concern about the possible corrosive effects of engine exhaust gas on metal. "As airplanes of simpler design come into general use," they allowed, "the employment of heat to prevent the formation of ice may become a simpler problem." At the present time, however, such a system was not feasible.[22]

NACA researchers were coming up with different results. Their wind tunnel tests showed that the ice accumulated only at or near the leading edge of an airfoil and did not melt and refreeze. Armed with this information, Theodore Theodorsen, the Langley's laboratory's talented physicist, and junior aeronautical engineer William C. Clay decided to conduct a flight test of the use of heat for deicing. They mounted a model wing of approximately full-scale dimensions on a Fairchild FC-2W2 monoplane and then inserted a small boiler in the engine's exhaust pipe. Steam produced by the boiler passed through conducting pipes and entered the leading edge of the model wing section by means of a steam distribution pipe that was equipped with small holes. Spraying jets were mounted four feet in front of the model wing. The pilot of the Fairchild flew in temperatures as low as 18° F as water sprayed onto the steam-heated wing. In no cases did ice form on the leading edge of the airfoil. "The most essential result obtained from this study," the researchers concluded, "is the fact that ample heat is available in the exhaust and in the cooling water for the purpose of ice prevention." The successful design of an airplane "immune from the dangers of ice accumulation," they predicted in June 1931, was "only a matter of technical development."[23]

Goodrich remained unconvinced. In the summer of 1935, Clay had a long discussion about the heat system with M. L. Taylor, the contact man for the aeronautics branch of the Goodrich company, during a visit to the Langley laboratory. Taylor indicated that he was acquainted with the NACA's research, Clay reported, but he was doubtful about the results because Goodrich had also made heat tests that indicated "that much more heat was necessary to prevent ice formation that we determined." Taylor explained that Goodrich had placed electrical heating units inside the leading edge of a wing and heated the air inside until it reached 400° F. Thermocouples attached to the surface of the wing recorded temperatures of 60° to 100° F, depending upon the water content of the airstream. "Mr. Taylor stated," Clay continued, "that ice could be made to form under these conditions."

Clay suggested that a vapor-heating system would offer more efficient distribution of heat and would prevent ice from forming at lower average surface temperatures. Clay believed that he managed to convince Taylor that there was ample heat in the exhaust and that the weight of a vapor-heating system would not be greater than any other available method of ice removal—that is, pneumatic boots. "He was apparently open-minded on the subject," Clay concluded, "although his acquiescence may have been influenced by courtesy rather than conviction."[24]

After Meyers's successful tests of the pneumatic system in March 1931, Goodrich engineers took the boots back to the tunnel for engineering modifications. They managed to reduce the weight of the system by 30 percent.

Also, the company adopted a lightweight pump and distribution valve combination that had been developed by the Eclipse Aviation Corporation. The modified system received its first major operational test during the winter of 1932–33 when Transcontinental and Western Air (TWA) equipped several of its Northrop Alpha 4-A mail planes with Goodrich boots on the wings and tail surfaces. The aircraft flew 45,000 miles on the eastern portion of the transcontinental route, frequently encountering ice. Pilots reported that the use of deicers for two or three hours was commonplace. Inflation intervals of thirty-five to fifty seconds were most effective in removing the ice. Overall, the pilots judged that the system worked extremely well.[25]

Encouraged by these results, TWA ordered its entire Northrop fleet equipped with the Goodrich boots for the winter of 1933–34. At the same time, United Air Lines ordered thirty sets of boots for its Boeing 247s, while American Airlines placed them on their DC-2s. The system now weighed fifty-eight pounds for the Northrops and seventy-five pounds for the larger transports.

The winter season for the boot-equipped aircraft got off to an unpromising start. During the early morning hours of December 12, 1933, two TWA Northrops flew into a winter storm over the Allegheny Mountains near Portage, Pennsylvania. The boots were unable to handle the severe icing conditions, and the pilots lost control of their aircraft. The airmen, a TWA official explained to the press, "used discretionary powers and leaped." They were uninjured. The loss of the two airplanes, the journal *Aviation* pointed out, "has been accepted as a lasting and permanent lesson in the formidability of the ice hazard by the entire air transport industry."[26]

Despite the loss of the two Northrops, pneumatic deicers continued to impress the airlines. Operational problems arose, but they were promptly overcome. For example, the method of attaching the boots to metal airfoil surfaces by means of lacing proved inadequate, for the boots tended to loosen in flight. Goodrich engineer Russell Colley solved this problem by devising a hollow threaded rivet—called a Rivnut—that could be installed from outside the wing and provide a mechanical attachment of the boot to the metal surface. When members of Canada's National Research Council visited Goodrich in the fall of 1935, a company representative was able to report that research had successfully overcome "the great menace of ice." After years of experimental use, the winter of 1935–36 would see the widespread introduction of pneumatic boots by the airlines.[27]

At about the same time that Goodrich was announcing the victory over ice, TWA was reporting, in private, that most of their pilots considered propeller icing to be far more serious than wing icing. Indeed, twenty-seven airmen who responded to a questionnaire believed that wing deicing alone was useless. Ice was slung off propeller blades, causing extreme vibration and loss of efficiency. Indeed, most TWA transports sported dents in their fuselages, attesting to the frequency of the hazard.[28]

Goodrich engineers had attempted to cope with the problem by covering the propeller blades with a thin sheet of rubber that had been saturated with castor oil. Applied before every flight, the oil reduced the adhesion of ice so that it could be removed by centrifugal force. Pilots, however, found that the oil would work into the cement that held the rubber to the blades and loosen the attachment so that the rubber flew off with the ice. In August 1935 the aeronautics branch of the Department of Commerce decided to become involved in finding a solution to the problem. Appropriating $8,000, the government contracted with Goodrich for a comprehensive wind tunnel and flight investigation of various means of preventing ice from forming on propellers. It was hoped that a solution might be discovered before the winter flying season of 1935–36.[29]

By the time the first tests took place in Goodrich's icing tunnel in August 1935, the company had devised a new method for distributing the ice-retardant liquid to the propeller blades, which were now covered with leather instead of rubber. Goodrich engineers fitted the hub of a propeller with a slinger ring that measured twelve inches in diameter and weighed 6¾ pounds. Tubes from the ring led around the base of the blade. Liquid was fed to the slinger ring through small copper tubing, from where it dropped into the ring slot. Centrifugal force distributed the liquid along the blade.

Using a mixture of alcohol and glycerin as an ice retardant, Goodrich engineers tested this arrangement at a temperature of 0° F in the tunnel. They found that the leather took fifteen minutes to become saturated with the liquid. When the covering became moist, ice would not form on it. The leather then dried out and required another fifteen minutes to remoisten. Before the first test was completed, the leather became loose and the propeller threw it off.

When Goodrich resumed the tests, someone—Jack Frye of TWA attributed the idea to I. R. Metcalf of the aeronautics branch[30]—suggested that the alcohol be allowed to flow over the bare blade. Ice was instantly removed. Impressed with the simplicity and effectiveness of these surprising results, Goodrich immediately transferred operations to Kansas City, where TWA would conduct flight tests.[31]

Although all the airlines had an interest in defeating the menace of ice, TWA clearly led the way. In the summer of 1935, TWA decided to convert its DC-1—the prototype for the DC-2 and DC-3—into a flying icing research laboratory. Under the direction of chief engineer D. W. Tomlinson, the airline set out to equip the aircraft with state-of-the-art deicing equipment. As part of this effort, TWA agreed to cooperate with Goodrich and the Department of Commerce in testing the newest propeller deicing system.

Work in Kansas City began with an investigation of the flow pattern of the alcohol and glycerin on a propeller blade. A standard Hamilton three-blade, variable-pitch propeller was mounted on a test stand. Observers watched as the antifreeze solution, dyed with methylene violet, left slinger ring tubes and then traveled out along the leading edge of the blades to the tip. At intervals of about ¼ inch along the blade, the liquid was carried from the leading toward the trailing edge. These initial tests showed an almost perfect distribution of the fluid.

They then attached the propeller to the DC-1. Set at low pitch, the propeller was accelerated to 900 revolutions per minute. Again, distribution pattern was excellent. Next came flight tests—and problems. The slingers proved structurally weak, too heavy, and poorly attached. With a redesigned slinger fasted to the propeller hub, the distribution pattern over the blades was so poor that ice likely could not have been removed. Also, the slinger rings were inefficient. Most of the liquid splashed out of the rings and was thrown over the engine because of the air turbulence in back of the spinner. The slinger went back to the factory for another redesign. It took almost three months before the system became flight-worthy.[32]

By the winter of 1935–36, the ice-protected DC-1 was ready to go. Whenever severe icing forced the cancellation of regularly scheduled flights, Tomlinson took off in the DC-1 in search of ice. "Encountered some very unusual icing conditions," he later recalled. The wing and tail pneumatic deicers worked well, as did the new slinger ring propeller deicer. His only major problem came while flying at 20,000 feet near Utica, New York. Without warning, both engines quit as a result of ice in the carburetor induction ducts. Tomlinson managed to start one engine at 15,000 feet and the other at 10,000 feet. He returned safely to Chicago through heavy icing.[33]

Walter R. Hamilton, superintendent of maintenance for TWA, came up with at least a temporary solution to the problem of carburetor icing by designing a system to heat the air entering the carburetor by means of intensifier tubes. The system worked, but it was costly to install and maintain, and it robbed the engine of power. Recognized as only a stop-gap measure, the airlines quickly adopted the system while work progressed on a more efficient alcohol-injection system and a redesign of the entire carburetor.[34]

Airline operations during the winter of 1935–36 saw great success in dealing with the icing problem. Jack Frye, president of TWA, announced that deicing equipment "has served virtually to eliminate ice formation as a danger to scheduled flight." TWA, United, Northwest, and Eastern planned to have their entire fleets equipped with pneumatic deicers for the winter of 1936–37, while American would outfit their DC-2s and DC-3s with the gear. Goodrich had orders for 360 deicers. "There is not a production line as such," NACA engineer William Clay observed during a visit to the Akron factory, "as all the work is done strictly by hand and by skilled workers." Goodrich charged an average of $1,000 for the equipment, which represented about 1 percent of the total cost of the average transport.[35]

All went well during the 1936–37 winter season until the early evening of March 25, 1937. A TWA DC-2, Flight 15-A, was en route from Newark to Chicago via Pittsburgh with ten passengers. As it approached the Allegheny County Airport, flying 800 feet above the ground, all seemed well for DC-2 registration number 320. Captain Frederick Bohnet made a routine radio report to the ground operator: "Okay—Bohnet in 320 to Pittsburgh. Okay." Captain A. M. Wilkins, who was flying the westbound route from Newark to Chicago, saw Bohnet's DC-2, flying near stalling speed and about one mile ahead, as the two aircraft made preparations to land at Pittsburgh. As Wilkins closed to one-half mile of Bohnet, he saw the DC-2 begin to make a left turn. The nose of the aircraft suddenly pointed downward, and the DC-2 plummeted into the ground. All ten passengers and three crew members died on impact.[36]

It did not take long to establish the cause of the accident. Observers who reached the scene of the crash reported that 1½ inches of ice remained on the leading edge of the ailerons and on the wing tips of the shattered airplane, which had not burned. A dramatic photograph of the ice-covered ailerons appeared in the New York Times on March 27. Investigators also found a postcard in the debris that had been written shortly before the crash by Pauline Trask, a schoolteacher from Germantown, Pennsylvania. It read: "Arrived safely—6:15 P.M.—pulling out of ice storm." The Accident Board of the Bureau of Air Commerce made it official in early May: ice had brought down the TWA transport.[37]

The optimism that TWA had displayed after the winter of 1935–36 vanished. On April 8, 1937, two weeks after the loss of the DC-2, Paul E. Richter, TWA's vice president in charge of operations, sent a plea for help to senior officers in the army and navy. "TWA has done everything in its power," he wrote, "to investigate, test and apply to its transports, every protection against ice. . . . Still, we know that ice formation involves unsolved variables that present the gravest hazard to aircraft. No transport company has the personnel or facilities to undertake the scientific program necessary to

A NASA OV-10A Bronco, built by North American Rockwell, used for an icing test program in 1973. (NASA photo no. C1973-948)

solve this vital problem." He asked the assistance of the military authorities in bringing the matter before the NACA. The NACA, he urged, should proceed "at the earliest instant with an investigation of ice formation which must produce solutions to every aspect of the problem."[38]

Richter's plaintive letter marked the end of round one in the battle against ice. At this point, the struggle might be viewed as a draw. Much had been accomplished to defeat the icing menace. The development of pneumatic boots had given airmen a valuable weapon in the struggle. Protection against propeller and carburetor icing had made substantial progress. But the TWA crash revealed that much remained to be done before the skies could be considered safe, as far as icing was concerned.

Time would show that the battle against icing would continue throughout the remainder of the twentieth century and into the twenty-first century. In the years that followed the TWA accident, it often seemed that victory was at hand. The NACA responded to Richter's request and launched an extensive investigation of the icing problem. The aerodynamicists at the Langley laboratory had always favored the development of a thermal deicing system. The pneumatic boots, they believed, caused drag. In 1938, tests of the boots in the NACA's eight-foot high-speed wind tunnel confirmed this judgment. Deflated boots, the tests showed, added 15 percent to smooth-wing drag, while inflated boots increased drag by 100 percent.[39]

Under the driving leadership of Lewis A. Rodert, the NACA embarked on the program of flight research to develop a practical thermal system. This work earned Rodert the prestigious Collier Trophy for 1946. When combined with research conducted in a large refrigerated wind tunnel at the NACA's Cleveland engine laboratory, a solution seemed at hand for the new generation of jet-powered transports. In 1955 NACA researcher Uwe von Glahn announced that "aircraft are now capable of flying in icing clouds without difficulty . . . because research by the NACA and others has provided the engineering basis for ice protection systems." Shortly thereafter, the NACA ended all icing research programs.[40]

The NACA's declaration of victory once again proved premature. Icing remained a menace to aviation, especially for helicopters and for the smaller turboprop commuter aircraft that began to appear in the 1970s. Ground deicing problems caused the spectacular crash to an Air Florida Boeing 737 into the Potomac River in 1982, while tailplane icing brought down several aircraft, including an American Eagle ATR-72 near Roselawn, Indiana, in 1994. Senator Paul Wellstone died in an icing-related crash of his private jet during his reelection campaign in the fall of 2002. Indeed, the Flight Safety Foundation identified ninety-one icing related commercial aviation accidents throughout the world between 1956 and 1996, with 776 fatalities.[41]

The National Aeronautics and Space Administration (NASA), the NACA's successor, returned to icing research in 1978. A great deal would be accomplished over the next twenty-three years, but the most recent generation of icing researchers have been hesitant to issue any victory announcements. On the contrary, they have developed a healthy respect for the icing foe, and they expect the struggle to go on for many years to come.[42]

NOTES

1. For a history of efforts to combat ice on aircraft, see William M. Leary, *"We Freeze to Please": A History of NASA's Icing Research Tunnel and the Quest for Flight Safety*, NASA Special Publication 2002-4226 (Washington, D.C.: U.S. Government Printing Office, 2002).

2. For a history of the Air Mail Service, see William M. Leary, *Aerial Pioneers: The U.S. Air Mail Service, 1918–1927* (Washington, D.C.: Smithsonian Institution Press, 1985).

3. Jack Knight to A. R. Dunphy, Nov. 19, 1921, copy in file 247, Langley Historical Archive, Floyd L. Thompson Technical Library, Langley Research Center, Hampton, Va.

4. Carl Egge to D. B. Colyer, Dec. 27, 1926, quoting report by Williams, Personnel File of Warren B. Williams, Records of the Post Office Department, Bureau of the Second

Assistant Postmaster General, Record Group 28, National Archives, Washington, D.C. [hereafter Post Office Records].

5. Luther Harris to Colyer, Apr. 23, 1927, Personnel File of John F. Milatzo, Post Office Records.

6. Bradley Smith, "Icing Wings," *U.S. Air Services* 15 (Apr. 1930): 22–25.

7. "Conference at Department of Commerce Regarding Formation of Ice on Aircraft," Mar. 28, 1928, Research Authorization (RA) 247, Langley Library, Langley, Va.

8. Thomas Carroll to Henry J. E. Reid, Mar. 17, 1928, RA 247, Langley Library. On the NACA's flight research, see Michael H. Gorn, *Expanding the Envelope: Flight Research at NACA and NASA* (Lexington: University Press of Kentucky, 2001).

9. Quote is from Thomas Carroll and William H. McAvoy, "The Formation of Ice upon Exposed Parts of an Airplane," NACA Technical Note 293 (1928), and "The Formation of Ice upon Airplanes in Flight," NACA Technical Note 313 (1929). See also Bradley Jones, "Icy Wings," *U.S. Air Services* 15 (Apr. 1930): 22–25.

10. William C. Geer, "The Ice Hazard on Airplanes," *Aeronautical Engineering* 4 (1932): 33–36.

11. Ibid.; Richard P. Hallion, *Legacy of Flight: The Guggenheim Contribution to American Aviation* (Seattle: University of Washington Press, 1977), 109–10.

12. William C. Geer and Merritt Scott, "The Prevention of the Ice Hazard on Airplanes," NACA Technical Note 345 (1930); Henry J. E. Reid, "Memorandum for the Files," Apr. 12, 1930, RA 247, Langley Library.

13. Smith, "Weather Problems Peculiar to the New York–Chicago Airway."

14. William C. Geer to George W. Lewis, Jan. 4, 1930, RA 247, Langley Library.

15. Quote is from William C. Geer to Harry F. Guggenheim, Mar. 26, 1930, copy in RA 247, Langley Library; "Russell S. Colley, Inventor," *Rubber World* (June 1982): 38–40. See also *NAT Bulletin Board*, no. 38 (Apr./May 1930), copy in the library of the National Air and Space Museum, Washington, D.C.

16. Reid, "Memorandum for the Files," Apr. 12, 1930; William C. Geer to George W. Lewis, May 6, 1930; George W. Lewis to Harry F. Guggenheim, May 8, 1930, all in RA 247, Langley Library.

17. Geer to Lewis, May 6, 1930; Henry J. E. Reid to the NACA, May 12, 1930; C. W. Leguillon to George W. Lewis, June 7, 1930, all in RA 247, Langley Library.

18. Russell Colley, "Problem #1825: Goodrich Refrigerated Wind Tunnel," Sept. 13, 1930, copy in author's possession, courtesy of David Sweet of BF Goodrich Aerospace, Akron, Ohio.

19. Ibid.

20. C. W. Leguillon to George W. Lewis, Sept. 13, 1930, RA 247, Langley Library; William C. Geer, "The Ice Hazard on Airplanes," *Aeronautical Engineering* 4 (1932): 33–36.

21. "Goodrich Airplane De-Icers," *Aero Digest* 18 (May 1931): 66; *New York Times*, Mar. 31, 1930.

22. Geer and Scott, "Prevention of Ice Hazard."

23. Theodore Theodorsen and William C. Clay, "Ice Prevention on Aircraft by Means of Engine Exhaust Heat and a Technical Study of Heat Transmission for a Clark Y Air-

foil," NACA Report 403 (1933). Although not published until 1933, the report is dated June 12, 1931. It is available at the Langley Library.

24. William C. Clay to the chief of the Aerodynamics Division, Aug. 24, 1935, RA 247, Langley Library.

25. *New York Times*, Oct. 29, 1933; "Ice," *Aviation* 33 (Nov. 1934): 353–54.

26. *New York Times*, Dec. 12, 1933; "Ice," 353–54.

27. "Colley," *Rubber World*; quote is from Clay to the chief of the Aerodynamic Division, Aug. 24, 1935, RA 247, Langley Library; *New York Times*, Oct. 23, 1935.

28. William C. Clay to Henry J. E. Reid, Nov. 12, 1935, RA 247, Langley Library.

29. Ibid.; S. Paul Johnson, "Ice," *Aviation* 35 (May 1936): 15–19.

30. Jack Frye, "No Ice Today," *U.S. Air Services* 21 (Aug. 1936): 13–14.

31. Clay to Reid, Nov. 12, 1935.

32. Ibid.; Johnson, "Ice."

33. D. W. Tomlinson to William M. Leary, Aug. 20, 1983; D. W. Tomlinson, "On Top," *Aviation* 36 (Dec. 1936): 21–24.

34. Johnson, "Ice."

35. Frye, "No Ice Today," 13–14; J. S. Pedler, Aeronautical Sales Division, Goodrich, to George W. Lewis, Sept. 4, 1936, and William C. Clay to Henry J. E. Reid, Sept. 29, 1936, both in RA 247, Langley Library.

36. *New York Times*, Mar. 26, 1937.

37. *New York Times*, Mar. 27, May 6, 1937.

38. Paul E. Richter to Admiral A. B. Cook, Apr. 8, 1937, RA 247, Langley Library.

39. Russell G. Robinson, "The Drag of Inflatable Rubber De-Icers," NACA Technical Note 669 (1938).

40. For Rodert's career, see Glenn E. Bugos, "Lew Rodert, Epistemological Liaison, and Thermal De-Icing at Ames," in Pamela E. Mack, ed., *From Engineering Science to Big Science: The NACA and NASA Collier Trophy Research Project Winners*, NASA Special Publication 4219 (Washington, D.C.: U.S. Government Printing Office, 1998), 29–58. Von Glahn, who was head of the icing branch at the Cleveland laboratory, delivered his comments at a conference in Ottawa, Canada, in June 1955. See Uwe von Glahn, "The Icing Problem—Current Status of NACA Techniques and Research," reproduced in NASA Technical Memorandum 81651 (1981), 1–10.

41. Flight Safety Foundation, "Protection against Icing: A Comprehensive Overview," special issue of *Flight Safety Digest* (June–Sept. 1997).

42. Leary, *"We Freeze to Please,"* 146–73.

Part 3

AERIAL WARFARE

CHAPTER 7
THE WRIGHT BROTHERS AND THE
U.S. ARMY SIGNAL CORPS, 1905–1915

On December 17, 1903, Wilbur and Orville Wright made the first successful flight of a heavier-than-air machine. Over the next two years, the brothers continued to refine their design. Unlike all other would-be inventors, they flew their aircraft under complete, although tenuous, control along all three axes—roll, pitch, and yaw. This control would be the basis on which they applied for a patent and the basis on which the patent would be upheld. Recognizing the airplane's potential as a military instrument, the Wrights decided to offer it first to the U.S. Army. In January 1905 Wilbur Wright made the first of several attempts to interest the War Department.[1]

Over the next ten years, the Wright Brothers would build an official relationship with the U.S. Army Signal Corps while developing personal friendships with many U.S. Army officers, particularly those who became pilots. The official relationship reached a zenith in 1912–13 but declined to a permanent nadir by 1915, yet many of the personal friendships survived. This chapter addresses how the Wrights established an official association with the U.S. Army Signal Corps and the immediate and long-term results. It also explores the personal friendships the brothers made with pioneer army aviators and how those friendships enhanced and otherwise affected official contacts. The army aviators had a definite impact on developing the technology of Wright aircraft. On the other hand, the Wrights, particularly Orville Wright, resisted technological advances. Their resistance to changing technology and the growing obsolescence of Wright airplanes adversely affected their business relationship with the army. Still, the personal relationships that army aviators developed with the Wrights and other early aircraft manufacturers proved long-lasting and influential.

The U.S. Army Board of Ordnance and Fortification viewed the first correspondence from the brothers in 1905 as just another appeal for money to build some theoretical flying machine and showed little interest. Eventually, at the instigation of President Theodore Roosevelt, on May 11, 1907, the board expressed interest in the Wright Brothers' flying machine. In reply, the Wrights offered for $100,000 an aircraft capable of carrying two people

200 kilometers at a speed of 50 kilometers per hour. When the U.S. Army demanded exclusive rights, the brothers refused to guarantee them, having already sold aircraft in Europe. Subsequently, the board, which did not have the money anyway, dropped the matter.[2]

The official contacts between the Wrights and the U.S. Army were in abeyance, but the opportunity for establishing a personal friendship with an army officer soon arose. Wilbur and Orville Wright spent the summer of 1907 laying the groundwork for a European Wright Company to manufacture and sell their flying machines. While in France, they became acquainted with U.S. Army Lt. Frank P. Lahm, who was detailed to the Cavalry School at Saumur, France. He became ill with typhoid, and while he was recuperating, his father visited him, bringing along Wilbur and Orville. This visit initiated a close and lasting friendship between Lieutenant Lahm and the Wrights. Shortly afterward, on August 1, 1907, the U.S. Army Signal Corps established a small Aeronautical Division to take "charge of all matters pertaining to military ballooning, air machines, and all kindred subjects." When the lieutenant received orders assigning him to the new Aeronautical Division, he wrote to the chief signal officer, Brig. Gen. James Allen, urging him to reconsider acquisition of the Wright Brothers' flying machine. He argued that the Flyer had considerable military value and the U.S. Army should be the first to purchase it, since it was an American invention. Eventually, the Board of Ordnance and Fortification contacted the Wrights, inviting them to meet with army officials. At the December board meeting, Wilbur explained the capabilities of the Wright Flyer and offered it to the army for $25,000.[3]

On December 23, 1907, the Signal Corps issued Specification No. 486, seeking bids for a heavier-than-air flying machine. The requirements establishing flight capabilities were based on the Wrights' contribution at the Board of Ordnance and Fortification meeting held earlier in the month. The airplane had to carry two people, up to 350 pounds combined weight, and sufficient fuel for a 125-mile flight. Speed was set at forty miles per hour with a bonus for more and a penalty for less speed. During trials, the machine would undergo an endurance flight of at least an hour. It had to be able to land in an unprepared field and to descend safely in case of motor failure. The successful bidder would have to train two individuals in handling and operating the machine. Delivery and trials were to be at Fort Myer, Virginia, just outside Washington. The Signal Corps accepted three bids on the specification, of which one was submitted by the Wrights. The brothers signed a formal contract on February 10, 1908, and immediately began to modify the 1905 Flyer to meet the army's requirements. In the spring they returned to Kitty Hawk, North Carolina, to test the new machine and sharpen their rusty flying skills. On May 14, 1908, Wilbur carried aloft the first aircraft passenger ever, an employee named Charles Furnas. By this time, one of the other two bidders had withdrawn his bid, and eventually the Signal Corps canceled the remaining bid because the bidder failed to deliver a flyable airplane.[4]

The Fort Myer Trials

In early May Orville traveled to Washington to inspect the drill field at Fort Myer. He found the small area barely adequate for flight. The good news was that Lieutenant Lahm had been detailed to assist in the trials. He assured Orville that he would have eight or ten soldiers as ground crew. Orville delivered the flying machine to Fort Myer on August 20 to meet the contracted time. He worked daily with Lieutenant Lahm and the enlisted ground crew to prepare the aircraft for flight. Orville began flying on September 3. On the 9th, Orville carried up Lieutenant Lahm, the first U.S. Army person to fly as a passenger in an airplane, for almost six and a half minutes. Three days later, he gave Maj. George O. Squier, acting chief signal officer, a flight lasting more than nine minutes. Each flight broke another world endurance record for powered heavier-than-air machines. Tragedy struck with the third army passenger, Lt. Thomas E. Selfridge, whom Orville carried aloft for the final preliminary flight on September 17. They had been in the air less than four minutes when a crack in a propeller caused it to foul a rudder control wire. The pilot lost control, and the airplane crashed, killing Selfridge and seriously injuring Orville, who spent the next three months in the Fort Myer Hospital. Lieutenant Lahm frequently visited Orville, cementing his friendship, and becoming acquainted with Katherine Wright, the sister and the "third member of the team." The army investigated but found no inherent problem in the airplane design or construction and extended the Wright Brothers' contract to the summer of 1909.[5]

By the summer of 1909, the achievement of the Wright Brothers was widely recognized in Europe and the United States. General Allen on June 18 at Dayton, Ohio, presented the brothers with a Congressional Medal for civilian accomplishments. Lieutenant Lahm served as the general's aide on this occasion. Two days later, Wilbur and Orville arrived in Washington to resume the army trials. Nine days later at Fort Myer, Orville made the first flight in the new airplane. The 1909 Wright Flyer was very similar to the 1908 model. The official tests began on July 27, when Orville flew, with Lieutenant Lahm as a passenger, on a record breaking one hour and twelve minute flight to meet the requirement for endurance carrying two people. Three days later, Lt. Benjamin D. Foulois was the passenger, official observer, and navigator for the speed and cross country tests. He had laid out the ten-mile course, with the airplane taking off from the center of the Fort Myer parade ground, flying south against the prevailing wind over Arlington Cemetery for five miles to Shuter's Hill near Alexandria, Virginia, coming around a tethered balloon on the hill, and returning to Fort Myer. Flying at an average speed of 42.5 miles per hour, the Wright machine earned its inventors a $5,000 bonus over the $25,000 price. On August 2, the Board of Officers certified that the Wright airplane had met all specifications and accepted it on behalf of the War Department. The army Wright Flyer was the first airplane purchased by any nation for military purposes.[6] The army's procurement of the revolutionary machine was the immediate result of the Wrights' official association with the U.S. Army Signal Corps, but the emphasis now shifted to a more personal aspect in the contract; i.e., training two army officers to fly the aircraft.

The Wrights required a better location than the Fort Myer drill field to conduct the training. Lieutenant Lahm chose a field at College Park, Maryland. The Signal Corps selected for instruction Lahm and Lt. Fredric E. Humphreys. Foulois, much to his disgust, was ordered to attend the International Congress of Aeronautics in France. Wilbur Wright began to teach the army officers on October 8, and after some three hours of dual flight both soloed on October 26. Years later, Lahm remembered, "Wilbur was a patient and understanding instructor, always ready to explain anything we did not understand, always ready to help us, to make easy our venture into this new field which was not too well understood by anyone at the time. Between flights and in the evening, we had long talks on aviation in general, on the Wright machine in particular, and on the future of aviation and flying."[7] The training period apparently served to deepen the friendship and respect that Lahm had for the Wright Brothers.

Returning in late October, Lieutenant Foulois made three flights with Wilbur before he departed. By November 5, Foulois received about three hours of dual instruction from Humphreys. But he did not solo at College Park. During the trials and flight training, Foulois developed great respect for the Wright Brothers. Like himself, they were "Honor Graduates of the University of Hard Knocks," who believed in practical experience as the best teacher. He never lost this respect, citing many years later the influence that the Wright Brothers had on him and on the development of aviation. At College Park, Foulois learned the basics of maintaining and operating the Wright Flyer. That experience would stand him in good stead in the coming months.[8]

Now that the army had two completely trained pilots, it promptly assigned them to other duties, leaving Lieutenant Foulois, who was only partially trained, as its sole aviator. At the end of November 1909, one partially trained pilot, eight enlisted mechanics, one civilian mechanic, and a damaged airplane constituted the U.S. Army's entire air arm. General Allen sent his motley crew and the Wright machine to Fort Sam Houston, Texas, where the weather was more favorable to winter flying. Foulois and his enlisted "Combat Air Force," as he called them, finally arrived on the post in February 1910. There he continued his flying lessons on his own. He frequently wrote the Wrights for instructions on the operation, maintenance, and modification of the airplane. Lieutenant Foulois encountered several problems in flying the Wright airplane. It tended to buck in the gusty wind, threatening to throw the pilot from his seat to the ground. Foulois, recalling Wilbur Wright's advice to put the nose of the airplane down and stay with his craft, managed to control it. To avoid being thrown out, he had a leather strap made to hold him in the airplane. Thus he could claim to have conceived and used the first safety belt in an airplane. He also altered the forward elevators on the advice of the Wright Brothers, moving one of the two to the rear of the airplane. This helped but did not totally solve the bucking problem. Another problem derived from the requirement to lay a 60-foot rail oriented into the wind for takeoff. Each time the wind changed, the crew had to move the rail. Obviously, a set of wheels in place of the skids would eliminate this problem and perhaps make for safer landings as well. In April, Foulois sent proposed blueprints of wheels to the Wrights, who made only one suggestion—the use of rubber bands to absorb the landing shock. Working through July and into August, the small group of mechanics devised wheels for the machine, and the pilot made service tests on the ground at first and then in the air. The wheels did away with the drudgery of launching the airplane from the rail. In late fall Foulois wrote that the chains frequently became fouled, interfering with the powering of the propeller. In December Orville replied that they should be boiled in grease mixed with graphite and tallow. On September 8, 1910, during windy conditions, Foulois wrecked the airplane. He and his crew rebuilt it, and he flew the remodeled airplane a few times in January and February 1911.[9]

While Lieutenant Foulois tried to keep the U.S. Army in the air in 1910 and 1911, the Wright Brothers turned their attention to setting up and running a business. The first product was the Model B. In comparison to the Wright army airplane, the Model B had wheels, the horizontal stabilizer in the rear, and a more powerful engine. Foulois, who on the basis of his experience obviously contributed ideas for the improvements, was very impressed with this model. Apparently, at this time, the Wrights were not as averse to input from fliers of its machines as they became a couple of years later.

Foulois would comment years later that the Wright B was the steadiest small airplane that he had ever seen. On March 3, 1911, Congress made its first appropriation, $125,000, to purchase airplanes. With the $25,000 made immediately available, the Signal Corps bought three Wright B and two Curtiss airplanes. On April 27 the Signal Corps began receiving its new airplanes, accepting a Wright B Flyer designated as Signal Corps (S.C.) No. 3. In May, at the suggestion of the Wrights, the War Department donated the original army Flyer, Signal Corps Airplane No. 1, to the Smithsonian Institution. The brothers reconditioned it at Dayton and sent it to College Park, Maryland, where Lt. Frank M. Kennedy turned it over to the Smithsonian on October 2.[10]

Training Military Pilots

Naturally, the Wrights were deeply involved in training pilots to fly the machines they sold. Orville Wright personally instructed or oversaw the training on Huffman Prairie near Dayton, Ohio, of at least 115 individuals, including U.S. Army pioneer aviators Henry H. "Hap" Arnold and Thomas DeWitt Milling. In late April 1911, Lieutenants Milling and Arnold reported to the Wright factory at Dayton, where Orville warmly greeted them. On May 2, Milling received his first flight as a passenger. A few days later he began lessons, completing fifteen dual flights for two hours and one minute before making his first solo. Orville witnessed this flight and subsequently flew with Lieutenant Milling. In 1956 Milling stated, "I have always felt that it was the instruction I received from Orville Wright that carried me through my flying career without being killed."[11] Orville had made another lifelong friend of an army aviator. Within eleven days, Lieutenant Arnold completed his flight training in twenty-nine flights totaling 3 hours and 48 minutes. This was about average for learning to fly the Wright airplane. However, Arnold found the control system hard to master.[12]

Of all foreign and domestic airplanes being flown at that time, the Wright was most difficult to control. The pilot's natural tendency when entering a stall would be to pull back on the control. But the Wright system required the pilot to push forward to regain flying speed. In addition to the difficult control system, the foot throttle on the Wright B airplane required the pilot to push to power down the engine. When a pilot throttled down the engine to land, it continued to pump gas. This excess gas spilled into a metal pan and frequently caught fire when the pilot released the foot throttle to obtain power to taxi. On November 6, 1911, and on January 10, 1912, Arnold wrote to the Wrights concerning a gradual decrease in No. 4's engine power. Captain Charles D. Chandler, another army aviator, also complained about the slow climbing rate. Orville replied that no other Wright airplanes seemed to have difficulty with engine power.[13]

After flight training, Milling and Arnold were assigned to the Signal Corps Aviation School at College Park. They arrived on June 15, 1911, the day before Foulois delivered S.C. No. 3 from San Antonio, Texas. The Signal Corps received its third Wright airplane, S.C. No. 4, on June 19 at College Park and another, S.C. No. 5, on August 7. In late November, the Signal Corps moved the flying school from College Park to Augusta, Georgia. The four pilots initially sent to Georgia were all Wright-trained, but only one Wright-manufactured airplane was sent. On December 9, 1911, Orville Wright, apparently in response to an inquiry, wrote to Lieutenant Arnold at Augusta, explaining how to measure propeller pitch and discussing ways to increase engine power. In early 1912, cross-country flights in the Wright airplane became impossible because of the loss of power in its engine. Wilbur Wright, visiting the school on January 20, estimated that the engine had lost five horsepower. In addition, inclement weather limited flying to 58 of the 124 days that the aviation school remained at Augusta.[14]

On December 11, 1911, the army shipped Wright B Flyer, S.C. No. 7, to the Philippines. Lieutenant Lahm, now with the 7th Cavalry in the Philippines, was detailed to temporary duty for aviation. Fortunately, on his way to the Philippines, he had visited with the Wrights in Dayton, where he received instruction in flying the Wright B equipped with wheels instead of the skids that he had become accustomed to using. With the help of two experienced enlisted mechanics sent with the airplane, he assembled it at Fort William McKinley. On March 21, 1912, Lieutenant Lahm made the first flight in the new machine. Relying on the refresher training the Wrights had given him the year before, he trained two army students, Lt. Moss L. Love and Corp. Vernon Burge, to fly. Corporal Burge, an enlisted mechanic, became the first enlisted man to receive a rating as a pilot. Lahm ended flying for 1912 with the onset of the rainy season. On March 10, 1913, he began the next season's flying lessons. On August 28, one of the new pilots, Lt. Herbert A. Dargue, was flying the airplane with pontoons. When the engine failed, he landed the airplane on Manila Bay. Before it could be towed back to the beach hangar, a squall dashed it against the rocky shore, damaging it beyond repair.[15]

Back in the Unites States, Captain Chandler consulted army aviators at College Park and Orville Wright at Dayton. Then, in September 1911, he submitted to the chief signal officer specifications for two new types of airplanes, a "speed scout" for long-range strategic reconnaissance and a "scout" to carry more weight for tactical reconnaissance. Early in 1912, the army re-

Wreck of Wright C, S.C. No. 10, in which Welsh and Hazelhurst were killed. (Courtesy of the Louis E. Goodier Jr. Collection, 168.7105, USAF Collection, AFHRA, Maxwell AFB, Ala.)

leased the new specifications and quickly purchased two scouts from the Wright Company. The army received its first scout airplane, Wright Model C, S.C. No. 10, at College Park in May. On June 11, the Wright pilot, A. L. Welsh, was flying the last acceptance test. He crashed the machine, killing himself and his passenger, Lt. Leighton W. Hazelhurst Jr. The Wright Company replaced the machine with another Model C, to which the army transferred the number 10. The Signal Corps accepted the second Wright C, S.C. No. 11, at College Park on October 3.[16]

The number of crashes and near crashes in the Wright airplanes dramatically increased in 1912 and 1913. On September 28, 1912, a newly certified pilot, Lt. Lewis C. Rockwell, with Corporal Frank S. Scott as passenger, took up S.C. No. 4, only to crash it from an altitude of about 25 feet while in a glide, killing both men. Rockwell apparently applied power to pull out of the glide, but either the elevator failed or the airplane stalled. Orville Wright thought that most likely the pilot glided down at too great an angle, although he admitted that the Wright airplanes would dive uncontrollably if the airspeed became too low. Milling later noted, "the difference between high speed and stalling was only about 8 miles [per hour]. Consequently, if the motor should stop, one had to be very alert in reaction to effect a landing without stalling the machine."[17] In November Lieutenant Arnold, while flying artillery support at Fort Riley, Kansas, with an artillery officer as a passenger, narrowly escaped crashing S.C. No. 10. A whirlwind had stalled his airplane, causing it to fall into a dive. He instinctively pulled on the controls, instead of pushing to gain airspeed and control. Arnold barely regained enough control to land the airplane. Badly frightened, he requested relief from flying duties and would not return to aviation until May 1916. However, Lieutenant Arnold continued to correspond with Orville Wright. In January 1913, responding to Arnold's inquiry, Orville gave a detailed explanation of how to obtain load test data by using sandbags on the wings and offering a chart to show the travel of the center of pressure at different speeds and weights. Later, in March, he defended in another letter to Lieutenant Arnold the Wright C scout, which had come under increasing criticism by army pilots for its performance. In this letter, Wright discussed the merits of the scout model and claimed that the high landing speed was its only drawback but that it was a very strong airplane and safer than other Wright models.[18]

In November 1912 the army had introduced a more stringent performance trial for military aircraft. The Wright Flyer Models C (scout) and D (speed scout) had great difficulty in meeting the test. The new six-cylinder engines that the Wright Company used with Models C and D were subject to frequent breakdowns and losses of power. Finally, after seven months of testing, the Signal Corps accepted the first Wright D speed scout as S.C. No. 19 on May 3, 1913. It accepted the second Wright D, S.C. No. 20, on June 6, but neither airplane was flown very much.[19]

In May 1913, Wright C scout S.C. No. 13 arrived at Fort McKinley, the Philippines. Lieutenant Lahm quickly made several successful flights from Fort McKinley and from the Manila Bay beach at Pasay. Then, in September, he tried flying with the pontoons that came with the airplane. He crashed into the water immediately after becoming airborne, damaging the airplane beyond repair. Fortunately, Lahm was unhurt. A replacement, Wright C scout S.C. No. 12 arrived on October 2. Lahm successfully trained two students to fly it. A third student, Lt. C. Perry Rich of the Philippine Scouts, crashed the airplane in Manila Bay on November 14. The engine broke loose, hitting Rich in the skull and killing him immediately. This accident ended the use of Wright Flyers in the Philippines.[20]

Back in the Unites States, in June 1913 the Signal Corps sent several Curtiss and Wright airplanes and pilots to its recently established Aviation School at North Island near San Diego, California. The mix of aircraft types led pilots to compare the strong points of each. Generally, a pilot was trained

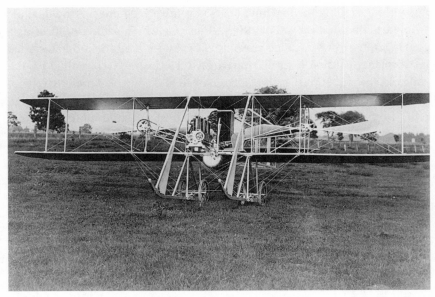

The Wright Model D in 1912 at the Simms Station near Dayton, Ohio. (Courtesy of Wright Aeronautical Corporation History and Figures Plus Photographs 168.651-5, 1900–1915, USAF Collection, AFHRA, Maxwell AFB, Ala.)

to fly either a Wright or a Curtiss airplane, but not both. Those who did fly both usually found features to admire in each, but the Curtiss control system was strongly favored over the Wright. The former with a single column, a shoulder yoke, and a foot throttle that the pilot pressed for more engine power encouraged instinctive motions of control. The Wright controls of a pair of levers and a foot throttle that the pilot had to press for less power was much more difficult to master. In August, Milling, Foulois, and Arnold pushed for a universal standard airplane control system. Orville Wright opposed the adoption of any system similar to the Curtiss one because of patent concerns. In general, the Wright Company's patent litigation retarded not only the development of the Wright aircraft but of all U.S. airplanes. This factor was overlooked by Orville Wright, who argued that responsibility for the general backwardness of the U.S. aircraft industry in comparison to that of Europe rested in a parsimonious Congress.[21]

Congress had good reason to limit appropriations for aviation. Between June 1912 and February 1914, Wright airplanes suffered six fatal accidents that killed eight army men and a civilian. Well over half of the fourteen army fatalities in airplane crashes had occurred in Wright machines. In the wake of so many fatalities, congressional appropriations for aviation declined by

A Wright Model C at Pasay, Manila Bay, 1913. (Courtesy of the Louis E. Goodier Jr. Collection, 168.7105, USAF Collection, AFHRA, Maxwell AFB, Ala.)

$25,000 in 1913. On February 16, 1914, the chief signal officer ordered that the Wright airplanes still in service should not be flown again until an investigation had been completed. The panel of pilots making the investigation generally condemned all pusher-type airplanes, particularly the Wright C, which it labeled as dynamically unsuited for flying. The panel recommended that only tractor machines be flown. From the viewpoint of the U.S. Army and particularly the pilots who flew the Wright airplanes, the machines were extremely dangerous. The Wright C, already obsolete when the U.S. Army purchased it in 1913, was slow, unstable, and difficult to control with the cumbersome dual-stick system. The center of gravity was too far to the rear. Also, the elevator was likely to fail as flight speed increased, usually in a dive. The pilots believed that the Wright airplanes, being pusher types, were very likely to result in death in the event of a crash. The engine usually fell forward onto the pilot and passenger, crushing them with its weight and momentum. Subsequently, the Signal Corps condemned the three remaining Wright airplanes of the fourteen purchased by the U.S. Army between 1908 and 1913.[22] The official relationship of the Wright Company with the Signal Corps had reached its nadir, never to be restored.

Nevertheless, in the spring of 1915 the Wright Company delivered to the U.S. Army one more airplane, S.C. No. 39. In March 1913, the Signal Corps had ordered the Model F, a pusher type with an armored fuselage and powered by a foreign manufactured engine. The Wright Company had trouble in obtaining the Austro Daimler engine, thus delaying aircraft delivery to the army. Lieutenant Dargue flew the airplane a total of seven times between May 27 and June 4, 1915, before the army dropped it from the inventory on June 13.[23]

A Parting of the Ways

Several factors led to the demise of the relationship between the army and Orville Wright. Obviously, a concern for safety in flying was an important factor. Also, as observed by the U.S. Army inspector general in a 1914 report, the Wright Company had failed to keep current in aviation technology. It left to others the development and perfection of hydro-airplanes, the use of tractor propulsion, the initial adoption of wheels in place of skids, the use of a closed-in cockpit and fuselage, a simpler control system, and an advanced engine with carburetor. Especially damaging was the Wright opposition to ailerons in place of wing warping, even though the ailerons were obviously superior. This neglect in advancing technology rested to a great extent in the Wrights' overriding efforts to receive recognition, fame, and financial rewards.[24]

After the successful delivery of the first U.S. Army airplane in 1909, Orville became deeply involved in the production and sale of airplanes while Wilbur focused on resolving legal issues, particularly the problems of patent infringement. On May 2, 1912, Wilbur returned home exhausted from a lengthy trip and ill with what was eventually diagnosed as typhoid. He died on May 30. After his older brother's death, Orville struggled for two years with running the Wright Company but soon realized his dislike for the job. In January 1914, he won the Wrights' patent suit against Curtiss. In early spring, he turned his attention to divesting himself of the airplane manufacturing business. He quickly obtained full control of the Wright Company and put it up for sale. He consummated the sale for a small fortune in October 1915. The Wright Company severed all ties with Orville Wright within a year of the sale and eventually became a major manufacturer of aircraft engines. Orville felt that he had accomplished the major goals that he and Wilbur had first set in 1905. He basically spent the rest of his life as a national hero and the elderly statesman of aviation.[25]

Even though Orville abandoned aircraft manufacturing, he maintained many of the personal friendships he had established during his dealings with U.S. Army Signal Corps officers. In fact, the relationship of the Wright Brothers with the Air Corps was characterized by frequent contact and the growth of personal friendships among many of the young aviators and the inventors. Interestingly, development of friendships with the army aviators was typical for early successful aircraft manufacturers, such as W. Starling Burgess, Glenn H. Curtiss, and Glenn L. Martin. When the army pilots encountered problems with the aircraft, they corresponded with the manufacturers, seeking solutions. In turn, often the Wrights and other manufacturers would visit the aviation schools or send representatives to help the pilots and mechanics. Foulois's experience in learning to fly by correspondence, while somewhat extreme, was not strikingly different from that of other army fliers. The major difference between the Wrights and other manufacturers is that increasingly the Wrights, and especially Orville, became defensive, stating on several occasions that the fault lay with the pilots, not with the aircraft. This obstinate position contributed to the army's crossing the Wright Company off its aircraft procurement list.

In general, the close relationships between manufacturers and the young army aviators continued through the interwar period and proved fruitful for both parties immediately before and during World War II. Government and business cooperated closely to achieve a vastly increased production of aircraft, initially for the Allies and eventually for the U.S. Army Air Forces. Particularly, General Arnold, first as chief of the Air Corps and later as commander of the Army Air Forces, relied heavily on his longstanding friendships with several of the aircraft manufacturers to expedite the manufacture of aircraft and to develop advanced aircraft types. Consequently, the Army Air Forces were able to build a strong air power base more rapidly than would have been possible otherwise.[26] Hence, in the long run the personal friendships proved important even beyond the pioneering years and the business association of the Army Signal Corps with the Wright Brothers.

NOTES

1. Richard P. Hallion, ed., *The Wright Brothers: Heirs of Prometheus* (Washington, D.C.: National Air and Space Museum, 1978), xii; Wilbur and Orville Wright, "The Wright Brothers' Aeroplane," *Century Magazine*, Sept. 1908, excerpted in Peter L. Jakab and Rick Young, eds., *The Published Writings of Wilbur and Orville Wright* (Washington, D.C.: Smithsonian Institution Press, 2000), 25–26; Tom D. Crouch, *The Bishop's Boys: A Life of Wilbur and Orville Wright* (New York: W. W. Norton, 1989), 291–92; Juliette A. Hennessy, *The United States Army Air Arm, April 1861 to April 1917* (Washington, D.C.: Office of Air Force History, 1985), 25; Orville Wright, "Interview by Fred C. Kelly," in Jakab and Young, *Published Writings of Wilbur and Orville Wright*, 262.

2. Hennessy, *United States Army Air Arm*, 25–26; Crouch, *Bishop's Boys*, 331–32; Law-

son W. Fuller, recorder of the board, to Wilbur and Orville Wright, May 11, 1907; Wrights to the board, May 17, 1907; the board to Wrights, May 22, 1907; Wrights to the board, May 31, 1907; the board to Wrights, June 8, 1907; Wrights to the board, June 15, 1907; the board to Wrights, July 16, Oct. 5, 1907, copies in Board of Ordnance and Fortification, *Letters and Proceedings*, 167.1-1, USAF Collection, Air Force Historical Research Agency (AFHRA), Maxwell AFB, Ala. [hereafter Board Letters and Proceedings].

3. Crouch, *Bishop's Boys*, 333–41, 346–47; Orville Wright to Wilbur Wright, Sept. 21, 1907, in Fred C. Kelly, ed., *Miracle at Kitty Hawk: The Letters of Wilbur and Orville Wright* (New York: Farrar, Straus and Young, 1951), 235; Frank P. Lahm, "Early Flying Experiences," *Air Power Historian*, Jan. 1955, 3; quote is from "The Birth of the United States Air Force," at <http://afhra.maxwell.af.mil>; Frank P. Lahm to Chief Signal Officer, Oct. 24, 1907, 168.651-6, 1950, USAF Collection, AFHRA, Maxwell AFB, Ala.; Generals Benjamin D. Foulois, Frank P. Lahm, and T. DeWitt Milling, Round Table Discussion, moderated by Gen. Carl Spaatz, June 29, 1954, K239.0512-767 in USAF Collection, AFHRA, Maxwell AFB, Ala. [hereafter Spaatz Round Table], 3; Ordnance and Fortification Board Meeting, Dec. 5, 1907, *Board Letters and Proceedings*; Hennessy, *United States Army Air Arm*, 26.

4. Wilbur and Orville Wright to General James Allen, Dec. 18, 1907, Jan. 27, 1908, in Marvin W. McFarland, ed., *The Papers of Wilbur and Orville Wright*, vol. 2, *1906–1948* (New York: McGraw-Hill, 1953), 843, 856; Spaatz Round Table, 3; Hennessy, *United States Army Air Arm*, 26–28, 225; Hallion, *Wright Brothers*, 116–17; Board to Wright Brothers, Jan. 3, 1908, in *Board Letters and Proceedings*; Crouch, *Bishop's Boys*, 354–59; Ernest L. Jones, Chronology of Military Aeronautics (1793–1948), unpublished, 168.6501 in USAF Collection, AFHRA, Maxwell AFB, Ala., Feb. 10, May 6–14, 1908.

5. Orville Wright to Wilbur Wright, June 3, 1908, in Kelly, *Miracle at Kitty Hawk*, 268–69; Spaatz Round Table, 4; Orville Wright, "My Narrowest Escape in the Air," in Jakab and Young, *Published Writings of Wilbur and Orville Wright*, 49–51; Hennessy, *United States Army Air Arm*, 28, 33; Crouch, *Bishop's Boys*, 353, 361, 374–78; Jones Chronology, Sept. 17, 1908; "Wilbur Wright to Octave Chanute, Dayton, June 6, 1909," in McFarland, *Papers of Wilbur and Orville Wright*, 2:953–54; quote is from Frank P. Lahm, "The Wright Brothers As I Knew Them," *Air Corps News Letter*, Mar. 6, 1939, 2, 167.63 in USAF Collection, AFHRA, Maxwell AFB, Ala.

6. Benjamin D. Foulois, "Early Flying Experiences: Why Write a Book, Part I?" *Air Power Historian*, Apr. 1955, 24–25; Maj. Gen. Benjamin D. Foulois, Oral History Interview by Alfred Goldberg, n.d., transcript K239.0512-766 in USAF Collection, AFHRA, Maxwell AFB, Ala. [hereafter Foulois oral history], 13–14; Spaatz Round Table, 26–27; Proceedings of the Board of Officers Convened . . . for the Purpose of Observing Trials of Aeronautical Devices, etc., Aug. 2, 1909, copy in *Board Letters and Proceedings*; Hennessy, *United States Army Air Arm*, 33–34; Crouch, *Bishop's Boys*, 397–99; Jones Chronology, June 17–18, Aug. 2, 1909; T. DeWitt Milling, "Early Flying Experiences," *Air Power Historian*, Jan. 1956, 93.

7. Lahm, "Wright Brothers," 3; Spaatz Round Table, 5; Hennessy, *United States Army Air Arm*, 34, 37; Crouch, *Bishop's Boys*, 408.

8. Foulois, "Early Flying Experiences," 23–24, 27 (quote); Foulois oral history, 64–65.

9. Hennessy, *United States Army Air Arm*, 39–40; Crouch, *Bishop's Boys*, 446; Foulois

oral history, 16–19; Foulois, "Early Flying Experiences," 27–28, 30–31, 33–35; Spaatz Round Table, 29–31, 61–65; "Airplane Remodelled," Jones Chronology, Dec. 22, 1909, Mar. 2, July 20, 1910, added entries.

10. Crouch, *Bishop's Boys*, 424, 435, 447; Hennessey, *United States Army Air Arm*, 40, 42, 252; Foulois, "Early Flying Experiences," 34–35; Jones Chronology, Mar. 3, Apr. 27, 1911.

11. Milling, "Early Flying Experiences," 96.

12. Crouch, *Bishop's Boys*, 435–37; Jones Chronology, Apr. 27, Aug. 22, 1911.

13. Hennessy, *United States Army Air Arm*, 47, 50, 53; Milling, "Early Flying Experiences," 95; Crouch, *Bishop's Boys*, 436; Jones Chronology, Mar. 10, 1913, Apr. 11, 1911, Mar. 28, 1913, added entries.

14. Hennessy, *United States Army Air Arm*, 47, 50, 54, 57, 79; Milling, "Early Flying Experiences," 96; Jones Chronology, Apr. 11, 1911, added entry; Lt. Henry H. Arnold to Orville Wright, Dec. 9, 1911, in "Material for the Arnold Collection," 168.65-41 in USAF Collection, AFHRA, Maxwell AFB, Ala. [hereafter Arnold Material].

15. Hennessy, *United States Army Air Arm*, 79–80; Lahm, "Wright Brothers," 2; Lahm, "Early Flying Experiences," 7–8.

16. Hennessy, *United States Army Air Arm*, 58, 62; Milling, "Early Flying Experiences," 99; Jones Chronology, Feb. 8, June 11, 1912, added entry.

17. Milling, "Early Flying Experiences," 97.

18. Hennessy, *United States Army Air Arm*, 71–72; Jones Chronology, Sept. 28, 1912; Milling, "Early Flying Experiences," 100; Orville Wright to Lt. Henry H. Arnold, Jan. 30, 1913, and Arnold to Wright, Mar. 22, 1913, copies in Arnold Material.

19. Jones Chronology, Nov. 28, 1911; Hennessey, *United States Army Air Arm*, 74.

20. Hennessy, *United States Army Air Arm*, 80–81, 84; Lahm, "Early Flying Experiences," 8; Jones Chronology, Nov. 14, 1913.

21. Hennessy, *United States Army Air Arm*, 88, 92–93; Crouch, *Bishop's Boys*, 457; Jones Chronology, Feb. 5–27, 1914; Rebecca H. Cameron, *Training to Fly: Military Flight Training, 1907–1945* (Washington, D.C.: Air Force History and Museums Program, 1999), 28; "Orville Wright to Captain Charles DeF. Chandler, Dec. 29, 1911," in McFarland, *Papers of Wilbur and Orville Wright*, 2:1031–32.

22. "Colonel Dargue's Observations," and "Behind the Times," in Jones Chronology, Feb. 9, 24, 1914; Hennessy, *United States Army Air Arm*, 102–3, 232, 252; Crouch, *Bishop's Boys*, 457, 464.

23. Hennessy, *United States Army Air Arm*, 117; Jones Chronology, June 29, 1914.

24. Jones Chronology, Feb. 5–27, 1914; Marvin W. McFarland, "Wilbur and Orville Wright: Seventy-Five Years After," in Hallion, *Wright Brothers*, 27; Hennessy, *United States Army Air Arm*, 93.

25. Crouch, *Bishop's Boys*, 424, 447–49, 464–67; Jakab and Young, *Published Writings of Wilbur and Orville Wright*, 193–94.

26. John W. Huston, ed., *American Airpower Comes of Age: General Henry H. "Hap" Arnold's World War II Diaries* (Maxwell AFB, Ala.: Air University Press, 2002), 1:93–95.

JOHN H. MORROW

CHAPTER 8
BRAVE MEN FLYING
THE WRIGHT BROTHERS
AND MILITARY AVIATION
IN WORLD WAR I

Prewar

*Flying has been brought to a point where it can be of great use in . . . scouting and
carrying messages in time of war.* — Wright Brothers to Congressman Robert M.
Nevin, January 18, 1905

*Everything presently serves war, there is no invention whose military use the military
does not contemplate, no single invention that it will not endeavor to use for military
ends.* — Nicholas Fedorov, 1906

From the origin of flight, in the form of hot-air balloons, the military foresaw
the prospect of a new arena for warfare. The only significant heavier-than-
air flights before 1908 were the experiments of Wilbur and Orville Wright.
After their first flight at Kill Devil Hill on December 17, 1903, they remained
far ahead of European competition through 1908. Yet European skeptics,
particularly the French, believed either that the Wright Brothers had not
flown or that their exploits were not significant. French seaplane designer
Antoine Odier and aviator and inventor Albert Étévé acknowledged that the
Wrights had flown, but both Frenchmen were convinced that their achievements were ignored by the world until the Wrights appeared in Europe in
1908, by which time French aviators believed that they had technologically
surpassed the Wrights. To the contrary, however, the Wrights' exploits had
not been ignored by the world, and French aviators had not surpassed their
efforts as of 1908.[1]

Despite their achievements and the alleged interest of the military in inventions, the Wrights had no military contracts. Although they advised Congressman Robert Nevin on the flying machine's usefulness in January 1905,
the U.S. Army's Board of Ordnance and Fortification rejected their offer of
a "Flyer." The brothers then conducted intermittent negotiations with the
British, French, and German armies between 1905 and 1908. The British
War Office insisted on flight demonstrations with no prior commitment to

The air trials at Fort Myers, Virginia, represented the first time the Wright Brothers flew their aircraft for the U.S. Army. (NASA photo)

purchase, while the brothers, afraid of piracy, insisted on an advance guarantee of purchase if the plane met its advertised performance. French performance stipulations remained beyond attainment, and the Prussian War Ministry found the price too high. The Wrights were unwilling to haggle over their $200,000 price for the machine and patent rights because they were convinced that no one else could develop a practical airplane within the next five years. Confident of their achievement, yet confronted with these impasses and dogged by the fear of patent espionage, the Wrights dismantled their aircraft and did not fly from October 1905 until May 1908.[2]

Yet neither the interest nor the negotiations ceased. German General Staff Captain Hermann von der Lieth-Thomsen, who was destined to become chief of German aviation during World War I, advised his superiors that it was dangerous merely to observe foreign aerial progress, but General Staff negotiations with the Wrights foundered on the Prussian War Ministry's continual objections to price. German interest rekindled French ardor, but once again French stipulations—the ability to take off from any location, to clear uneven terrain, and to carry an observer to a minimum altitude of 300 meters—stymied the brothers. By 1907 British army balloon chief

Col. J. E. Capper, who had earlier been interested in buying a Wright Flyer, had become more interested in sponsoring the aviation experiments of Britain's resident Wild West showman Samuel "Buffalo Bill" Cody and British lieutenant John Dunne. The pressures of the arms race in the early twentieth century kept the European powers, particularly the French and the Germans, interested in the Wrights, but not sufficiently to meet the brothers' demands.[3]

The foreign interest in aviation did have some effect in the United States. In 1907 the U.S. War Department sent Maj. George Owen Squier to Europe to study aviation progress and established an aeronautical division in the office of the chief of the Signal Corps, which requested bids for an airplane. The Wright Brothers won the contract, and when Congress failed to appropriate the $200,000 that the Signal Corps requested for aviation in 1907, the U.S. Board of Ordnance and Fortification covered the airplane's cost. Although the Flyer surpassed the contract's performance requirements in trials at Fort Meyer, Virginia, it crashed on September 17, 1908, severely injuring pilot Orville Wright and killing his passenger, Lt. Thomas E. Selfridge. The army consequently delayed acceptance of its first airplane until August 1909.[4]

In 1908, while the Wrights negotiated, a small coterie of French inventors began to make significant strides in powered flight. Most critically, on January 13, 1908, Henri Farman had flown a biplane designed by the Voisin Brothers and powered by a fifty-horsepower Antoinette V8 engine designed by Léon Levavasseur over the first officially monitored closed-circuit one-kilometer course at Issy-les-Moulineaux outside Paris. The Wright Brothers' four-cylinder engine had delivered twelve and later thirty horsepower for a weight of 180 pounds. Levavasseur's Antoinette produced twenty-four and then fifty horsepower for a weight of 110 pounds, as Levavasseur, through trial and error, determined the minimum admissible engine weight by reducing the thickness of the parts until they broke under tests.[5] The official French history of military aviation proclaimed this event "the true birth of practical aviation" because Farman, flying the wheeled Voisin, was the first to take off under the plane's power, while the Wright biplane lacked wheels and required a launching apparatus.[6]

Farman had managed to stay airborne for thirty minutes by August 1908, when Wilbur Wright arrived in France to demonstrate the Flyer. By December he had astounded and electrified French spectators by easily outdistancing French competition and raising the international duration record to two hours and twenty minutes. French businessmen Henri Deutsch de la Meurthe and Lazare Weiller founded the Compagnie Générale de la navigation aérienne to exploit the Wright patent. Wilbur Wright's exploits kindled interest in aviation in Italy and in Russia, where army engineers wanted to buy Wright Flyers to start flight training, and in Austria-Hungary, where negotiations collapsed over price.[7]

Yet the French and German armies were more impressed when Henri Farman made the first cross-country flight of some thirty kilometers from Bouy to Reims on October 30, 1908. The Technical Section of the German General Staff reported: "With this flight of thirty kilometers, a new epoch in aviation has dawned. Farman is the first aviator to leave the maneuver field and undertake a flight to a distant destination."[8] Already in September 1908 the British War Office had issued aircraft specifications that included the ability to take off under the craft's own power, thus effectively eliminating the Wright Flyer from consideration.[9] These stipulations and assessments explain why the French history of military aviation suggests that Farman's exploit held more significance for military aviation than Orville Wright's flights.[10] As long as the airplane's takeoff was tied to a launching apparatus, which an army would have to transport in its line of march to use the Flyer, and its flight bound by a maneuver field, the military perceived it to be of little practical use. Overland flight in 1908 demonstrated the airplane's potential for reconnaissance and communications, and thus its military significance.

After 1909, a host of European aviation inventors, particularly French ones, surpassed the Wright Brothers, and even in the United States, where the army owned only one airplane, a Wright biplane, as of late 1909, the Wright Brothers found themselves equaled and then exceeded by their archrival Glenn H. Curtiss. Their creative hour had passed, but powered heavier-than-air flight launched by the brothers was on its way, and European military establishments had awakened to the significance of the airplane for their purposes. The link between military and the inventors of the nascent industry to develop aviation that the Wrights had sought to fashion, with limited success, would now be forged as the European powers first prepared for future war, and then fought a much larger and longer war than many had expected.

Yet the connection between the Wright Brothers and aviation in World War I was based not only on their relationship with the military but also on the fact that early aviation inventors sought out the military to fund their development of airplanes. It was also inherent in the very nature of early aviation inventors and entrepreneurs, for many of them not only conceived, designed, and built their planes, but also flew them. Early powered heavier-than-air flight was a highly dangerous business, as Orville Wright's crash and the death of his passenger in 1909 indicate. Louis Blériot, the French-

Many American pilots during World War I trained on the Curtiss JNS-1 Jenny. Here it is in flight with a trailing pitot-static tube for airspeed calibration in August 1922. (NASA photo no. L-343)

man who flew the English Channel in 1909, was prone to crashes and injuries from serious burns to broken bones. Gabriel Voisin's brother Charles would be killed in a flying accident.

The Wright Brothers and Blériot epitomized the early aviation inventor/ entrepreneur, staid, middle class, upright in moral rectitude, and methodical in approach to the conquest of the air. Yet they were not only creative, ingenious, and inventive; they were also courageous and daring, perfectly prepared to risk life and limb to test their own inventions. While other inventors in Europe and the United States might be less upright or staid, almost all came from the middle class, which provided them the material wealth and education that enabled their pursuit of the fanciful goal of flight.

Documentary films from this early era occasionally show rather hilarious sequences of early birdmen running determinedly, flapping wings attached to their arms, and leaping off hills, or incredible ten-winged contraptions collapsing in a heap of wood splinters and cloth on their intrepid pilots. Inventors were killed even in glider crashes; when engines and fuel were added to those wood-and-fabric contraptions, the result was a highly combustible or inflammable mix.

The public was entranced with these new machines and turned out in the hundreds of thousands, particularly in Europe, to witness early air shows. The public and the aviation press fêted the early inventor-fliers as heroes conquering the heavens. In prewar Europe and America, aviation displaced auto racing—mere child's play in comparison to flight competitions—as the most popular and dangerous sport, and aviators replaced automobile racers as the daredevils of the day. Aviators became popular heroes and sportsmen, conquerors of speed, height, and distance, masters of the use of technology in the conquest of nature.

But a morbid side lurked behind the fascination. That flight was such a hazardous undertaking enhanced its appeal. Prewar aviation journals enthusiastically printed graphic accounts of bloody crashes in which fliers and passengers were killed, their bodies and spines crushed to a pulp, their brains and bodies impaled on broken gasoline pipes or chassis struts, and their bodies occasionally so burned and crisply grilled in flaming wreckage that "nothing . . . was left but ashes and two unrecognizable trunks of burnt flesh."[11] Some reporters observed that the possibility of witnessing bloodshed attracted large crowds. Occasionally, even the spectators were at risk. On May 21, 1911, at Issy-les-Moulineaux, aviator Émile Train lost control of his plane, which plunged into the crowd, killing the French minister of war and injuring the French prime minister,[12] whose presence was proof of governmental interest in flight for military purposes.

Prewar Europe from 1908 to 1914 witnessed the militarization of flight. In the absence of substantial sport or commercial markets, and in an atmosphere of increasing nationalistic bellicosity, supporters of military aviation molded popular attitudes to benefit their cause. Individual authors, the press, and aviation organizations pressured government to develop military aviation. Public perceptions and expectations of air power's future importance for aggression or defense fueled its growth and the development of small prewar aviation industries primarily dependent upon military contracts for their survival. Prewar literature foretold nearly every role that aircraft would actually play in World War I, including the bombing of civilians to destroy morale. From its very origins military aviation captured the rapt attention of civilians, who regarded aeronautical achievements as measures of the greatness of nations at the beginning of the twentieth century. Although the airplanes of the day were too small and fragile to factor into prewar politics and diplomacy, Germany's development of the giant dirigible, or Zeppelin, which had substantial range and seemed a potential bomber, did prompt some German aviation magazines to threaten a preemptive strike,[13] thereby heightening prewar tensions. Public involvement enhanced the social and cultural significance of military aviation in World War I.

The inventor/pilots and other aviators had provided the public with drama and spectacle, daredevils and heroes, and blood and gore. Prewar military aviators basked in the reflection of the heroic image of their civilian counterparts. In the small air services of all countries a different type of warrior arose, exemplified by the dashing and audacious "Lieutenant Daedalus Icarus Brown," Royal Flying Corps pilot of "fame and renown" proclaimed in British doggerel.[14] Wartime would provide the world stage for military aviation to deliver much, much more of the same, from drama to gore, as intrepid aviators—true heirs to the Wright Brothers and their peers, who had developed and often flown the airplanes that the military adopted—took to the air in their flying machines, the products of an age that had begun barely more than a decade before.

Wartime

Nothing but the supreme stimulus of war considerations, and nothing but the large and generous flood of money which the taxpayer can provide, will carry aviation forward to the foremost place in the world. — Winston Churchill speaking at a Royal Aero Club dinner at the Savoy, 1914

Men were going to die in the air as they had died for centuries on the ground and on the seas, by killing each other. The conquest of the air war truly accomplished. — French aviation historian René Chambe, 1955

The air war of World War I has become one of the most highly romanticized and mythologized subjects of military history. At the mention of World War I aviation, popular imagination invariably conjures up the aces, the high-scoring fighter pilots who became the ultimate heroes of the Great War. Aerial combat—individual, chivalrous, and deadly—dominates the popular conception of aviation in World War I. This conceptualization of World War I aviation remains, although recent sources and scholarship attempt to deromanticize the history of wartime aviation. Cultural studies of early airpower by Robert Wohl and Peter Fritzsche indicate why the popular notions possess such enduring strength.

Airmen were *the* heroes of the great war of 1914–18. Immortalized through their exploits, the air aces are probably far better remembered today than the political and military leaders of the time. After all, the Red Baron—the real one, not that scandalous impostor that graces the Red

Baron pizza commercials—still patrols the skies of the western front, somewhere north of Pont-à-Mousson, if challenged only by a devil-may-care beagle mounted on a Sopwith Camel doghouse. Who in America, after all, remembers Hindenburg and Ludendorff other than diehard World War I history buffs and students.

It was natural for aviators to become the heroes of that conflict. Dreams of flight embodied in the myths of Daedalus and Icarus long antedated the actual achievement of powered flight, just as visions of aerial warfare preceded the formation of air arms.

In wartime Europe, public fascination with these new warriors encouraged a mythologizing of the air war into a single image of individual combat, deadly but chivalrous. In the trenches mass slaughter on an unprecedented scale rendered individuals insignificant. Aerial heroes provided a much-needed affirmation of the importance of the individual and youth in a slaughter of both. The fighter pilots consequently became not only the symbols of aviation but also the ultimate heroes of World War I.

The great war fliers, particularly in France and Germany, were worshiped by the public. Oswald Boelcke, one of Germany's first and most famous aces, won Germany's highest award for valor, the *pour le Mérite*, early in 1916. Photographs of the handsome youth and jaunty verses about his victories flooded the press. When he visited Frankfurt-am-Main in the spring, crowds stared at him in the streets. During intermission at the opera, the audience crowded around him, and at the finale, instead of singing an encore, the Heldentenor sang a verse in Boelcke's honor. The audience went mad, clapping, shouting, and tramping their feet. Boelcke, imperturbable in aerial combat, was so startled that he fled the theater. He crashed to his death in October 1916, the victim of a collision in combat after forty victories. A German nation in mourning commemorated Boelcke in two elaborate funerals, sent condolences to the family, and composed eulogies to inspire German youth to protect the fatherland as their hero had.[15]

Manfred von Richthofen, Boelcke's pupil and heir, elicited the same worship during 1917 and 1918. The first two editions of his wartime memoirs, *Der Rote Kampfflieger*, which appeared in 1918, sold a half million copies. His funeral service in Berlin in May 1918 was even more spectacular than Boelcke's, as the Hohenzollern royal family joined the Richthofens in the pew.[16]

The legendary Georges Guynemer was France's greatest hero, and upon his death in the fall of 1917 after scoring fifty-three victories, teachers instructed schoolchildren that he had flown so high that he could not descend. In October the government enshrined "Capt. Guynemer, symbol of the aspirations and enthusiasm of the army of the nation," in France's memorial to its national heroes, the Panthéon, "whose cupola alone has sufficient span to shelter such wings." The frail youth embodied the victory of the spirit over the flesh, of France's will to endure despite her grave wounds.[17]

In England the Royal Flying Corps (RFC) characterized air combat as a sport, a notion that stemmed from the corps's composition early in the war —commissioned officers recruited mainly from the ranks of public school sportsmen attracted to military aviation for the adventure. The image of the air war passed down to us in most of their memoirs and histories is one of a clean and glorious struggle, far above the squalor of the western front below. In the most literary British memoirs, such as Cecil Lewis's *Sagittarius Rising*,[18] the war assumed the characteristics of sport and medieval tournament, a joust between heroes who bore only the utmost respect for one another, as, bound together in the brotherhood of the air, they rose daily to do battle. They fought and lived by unwritten codes. In squadron at the end of the day, for example, they were never to dwell on their losses except in absolute privacy. Instead, maintenance of a "stiff upper lip" was mandatory; these young aviators consequently released nerves, rage, and fears together in "rags," or brawls, in the mess or in bruising football games. Mess bills for broken furniture were common, and although no intrepid historian has yet studied the casualty rates for these "friendly" terrestrial struggles, at least one top British aviator, forty-victory ace Philip Fullard, suffered a seriously broken leg in one in 1917. Fullard, a tremendously gifted flier and marksman, was scoring at such a rapid rate that he seemed destined to become Britain's greatest ace. Yet he never returned to combat, as his nerves gave way when he was finally scheduled to rejoin his squadron at the front after a long and difficult recovery, his sense of invincibility apparently as shattered as his limb.[19]

British aviation magazines such as *Flight* and *The Aeroplane* romanticized the RFC and the sporting, chivalric, heroic, and sacrificial images of the air war. From RFC headquarters Philip Gibbs's column "Daily Chronicle" depicted the RFC as "Knights-Errant of the Air," recalling the Black Prince in Flanders during the Hundred Years' War. In a war with precious little romance, he found it in the "daily tourneys" in the air, as fearless British fliers fought with the ardor of schoolboys flinging themselves into a football scrimmage.[20] And why not? They were, in fact, like their counterparts in all countries, overwhelmingly youth in their late teens and early twenties, primarily volunteers from the middle class lured to aviation by the adventure, excitement, and risk, schoolboys transformed into warriors by the greatest war humankind had witnessed to that date.

The Aeroplane of May 30 carried a poem, "The Lament of the Broken Pilot," who bid farewell to France, "the land of adventure and knightly deeds, / where the pilot faces the foe / in single combat as was of yore— / giving him blow for blow." No longer among the "throng of chivalry, youth, and pride," where his comrades entered the "airy lists in the name of Freedom and Right," our broken pilot would now keep their "armour bright." Exclusive London stores advertised aviation clothing intended to dress the wealthy young sportsman-knight stylishly and appropriately for the airy lists.[21]

What glorious images! What a beautiful way to fight a war otherwise characterized by brutal and senseless mass slaughter. One discerns the origin of the romantic images immortalized in movies like *Hell's Angels*, *Wings*, and the *Dawn Patrol*, where aviators pursue one another in individual combat, fight tenaciously, win gallantly, or die heroically, their flaming craft plunging to earth like meteoric funeral pyres, extinguishing their equally meteoric careers with scorching finality. The greatest of these warriors— Albert Ball, Oswald Boelcke, William Bishop, Rene Fonck, Georges Guynemer, Edward "Mick" Mannock, Manfred von Richthofen, Ernst Udet—were legendary, lionized by adoring publics; their exploits, the material of myth. Yet their lives were often terribly short. It is sobering to reflect that of those eight aviators, six died in 1916, 1917, and 1918, waging an aerial conflict that became a mass war of attrition just like the struggle on the ground, thereby seriously eroding any notions of chivalry or sport that still lurked in the breasts of aerial combatants. These aviators are the symbols of the first war in the air, its heroes and victims, and the focus of most studies of the subject.

Anecdotes of wartime aviation that concentrate on the exploits of fighter pilots have given the impression that in this mass war of technology and industry, the air arm was merely an atavistic appurtenance in which a few exceptional aces were the dominant feature. This approach robs World War I air power of its genuine military and industrial significance. The air arms did more than provide the warring nations with individual heroes, for their individual exploits occurred within the context of an increasingly mass aerial effort in a war of the masses.

During World War I, aviation evolved from an instrument of reconnaissance used singly in 1914 by tiny air arms into a weapon for fighting, bombing, and strafing in 1918 employed by air forces with up to 300,000 men in service. Aviation played a significant role, first in rendering ground forces more effective through reconnaissance or artillery observation. Later, its effectiveness as a weapon for fighting, bombing, and strafing required its deployment en masse against the enemy. Air services that had begun the war with some 200 frontline airplanes had 2,000 to 3,000 airplanes at the front in 1918. National aviation industries that had a few thousand workers to deliver a hundred planes a month in 1914 employed hundreds of thousands of workers to deliver thousands of planes and engines monthly in 1918. French, German, and English wartime aircraft manufacture was 52,000, 48,000, and 43,000 respectively, and the French produced some 88,000 engines to English and German totals of 41,000 each. In 1918 the great powers were producing 2,000–3,000 planes and 2,000–4,000 engines a month.

In what by 1917 had become an aerial war of attrition above the one in the trenches, generals and politicians recognized the importance of mass. Both Gen. Henri Pétain, French commander in chief, and Winston Churchill, British minister of munitions, recognized the capital importance of aviation when deployed in mass, Churchill desiring to replace the attrition of men with a war of machines using "masses of guns, mountains of shells, clouds of aeroplanes."[22]

Concentration on the individual exploits of a few fighter pilots has thus given an archaic, anachronistic image to the most advanced and innovative technological arm of warfare, the one that epitomized the new total warfare in its requirement of meshing the military, political, technological, and industrial aspects of war—the front and the rear, the military and civilian.

Military and political leaders had to make crucial decisions to expand the tiny air arms of 1914 and mobilize the embryonic supporting industries, for in air power technological and industrial superiority essentially determined the outcome of the struggle. The race for aerial superiority had to be won first in design offices and then on factory floors, as the airplane evolved from an experimental vehicle into a weapon. Aircraft manufacturers like Albatros, SPAD, and Sopwith; engine manufacturers like Daimler, Hispano Suiza, and Rolls-Royce—their designers and skilled workers—were the essential backbone of their countries' aerial effort.

The aircraft companies evolved into large enterprises during the war, as they enlarged their plants to meet the demands for aircraft of the armies and navies of the fighting powers. As the struggle for aerial supremacy developed after 1915, the manufacturers had not only to build more airplanes, but also to develop more specialized aircraft types endowed with steadily improving performance. The aircraft manufacturers themselves evolved with the industry, or were left behind by aviation progress. The rapid progress of aeronautics left the Wright Brothers in its wake, and by 1917 their most significant participation in aviation was a patent suit against Glenn Curtiss over aircraft controls that may well have delayed the progress of the American aircraft industry. Louis Blériot's talents as designer were also no longer equal to the pace of aviation, but his entrepreneurial talents enabled him to direct aircraft manufacture throughout the war. The Caudron Brothers' prototype of the R4-5 series, a so-called "omnibus" battle plane for the French air service, crashed on December 15 with Gaston Caudron at the controls, leaving his brother René, assisted by a young designer of future fame, Henri Potez, to attempt modification of what was soon recognized as an abortive and doomed concept.[23] Gabriel Voisin, whose early wartime aircraft proved highly serviceable if mediocre in performance, by the end of the war had been driven from aviation to automobile manufacture, hounded by his lack of success in the former and the rude nickname *Bébé Grillard*, or "Baby Grillmaster," that aircrew had given him because of the tendency of his aircraft to burst into flame when hit.

In contrast, some manufacturers proved themselves capable of continuing to fly and design while directing the wartime expansion of their firms. French designer/manufacturer Louis Breguet was flying his own airplane as a volunteer military aviator as the German forces approached Paris at the end of August 1914. On September 2, Corporal Breguet returned from a mission to report that the German Army was turning to the east of Paris, information that later reconnaissance flights confirmed and that ultimately led to the Battle of the Marne, where the French Army halted the German advance.[24] Breguet's design and manufacture of bombers for the duration of the war would culminate in the superb Breguet 14 single-engine reconnaissance bomber of 1918.

Probably more widely known is the experience of the "Flying Dutchman," Anthony Fokker, prewar manufacturer of small, light, and maneuverable monoplanes. Fokker's fighter plane designs, from the first *Eindecker*, or monoplane, with a fixed forward firing machine gun, through the DrI *Dreidecker*, or triplane, of 1917–18, to the Fokker D7 of 1918, often considered the best production fighter of World War I, and his last parasol-winged monoplane fighter, the D8, probably merit his selection as the best fighter plane designer of the war. A key to the success of the young manufacturer was his superb ability as a pilot. His youth and skills enabled him to demonstrate his aircraft to frontline pilots and establish an immediate rapport with Germany's greatest fighter pilots, such as Oswald Boelcke, Max Immelmann, and Manfred von Richthofen. This rapport in turn enabled him to design his aircraft with intimate knowledge of their requirements in mind, while his talent as a pilot/designer meant that in test-flying his own designs, he could ascertain both their flaws and the modifications to correct the problem, as he did in the case of the D7.[25]

The airplane and its engine exemplified the harsh demands and enormous waste of modern industrial warfare, as the intensifying air war necessitated greater production to replace destroyed craft and to meet the front's incessant demands for more aircraft. They had to be sufficiently simple to lend themselves to serial production, yet of sufficient reliability and performance to remain effective under rapidly changing frontline conditions despite their limited combat life. Planes and engines demanded much higher standards of precision and reliability than the automobile did, and their rapid obsolescence in wartime rendered them unlike small arms or artillery, which was of standard types that changed infrequently and could be produced by state-run arsenals.

The sheer numbers of airplanes on the western front by 1918, more than 8,000 on all sides, indicate that the air war in general, and aerial combat in particular, was no longer an individual affair but instead a mass struggle of attrition. As the war had expanded in scope, the basic tactical unit, the French escadrille of six planes, the German Flieger-Abteilung of six, and the British squadron of twelve planes expanded in size to twelve, nine, and eighteen planes respectively. These units were subsumed under increasingly larger ones, like the German fighter circuses of sixty planes, as the attempt to achieve aerial superiority led to a concentration of forces. Under these circumstances, only a very few exceptional aviators, epitomized by the Frenchman René Fonck, could fly alone and survive in 1918. The ultimate unit was the French aerial division of 1918, with more than seven hundred bombers and fighters intended for tactical air raids over German lines.

A war of attrition meant death for aviators.

War flying was dangerous, as illustrated by the most accurate figures available of losses in aviation. Some 39 percent of France's more than 18,000 aircrew trained in the five years from 1914 to 1919 fell as casualties. More than 50 percent of 22,000 British pilots trained became casualties. In the absence of figures for aircrew trained in Germany, one can assume that their percentage of casualties was at least as high as that of the French and may have been higher than that of the British, because their force was smaller than their opponents while their total number of casualties apparently nearly equaled that of the British. While it is hard to compare these loss rates with those in the infantry, we do know that in the first six months of 1918, French infantry losses amounted to 51 percent of effectives, while the losses of French pilots at the front reached 71 percent. This toll is ample proof that aviators paid no less a price in wartime sacrifice than infantry did.[26]

The Royal Air Force was sufficiently concerned about fragmentary evidence of casualties to trace the careers to October 30, 1918, of nearly 1,500 pilots sent to France from July to December 1917. The results: 18 percent had been killed; 26 percent had been injured or sick and admitted to the hospital; 20 percent were missing over the lines; 25 percent had been transferred home; and 11 percent were still in France. Overall, then, 64 percent of those nearly 1,500 pilots were killed, wounded (or sick), or missing, and of the surviving 36 percent, about a quarter returned to England early in their tour. Only about one-fourth of all pilots completed a tour of duty of nine months, a chastening thought should one be tempted to minimize the toll of flying in World War I.[27]

One cannot evade the grim reality of the circumstances of their deaths. Take, for example, the case of the Red Baron, ace of aces. Entire books, countless chapters, if not yet verse, have contended over the circumstances of his death. The debate over whether Richthofen fell to another airman or to ground fire has a certain symbolic significance. Those who view the air war romantically would prefer the former conclusion, which currently seems more likely based on the evidence. The second likely cause of his death befits the capricious nature of total war in the air—the Baron, like the Irishman Mannock and Italian Baracca, the leading aces of the Royal Air Force and Italian air service respectively, was brought down by a bullet from an anonymous machine gunner on the ground. The manner of death does violence to the mythology of the knights of the air in the same way that, centuries before, the firearm's ability to kill impersonally and at long range first occasioned the resentment and ultimately the demise of the knight.

A further examination of casualties also indicates why authors, especially those preoccupied with the heroic nature of aerial combat, shy away from them. Accidents, termed by an American medical officer "the most important medical problem of aviation," were the greatest source of fatalities. In the U.S. Air Service, of 681 fatalities in flight personnel, 25 percent fell in combat and 75 percent in accidents, most of which occurred in flight school.[28] As slightly more than 2,000 American flying personnel arrived at the front during the war, for every four who survived to fight, one had not, that one symbolized by the short-lived Gary Cooper character in the movie *Wings*. In 1918 five Italian aviators were killed in accidents for every one in combat. But these were relatively small forces.

Among the major combatants, just over 36 percent of French fliers dead, or missing and presumed dead, perished in accidents in the rear areas. Of German losses, more than half were not attributable to enemy action. It is impossible to determine similar breakdowns of British losses, but training casualties were high. Adm. Mark Kerr, who commanded the southwest training area in England in 1918, lamented that nearly three hundred pilots were killed in his region in three months' time.[29]

It is certainly safe to say that aviators in all countries were more likely to die in accidents than in combat. The youthful volunteers knew that. Initially their irrepressibility and the callousness of wartime enabled them to cope with the situation. French aviation artist Marcel JeanJean recalled one exchange on the training school flight line while watching a crash: "Those poor fellows, they are going to kill themselves." "Too bad! That's war."[30] A British pilot recalled that fatal crashes on Sopwith Camels in training were so frequent that they stopped bothering to look up when they heard a Camel go into a spin and took for granted that they had lost another pilot trainee.

In most accounts there is a near-total absence of blood and gore, giving death in the air a certain cleanliness. Yet the following account from Bernard Lafont's *Au Ciel de Verdun* dispels that romantic illusion. Two aviators have fallen from their Farman when attacked during the battle of Verdun: "The first of the two men is impaled on the iron gate. There is the pierced body, a bloody rag. The wounds are enormous. Purple streams flow onto the clothes; drops hang and then fall one by one in a large puddle on the ground below."

"The second fell on the roof of the house. I clearly heard the dull sound of the body when it was crushed in a heap. Flouc! The body was recovered from the roof, entirely broken, shattered, shapeless, and without rigidity like a heap of slime. They filled a coffin with it." Contrary to the myth of Guynemer's death, that he had flown so high he could not descend, Lafont's account graphically reminds us that all aviators returned to earth, one way or another.[31]

This was no sport, no game. It was a deadly, ruthless, and capricious business, where a man's life depended not solely on his individual skills but on a combination of those skills, on luck, and on machines that were very far from perfect. The widespread incidence of occupational traumas indicated the stress involved in war flying. Compare photographs of Boelcke and Richthofen taken at the start of their careers to those taken a year or two later; they have aged greatly and no longer look like youth in their mid-twenties. Many men fell victim to flying fatigue, which caused sleeplessness, irritability, exhaustion, and shakiness after landing, and most dangerous, carelessness in combat. The great French ace Guynemer may have suffered from tuberculosis, which, exacerbated by his refusal to rest, meant that he was not only sick but increasingly nervous and irritable in the period before his death in combat in 1917. An American pilot complained that his nerves were shot; he knew that he would die sooner or later, but waiting for the moment was killing him.

Even for the survivors, the nerves did not necessarily end with the war. Elliott White Springs, a South Carolinian, Princetonian, and postwar textile magnate, spent seven months in combat with RAF Squadron 85 and American squadron 148 attached to the British air arm. He survived to write *The Diary of an Unknown Aviator*, published in book form under the title *War Birds* in 1927, a brawling, boozing, yet grim and moving novel of the American aviators who flew with the British. The persistent postwar anxiety and depression that gripped Springs culminated in a nervous breakdown in 1942, from which he never completely recovered. Its roots lay in his poor relationship with his father and "a genuine war neurosis after 1918," which he managed for a time with writing and drink. Springs had written in 1918 that the "best part of me will always remain" at the front.[32] In a sense, he was correct.

American Edwin Parsons attributed the consumption of liquor to the need for a sedative for strained nerves. Opinions varied on the value of liquor, as German ace Max Immelmann, a physical fitness fanatic and teetotaler, believed that liquor led to overstrained nerves. If that was the case, the accounts of some Americans suggest that they must have suffered a surfeit of nerves, and one British author humorously conceded that if the British and Americans had drunk as much as some memoirs declared, they would not have lasted very long at the front.

Perhaps the most graphic account of one pilot's struggle with nerves is contained in the diary of Edward "Mick" Mannock, the Irishman who was Britain's highest-scoring ace. According to his biographer Frederick Oughton, Mannock had two temporary nervous collapses and was often sick before patrols, much of his tension occasioned by the repeated breakdowns of his airplane, or "bus," as he called it. His diary recounts constant engine failures and gun jams; once during target practice his right bottom wing fell off. By the summer of 1918 the nervous strain was so great that his hands shook and he would burst into tears. He knew that he would die, but he feared burning to death, a hell to which he had gloatingly consigned many of the "Huns" that he passionately hated, so he carried a pistol to shoot himself. Shot down in flames from ground fire, Mannock's remains were never found, thus no one will ever know whether he had time to use the pistol.[33]

Beneath the veneer of glamour and chivalry, aerial combat was undeniably exhilarating and intoxicating for many of its participants, but also nerve-racking and frightening as well. Mannock's tearful outbursts after mechanical failures and witnessing the deaths of friends seem anomalous. Yet an incident cited in a recent book, *A Yankee Ace in the RAF: The World War I Letters of Captain Bogart Rogers,* indicates that the unwritten code of the prohibition of open mourning in common actually led to the suppression of recollections of incidents that violated the code. Rogers published an article in 1930 describing how Squadron 32 disintegrated into hysterical tears, rage, and mourning in the mess the evening of the death of its favorite pilot. The next day the dead flier's best friend took off, shot down a German in revenge, and collapsed upon his return to the field, to be invalided out of the service with nerves from which he had still not recovered nearly a year later. Some of Rogers's former squadron mates reproached him for disclosing such an event; others accused him of exaggerating it. The essential point was that Rogers had compounded their violation of the unwritten code by writing about it twelve years later.[34]

As grim and dangerous as it was, aerial fighting was only one aspect of air warfare. Ground attack, reconnaissance, and bombing were significant roles that directly intruded on the course of the war on the ground.

One of the most difficult, and most important, tasks of aviation as the war continued was ground attack, for which the Germans evolved special units of battle or storm fliers equipped with light, maneuverable two-seat biplanes. These infantry fliers became an effective offensive and defensive weapon in 1917, attacking enemy batteries, strong points, and infantry reserves with machine guns, grenades, and light fragmentation bombs. They suffered high losses in their dangerous work, as they ranged in squadron or group strength over the front at 2,000 feet, buffeted by the drafts of passing shells, and then descended to strafe troops from 300 feet above the trenches in the dead zone between the artillery fire from both sides. In these German ground attack units, only the commander was a commissioned officer; the crews were almost entirely NCOs and soldiers. Their fighting spirit was high, as they protected their infantry brothers below by flying above the "rue de merde," or "shit street," as they called the front, on days when heavy rain and low clouds grounded other units. This was the air war at its grittiest, and at the battle of Messines in June and then at Cambrai in November and December 1917, they effectively controlled British breakthroughs and led attacks with demoralizing battlefield strafes.[35]

British counterparts to these storm fliers in 1918 flew Sopwith Camels, which had won praise as the war's preeminent dogfighter in 1917, but whose essential task in 1918 was ground attack in high-risk assault squadrons. V. M. Yeates, author of the novel *Winged Victory,* and survivor of 248 hours and four crashes in Camels during 1918 before being discharged with tuberculosis in the summer, termed ground strafing "the last occupation on earth for longevity" and "the great casualty maker." Yeates considered it the most dangerous and valuable work that fighter pilots performed, though they received little credit for it.[36]

From March to November 1918, RAF Assault Squadron 80, with a strength of twenty-two officers, suffered 168 casualties from all causes, or about 75 percent monthly, with almost half killed.[37] In *War Birds*, an American pilot who flew with the British, after surviving training in which three pilots practicing on Camels were killed in one day, commented that fighting Fokker D7s in Camels during the summer was exhausting and caused high losses. He concluded that it was "only a matter of time until we all get it."[38] As another American explained, more humorously, "A Camel pilot had to shoot down every German plane in the sky in order to get home himself, as the Camel could neither outclimb nor outrun a Fokker."[39]

The essential task of aviation throughout the war was reconnaissance, and in the French and German air arms these army cooperation planes were the preponderant types. The crews of the two-seater army cooperation planes who routinely carried out these missions often flew in machines that left much to be desired. British Be2 biplanes, already obsolete in 1915, remained in service, cannon fodder for German fighters, into 1917, in part to complete production runs but also because many RFC pilots were not sufficiently well trained to fly higher-performance planes. The "Quirks," as fighter pilots named these two-seater crews, flew straight to their target and back at low altitude. A fighter pilot awed by such apparent indifference to evasive maneuvers presumed that they were so accustomed to being shot at by ground fire, at low altitude, that they had become fatalistic, like infantrymen. It never occurred to him that inadequate training may also have limited their ability to perform intricate maneuvers. In French aviation many army corps crews similarly struggled in the AR biplane, which was intended only as a stopgap when it appeared in 1916 but which served into 1918. The crews who manned these aircraft provided many of the victories for opposing fighter pilots.

The Germans tended to at least equip some of their reconnaissance crews with better planes in order to husband their dwindling manpower. They sent expert crews alone and at high altitude in high-performance machines, using their skill and the planes' ability to evade the enemy. By the end of 1917 their Rumpler biplanes were capable of 20,000-foot ceilings on these missions with little loss of altitude, thanks to their high-compression engines. British aces had a healthy respect for these crews, some of whom were formidable. In James McCudden's accounts of separate combats with four two-seaters at high altitude in December 1917, three of them escaped. Canadian ace Billy Bishop's patrol of six once jumped a lone German two-seater, who turned in a flash, attacked them head-on, hitting Bishop's plane and another member of his squadron, and escaped, earning Bishop's accolade "a very fine pilot and a very brave man."[40] Two-seat crews were usually the prey of fighter pilots, but occasionally the prey became the predator.

A final task of wartime aviation was bombing. On the Italian front by 1917, waves of thirty to forty Caproni trimotored biplane bombers supported infantry attacks by bombing Austro-Hungarian troops. The Capronis also staged long-distance raids across the Adriatic to bomb targets, sometimes flying as low as forty feet above the waves in their effort to strike by surprise and avoid antiaircraft defenses.

Over the western front in 1918 the French aerial division, whose nucleus was the superlative Breguet 14, a fast, sturdy biplane carrying twenty-four twenty-two-pound bombs and defended by gunners armed with twin Lewis guns, aggressively raided across the lines in massed formations. General Pétain, the French commander in chief, sought certain aerial superiority in 1918 and methodically attacked enemy lines of communication continually and in mass.

The culmination of these massed tactical raids was the aerial support of the American Expeditionary Force's attack on St. Mihiel in September 1918. There Col. Billy Mitchell commanded nearly 1,500 airplanes, half American and half French, the largest concentration of allied air forces during the war. This armada gained aerial control as the fighters penetrated over German airfields and the bombers struck targets on the battlefield and in the rear.

Tactical bombardment of enemy forces was one thing; the strategic bombing of enemy cities and civilians in no way comports with ideals of sport, chivalry, and individual combat. The German government launched Zeppelins in 1915 and 1916 and then large bombers in 1917 and 1918 to bomb England in an attempt to drive it from the war. The attempt failed, but the campaign indicated a willingness to strike at civilian morale. The French had waged an unsuccessful strategic campaign against west German industrial towns in 1915. The British, unable to retaliate against German civilians until 1918, wanted to start, in the words of Secretary of State for Air Lord William Weir, "a really big fire" in a German town, assuming that such attacks would undermine German morale.[41] The war ended with the British poised to begin bombing Berlin, and with the value of strategic bombing unproven, but with the notion that the bombing of civilians could undermine their morale. The postwar demobilization was so rapid that air forces and aviation industries shrank within two or three years to mere shadows of their wartime selves. The losers of the war, such as Germany, were forced to disarm and forbidden to possess an air force, but even the victors, confronting huge wartime debts, dismantled their air forces so rapidly that in some cases they did not give the industry sufficient time to transition into the civil aviation field, and many aircraft firms disappeared. Yet wartime exploits and expectations left an enduring legacy that would form the basis for the future evolution of military flight.

As opposed to the concentration on knights of the air, the myth of the potency of the bombing of civilians was the less often acknowledged and darker side of the legacy of the air war. Many of the studies of World War I aviation today still concentrate on aces and airplanes, the personal and technical aspects of the war, rather than on the use of the air weapon to strike at defenseless civilians in strategic bombing campaigns. Yet those early campaigns, however ineffective, remind us that the air weapon of World War I was truly the spawn of the era of total war, which conflated civilian and military targets and deemed the bombing of civilians, women and children included, an acceptable means of winning a war. Both of these powerful images—the romantic idealization of individual aerial combat rooted in the past and the brutal vision of massive civilian destruction foreshadowing the future—constituted World War I's dual legacy for airpower in the twentieth century.

This evolution began, in concrete form, first in the realization of powered heavier-than-air flight and then in the initiation of the relationship between the military and aircraft designers and entrepreneurs, with Orville and Wilbur Wright.

NOTES

1. Antoine Odier, *Souvenirs d'un vieille tige* (Paris: A. Fayard, 1955), 117; Albert Étévé, *Avant les Cocardes: Les Débuts de l'aéronautique militaire* (Paris: Lavauzelle, 1961), 21–23.

2. On the Wright Brothers, see Fred Howard, *Wilbur and Orville: A Biography of the Wright Brothers* (New York: Alfred A. Knopf, 1987).

3. John H. Morrow Jr., *Building German Air Power, 1909–1914* (Knoxville: University of Tennessee Press, 1976), 14; Alfred Gollin, *The Impact of Air Power on the British People and Their Government, 1909–1914* (Stanford, Calif.: Stanford University Press, 1989), 170–73.

4. Charles J. Gross, "George Owen Squier and the Origins of American Military Aviation," *Journal of Military History* 54 (July 1990): 281–305.

5. Charles Christienne and Pierre Lissarrague, *Histoire de l'aviation militaire française* (Paris: Lavauzelle, 1980), 31–34; Étévé, *Avant les Cocardes*, 15–16; Odier, *Souvenirs*, 128–29.

6. Christienne and Lissarrague, *Histoire*, 36–37.

7. Morrow, *Building German Air Power*, 103.

8. Kriegswissenschaftliche Abteilung der Luftwaffe (KAdL), *Die Militärluftfahrt bis zum Beginn des Weltkrieges 1914*, 3 vols., 2d rev. ed., edited by Militärgeschichtliches Forschungsamt (Frankfurt-am-Main: Mittler, 1965–66), 3:10.

9. Gollin, *Impact*, 309–11.

10. Christienne and Lissarrague, *Histoire*, 37.

11. Quoted in *Aeroplane* 1, no. 14 (Sept. 7, 1911): 323. See also *Aeroplane* 1, no. 16 (Sept. 21, 1911): 375–76; *Aeroplane* 1, no. 19 (Oct. 19, 1911): 462; *Aeroplane* 5, no. 23 (Apr. 4, 1912): 602.

12. Henry Serrano Villard, *Contact! The Story of the Early Birds* (Washington, D.C.: Smithsonian Institution Press, 1967), 127.

13. KAdL, *Militärluftfahrt*, 2:86; Jürgen Eichler, "Die Militärluftschifffahrt in Deutschland 1911–1914 und ihre Rolle in den Kriegsplänen des deutschen Imperialismus," *Zeitschrift für Militärgeschichte* 24, no. 4 (1985): 350–60 (part 1), no. 5 (1985): 403–12 (part 2).

14. *Aeroplane* 3, no. 14 (Oct. 2, 1913): 374.

15. Johannes Werner, *Knight of Germany: Oswald Boelcke, German Ace* (New York: Arno Press, 1972 [1932]), 145, 164, 172, 233–36.

16. Lee Kennett, *The First Air War, 1914–1918* (New York: Free Press, 1991), 160; Peter Kilduff, *Richthofen* (London: Arms and Armour, 1993), 165, 183–84, 219–20.

17. "Guynemer et les Cicognes," *Icare*, no. 122 (1987): 27, 74, 87.

18. Cecil Lewis, *Sagittarius Rising* (New York: Harcourt, Brace, 1936).

19. Peter H. Liddle, *The Airman's War, 1914–18* (New York: Blandford Press, 1987), 64–69.

20. *Flight* 8, no. 5 (Feb. 3, 1916): 97; *Flight* 8, no. 33 (Aug. 17, 1916): 705–6.

21. *Flight* 9, no. 16 (Apr. 19, 1917); *Aeroplane* 12, no. 22 (May 30, 1917).

22. Randolph S. Churchill and Martin Gilbert, *Winston S. Churchill* (Boston: Little, Brown, 1975), 4:61.

23. Albert Étévé, *La Victoire des Cocardes* (Paris: Lavauzelle, 1970), 158–59, 161–62, 167.

24. Christienne and Lissarrague, *Histoire*, 83–87.

25. John H. Morrow Jr., *German Air Power in World War I* (Lincoln: University of Nebraska Press, 1982), 41–42, 73–120. For more information on Fokker, see Marc Dierikx, *Fokker: A Transatlantic Biography* (Washington, D.C.: Smithsonian Institution Press, 1997).

26. *Statistisches Jahrbuch für das Deutschen Reich 1924–1925*, 30; Walter Raleigh and H. A. Jones, *The War in the Air* (Oxford, Eng.: Clarendon, 1922–37), 3:35; "Statistics Regarding French Aviation during the War," U.S. Naval Attaché to Office of Naval Intelligence, Jan. 14, 1920, Record Group 313, Naval Intelligence Office, 38.4.1, National Archives and Records Administration, Washington, D.C.

27. Folio 5, AIR 9/3, Public Record Office, London.

28. *Aviation Medicine in the AEF* (Washington, D.C.: U.S. Government Printing Office, 1920), 205. See also ibid., 217.

29. Mark Kerr, *Land, Sea, and Air: Reminiscences of Mark Kerr* (New York: Longmans, Green, 1927), 280.

30. Marcel JeanJean, *Des Ronds dans l'air: Souvenirs illustrés* (Paris: Aurillac Imprimerie Moderne, 1967); Marcel JeanJean, *Sous les Cocardes: Scènes de l'aviation militaire* (Paris: Serma, 1964 [1919]).

31. Bernard Lafont, *Au Ciel de Verdun: Notes d'un aviateur* (Paris: Berger-Levrault, 1918), 28–29.

32. Burke Davis, *War Bird: The Life and Times of Elliot White Springs* (Chapel Hill: University of North Carolina Press, 1987), 43–44.

33. Frederick Oughton and Vernon Smyth, *Ace with One Eye: The Life and Combats of Maj. Edward Mannock* (London: F. Muller, 1963).

34. John H. Morrow Jr. and Earl Rogers, *A Yankee Ace in the RAF: The World War I Letters of Captain Bogart Rogers* (Lawrence: University Press of Kansas, 1996).

35. Georg P. Neumann, ed., *In der Luft unbesiegt* (Munich: Lehmans, 1923), 79–91, 166–75; J. C. Nerney, "The Battle of Cambrai," AIR 1/678/21/13/1942, Public Record Office.

36. V. M. Yeates, *Winged Victory* (London: Buchan-Wright, 1985 [1934]).

37. John C. Slessor, *Air Power and Armies* (London: Oxford University Press, 1936), 100.

38. John M. Grider, *War Birds: Diary of an Unknown Aviator*, edited by Elliot W. Springs (Garden City, N.Y.: Doubleday, 1938), 221. See also ibid., 233–37.

39. Quoted in James J. Hudson, *Hostile Skies: A Combat History of the American Air Service in World War I* (Syracuse, N.Y.: Syracuse University Press, 1968), 202.

40. William A. Bishop, *Winged Warfare* (Garden City, N.Y.: Doubleday, 1967), 133.

41. William Weir to Hugh Trenchard, Sept. 10, 1918, MFC 76/1/94, Royal Air Force Museum, Herndon.

CHAPTER 9
STRATEGIC AIR WARFARE
AN ANALYSIS

The entry of airplanes into the arsenals of nations marked a profound and permanent change in the nature of war fighting. Indeed, airplanes quickly became key tools in the prosecution of twentieth-century conflict. In their most dramatic role, they delivered bombs onto the homeland territory of combatant states. What evolved into the theory of strategic bombing began with a simple assumption: aircraft, specifically long-range or "strategic" bombers, can avoid an enemy's army and navy and proceed directly to its "vital centers," where they can cause enough destruction and disruption to induce surrender on terms favorable to the attacker. It was a simple idea, but it would prove to be a powerful and tenacious one.

Where did this faith come from? On what was it based? Why was there a rather large gap between interwar expectations for strategic bombing and the reality of air warfare in the early years of World War II? And why do our expectations for strategic bombing still seem to be so high, even though there has been plenty of evidence to temper the bold claims of air forces over the years? The purpose of this essay is to provide a brief overview of strategic air warfare in the twentieth century, analyzing the record of this mode of war fighting in light of the claims and the rhetoric surrounding it. Because I cannot cover the entire sweep of its history in a brief essay, I shall concentrate mainly on the interwar and World War II period, and I shall focus mainly on Britain and the United States.

In the years leading up to World War I, expectations about air warfare developed in a particular context that later shaped the way the experience of aerial bombing was perceived and interpreted. In Britain, for instance, Edwardian society was wringing its hands about whether the working classes would be controllable in war. The urbanization that had been the inevitable accompaniment to industrialization nonetheless raised anxieties among the governing classes. City life was seen as an unnatural source of decline in the genetic health and robustness of the British people, and a potential wellspring of political radicalism. These generalized anxieties were furthered by ongoing, bitter labor-management disputes (especially in mining and railways) that peaked on the eve of World War I.[1]

A rash of alarmist popular literature had appeared in the first years of the century; it speculated about whether Britain would be invaded and whether the nation could defend itself adequately. In William LeQueux's book, *The Invasion of 1910* (1906), England's problems stemmed from the fact that her old, strong aristocratic government had been replaced by a weaker one "swayed by every breath of popular impulse."[2] Issues of how to maintain civil order in wartime were raised in lectures before Britain's Royal United Services Institution. In November 1913, for instance, Maj. Stewart L. Murray concerned himself with potential civil strife generated by working-class radicalism. A war, he argued, might see an "explosion of those volcanic forces which underlie every modern democracy. . . . Unless steps are taken to prevent the hardships of war pressing intolerably upon the new working classes, the whole organized political power of labour may be used to demand the cessation of the war, even at the price of submission to our enemies."[3]

In commencing their air campaigns against Great Britain (with Zeppelins in 1915 and Gotha bombers in 1917), the Germans sought to exploit both the material and the psychological or "moral" effects of long-distance bombing against home front economies and populations.[4] German Zeppelins and bombers, which dropped their loads disproportionately on the London poor, began to look like the instruments that might impose the "hardships of war" on the "new working classes." If the British reaction to bombing was not exactly civil conflict and passionate disorder, the fear of such things was not far from government minds. In fact, public behavior in Britain under the fall of bombs was generally admirable. Certainly there were instances of shop fronts being bashed in by crowds assuming that those shops were owned by Germans, and instances of people trekking out of cities to escape danger. Mostly, though, people behaved pretty well under the circumstances. But the government was nervously fearful that the heavy pressures on the domestic population might move them toward some dangerous brink.

British government debates reflected worries about the course of the war and anxiety about the future. Of overwhelming concern were the Russian Revolution and the reemergence of industrial action in 1917. Concerned about German efforts to terrorize the British population, the War Cabinet discussed the possibility of large-scale air attacks on London that might overwhelm the capacity of the fire brigades. So profound were concerns about the stability of the home front that government planners would develop (in the spring of 1918) Emergency Scheme L. This highly secret plan aligned military districts with police districts and replaced civil authority with military control. It required government authorities to maintain in Britain the equivalent of about eight divisions otherwise capable of overseas service. Though its purpose was not stated explicitly and must be deduced from the structure and requirements of the scheme, the purpose was almost certainly to maintain control in a domestic emergency, and to counteract the emergence of revolutionary dissent.[5] These many worries ultimately served to catalyze the creation of a new service, the Royal Air Force (RAF)—independent of both the army and navy.[6]

This nervousness carried over into the interwar years and was fueled by many sources. Captain B. H. Liddell Hart's book, *Paris; or, The Future of War* (1925), was written when Europe was still trying to come to terms with the Russian Revolution. Liddell Hart assumed that the disruption caused by bombing might lead to chaos, particularly among the lower classes: he envisioned "the slum districts maddened into the impulse to break loose and maraud." And such chaos would lead to the loss of government control.[7] This sort of daunting prediction was reinforced by the translated work of Italian air theorist Giulio Douhet, who wrote graphically about the prospect of such things as aerial gas attacks on cities.[8]

Helping to shape the memory of World War I air war was the perception that the indirect effects of bombing had been terribly important and consequential. In a 1924 lecture to the newly opened RAF Staff College, the new commandant argued that the Germans had been able to impose a very high defensive cost on the British, including fighters, antiaircraft guns, searchlights, and manpower, in exchange for what was actually a very limited offensive effort. This was dangerous and could not be allowed to happen again, since the human and material resources diverted into defense represented a significant loss to the crucially important offensive aspect of the war effort.[9]

In fact, wartime experience had in fact traced only the very steep, initial portion of a diminishing marginal returns curve, but those studying it tended to extrapolate linearly. The lesson they drew was that under no circumstances could one surrender the strategic initiative in the air, since doing so would mean starting down the slippery slope to defeat. The solution, the aviation experts said, was an offensive air force capable of bombing enemy territory right from the outset of a conflict.[10] It would deter war. But if deterrence failed, it would offer the quickest and surest means to victory. In large part, the politicians bought the argument. It had appeal because, in addition to promising economies, it seemed to provide a possible escape from the nightmare of World War I–style mud and trench warfare. Those proposing the theory of the aerial offensive assumed, implicitly, that their own population would have greater determination and grit than the enemy, whose air force would be attempting to do the same thing.

Through the 1920s there was a large gap between RAF declaratory policy (public claims) and actual capability. At the time, though, it did not matter too much, since there was no enemy on the immediate horizon, and thus brave talk was cheap. Besides, problems with the other services could be left largely unresolved. In the United States, the emergent air force remained very much a part of the army. Without significant experience of independent operations during the war, air advocates had little chance of separating themselves from their parent service. Some, like Gen. Billy Mitchell, chafed at this; he took his case to the public, dramatically and repeatedly, but did not obtain the result he desired. While the American people watched his aerial stunts with interest, they did not feel so imminently threatened as to call for a separate service. The navy, long accustomed to getting its way in defense matters, dubbed Mitchell "General of the Hot Air Force."[11]

Closely watched and under the thumb of the army, American air service personnel could not easily advocate futurist theories about the use of air power in war. The army was suspicious of conceptions of "independent air power," and the navy perceived the idea as a threat. Also, the larger ethical implications of strategic bombing had not sat well with American politicians like Secretary of War Newton Baker. In light of this, the most influential theorizing about the future took place quietly. Those who had observed World War I bombing wrote their conclusions into Air Service (and later Air Corps) manuals. William Sherman, an instructor at the Air Service Field Officer's School, wrote some of those manuals and then articulated an early version of what would come to be called the American "key-node" or "industrial fabric" theory of targeting.[12] In a 1926 book called *Air Warfare* he argued: "Industry consists . . . of a complex system of interlocking factories, each of which makes only its allotted part of the whole. . . . Accordingly, in the majority of industries, it is necessary to destroy certain elements of the industry only, in order to cripple the whole. These elements may be called the key plants. These will be carefully determined, usually before the outbreak of war. . . . On the declaration of war, these key plants should be made the objective of a systematic bombardment, both by day and by night, until their destruction has been assured, or at least until they have been sufficiently crippled."[13] Similar notions filled the manuals of the army's Air Corps Tactical School (ACTS), located in Montgomery, Alabama, during the 1930s. ACTS became the home of forward thinking about air power in the interwar years.

The "industrial fabric" theory of targeting appealed to the Americans from the outset. Geography had allowed the Americans to take a more economic-oriented, antiseptic approach to the issue of air war than did the Europeans, whose cities were under threat of direct attack. The industrial fabric idea was intuitive and elegant—attractive to a nation interested in industrial efficiency and the scientific management principles of Frederick Winslow Taylor. Also, it seemed to be borne out by the 1929 collapse, which highlighted the inherent interconnectedness of modern economic systems and the structures resting on them.

In 1930s Europe, by contrast, many popular writers elaborated upon the dark dread of aerial bombardment. Titles of this sensationalist fiction included *The Gas War of 1940*, *The Poison War*, *The Black Death*, *Empty Victory*, *War upon Women*, and *Invasion from the Air*.[14] Earlier, the British general strike of 1926 and the financial collapse of 1929 had ensured that governments would continue to be uneasy about the behavior of their populations. Seeking to protect service autonomy and guarantee funding for bombers, the British Air Staff itself contributed to public anxiety about future warfare. Its focus on the offensive in wartime had fueled the notion that the bomber would always get through, as once-and-future prime minister Stanley Baldwin argued dramatically in 1932. Drawing on figures from World War I, the Air Staff had estimated that in a future war there would be fifty casualties per ton of bombs dropped in England. Though the multiplier was flawed for all sorts of reasons, it was used by the Ministry of Health and other civil agencies for planning purposes, beginning in the 1920s. Later, as the Luftwaffe expanded under Adolf Hitler, the scenarios became more dire.[15] A July 1938 report by the Advisory Committee on London Casualty Organization anticipated 30,000 casualties per day requiring hospital treatment, continuing over several weeks.[16] In October 1938 the Ministry of Health received, from eminent psychologists in London, a report indicating that psychiatric casualties might outnumber physical casualties by three to one. The previous month the director of one highly regarded London clinic predicted widespread outbreaks of neurosis upon a declaration of war, especially after the first air raids. In April 1939 the Ministry of Health issued one million burial forms to local authorities.[17] Not only would these figures prove to be grossly inflated in the early stages of World War II, but the predictions of psychological casualties would prove to be quite wrong. Despondency and depression during the Blitz was almost always symptomatic of the inadequacies of local relief efforts. Contrary to prewar prognostications, hospital admissions for neurosis declined, suicide rates fell, and drunkenness declined by half relative to peacetime.[18]

During the tense years of the late 1930s, as Hitler's Luftwaffe forced Britain to rearm, the RAF had to face up to the gaps that had grown between its declared policy and its actual capabilities. Sir Hugh Trenchard, the chief of Air Staff through the 1920s, had not indoctrinated his service with an analytical mindset. His failure to do so would prove to be a problem for a service born of technology and wholly dependent upon it. The RAF would go into war unprepared and unable to carry out its publicly declared mission. Most of the types of planning and analysis needed to support a modern air offensive had not been undertaken. Navigation, in particular, had been allowed to languish (even though navigation in civilian airlines was making great strides). All kinds of urgent work needed to be done, and—as it would turn out—there would not be enough time to complete it all before the outbreak of war.[19] As he surveyed his forces on the eve of war, bomber command chief Sir Edgar Ludlow Hewitt admitted that 40 percent of his force could not find a target in a friendly city in broad daylight. When war planning began in earnest, he argued that sending bombers over Germany might end in a major disaster.[20]

Under the circumstances, Air Staff plans for the outset of the war were very different from the bold 1920s conceptions of an immediate all-out aerial offensive. Britain would engage in only a very restricted air effort, designed to do nothing to provoke the Germans into heavy attacks on London. In the meantime, the nation would rely on the navy for defense and economic warfare and would build fighters and bombers as rapidly as possible. Targeting was very restricted early on: for the most part, bomber command crews confined themselves to dropping propaganda leaflets and striking at self-contained naval targets, such as ships at sea. Targeting guidance expanded at the time of the Battle of France, although bomber command was able to offer little assistance to the ground armies. It is interesting that German failure in the Battle of Britain did not blunt British enthusiasm for an air offensive against Germany. The best explanation is a simple one: The British were in no position to admit to themselves, in 1940, that their one offensive tool against Hitler might not work. Winston Churchill's decision to fight Hitler rather than negotiate had rested in large part on a faith in the role that bomber command would play in war. Alone, and needing to take some kind of offensive action against the Third Reich (as much for home front morale as for any damage it might do in Germany), the British could not afford doubts about their air force and its potential for successful war fighting.

The strength of German defenses forced the British to fly raids almost exclusively at night—and at night targets were hard to find and hit. Sir Charles

Portal, who headed bomber command in 1940 before being named chief of Air Staff, began to argue for an air campaign designed principally to undermine German morale. The argument gained momentum all through 1941 and was given a substantial push in August and September by the results of the Butt Report, the first thoroughgoing examination of British bombing accuracy up to that point in the war. D. M. Butt, of the War Cabinet Secretariat, reviewed more than six hundred operational photos, comparing them with crew claims and bomber command assessments. It revealed, shockingly, that only about one in five of Britain's bombers was getting within five miles of its target.[21] This statistic made clear, although some on the air staff were hesitant to believe it, that if bomber command was to contribute to the war effort at all, it would have to concentrate on the one target it could find and hit reliably: cities. A new bombing directive of February 14, 1942, explicitly told crews to strike area targets with the aim of undermining the "morale of the enemy civil population and in particular, of the industrial workers."[22] A week after this directive had been issued, Sir Arthur Harris took the helm of bomber command. He had not been responsible for the policy, but he supported it fully, believing that concentrated attacks on cities would offer the most efficient means of breaking the German ability and will to fight.

As the British prepared to wage what would be called the "area bombing" campaign, the Americans prepared to go in a very different direction. In the United States in the 1930s the first B-17 bomber prototypes were authorized, by a cautious, defensively oriented Congress; they were intended to attack ships at sea threatening the American coastline. The new plane, equipped with special bombsights, looked like the instrument to carry out a doctrine already worked out in detail at ACTS. In the early 1930s, instructors at ACTS had advanced several main propositions:

- Strategic bombers would play an important part in future wars between industrialized states.
- High-altitude, self-defending bombers flying in groups could attack their targets in daylight without suffering prohibitive losses.
- Targeting must focus on "bottleneck" industries in the enemy war economy, identified by careful analysis.

The Americans placed great faith in their industrial fabric theory, to be implemented through high-altitude precision bombing. Like the British, however, the Americans fell victim to mistaken assumptions that would undermine their conclusions. The idea that the bomber would always get through came out of a problematic reading of World War I experience and a belief (correct for a time) that bomber technology was ahead of fighter technology. As defensive techniques made strong advances in the late 1930s, bomber advocates could not bring themselves to admit that these advances called into question the foundation on which their ideas rested. Both the British and the Americans struggled with the idea of long-range fighter escorts: They could not see how to design a plane that could both keep up with the bombers over long distances and then compete on equal terms with enemy short-range fighters. In addition, an operational war plan for the industrial fabric theory would have to rest on sophisticated intelligence, and this was simply not available before the war. Even though the Americans tried to be analytical in their approach, they suffered from many of the same oversights and mistaken assumptions about bombing as the British.[23]

When American entry into war started to look imminent, President Franklin D. Roosevelt turned to the Air Corps. It is no coincidence that both Britain and the United States ended up relying heavily on air power in World War II. Neither nation had a history of maintaining large standing armies. Both had been, and continued to be, naval powers interested in advanced technology. Both were democracies and casualty-averse. Air power promised a means of offensive thrust against the enemy that would be more powerful and less costly than other, more traditional means. This remains the case, as air power continues to have great political appeal. Since World War II American politicians have turned to it repeatedly, with high hopes every time, to solve intractable dilemmas without incurring the formidable costs of employing ground forces.

A new advocate of air power, FDR had high hopes. But the American air campaign (like the British campaign before it) got off to a slow start. It took nearly all 1942 to simply build up the aircraft and infrastructure needed to wage a strategic bombing campaign from Great Britain. Despite heavy pressure from the British, the Americans declined to join in the nighttime area offensive. They were determined to try out their own theory even if the earlier British attempts at daylight bombing had met with failure. They therefore undertook high-altitude precision bombing of Axis targets, beginning cautiously over occupied France in August 1942 and then commencing attacks on Germany proper in early 1943. Allied grand strategy accommodated the difference of views. At the Casablanca conference of January 1943 the Allies agreed to disagree on air tactics and created a directive that let both parties go their own ways.[24]

Sir Arthur Harris believed, emphatically, that he could bring Germany to its knees by bombing its cities.[25] In a short period of time Harris had made bomber command a much more effective force operationally than it had been in 1940–41. By 1943 he had undertaken some unprecedented attacks on

A B-29 flying over water. This was one of the great bombers of World War II, used extensively in the Pacific theater and later modified to deliver the first nuclear weapons to Japan in 1945. (NASA photo no. L-49692)

Germany, including the infamous July 1943 attack on Hamburg which culminated in a massive firestorm. But by the end of 1943 both the British and American strategic air campaigns were in crisis. Bomber command's year-end Battle of Berlin—in which Harris had placed great hopes—brought Britain no closer to victory. Instead, German night fighters were beginning to win the battle against bomber command. For the Americans, the very costly Schweinfurt raids against German ball bearing factories (flown in August and October 1943) finally laid to rest the much-vaunted theory of the self-defending bomber.[26]

The Americans chose to change tactics rather than targets. Once it had become clear that bombers needed escorts, the Americans brought them into the theater in great numbers. American bombers flew to targets that the Germans felt compelled to defend, and the American fighters took on the Luftwaffe aircraft sent up to meet the bombers, resulting in vast battles of attrition. This was not what prewar theory had envisioned, but it proved to be a breakthrough in the Allied war effort. The resulting defeat of the Luftwaffe facilitated the Normandy landing, opened Germany up to a more effective Anglo-American selective targeting campaign, and aided the British directly because of the toll it took on German fighter pilots.

The Americans also changed their approach to bombing. In early November 1943 U.S. chief of air staff Gen. Henry "Hap" Arnold permitted blind bombing attacks on days when the weather was poor. Since American visual sighting methods could not be relied upon to penetrate the heavy cloud cover that frequently prevailed over the European theater, the Americans routinely had to fall back on blind bombing methods. This had the effect of causing British and American bombing practices to converge. Indeed, in poor weather in the autumn of 1943, American accuracy rates were no better than those recorded by the Butt Report in 1941.[27] While the Americans claimed throughout the war that their policy in Europe was one of precision bombing against selected military/industrial targets, the practical effects of British area bombing and American "precision" bombing were quite often similar. In poor weather American bombers typically would strike marshaling yards because they were relatively easy to find and hit. These raids, which usually included significant percentages of incendiary bombs in the bomber's ordnance mix so as to increase the collateral effects of the strike, tended to cause a great deal of damage, not to mention casualties, all around the target area.[28] And, frustrated by their inability to carry out effective precision strikes against Japanese industry, the Americans in 1945 adopted fire-bombing methods that had little to distinguish them from British practices in Europe. The many problems that both air forces faced demanded that they engage in almost constant, real-time modification of their tactics and techniques in order to make aerial bombing effective.

Both Sir Arthur Harris and Gen. Carl Spaatz, commander of the U.S. Strategic Air Forces in Europe (USSTAF) had hoped that an air campaign might obviate the need for an amphibious landing in France. When this hope foundered, they sought at least to tie the pre-Overlord preparations to their preferred targets lists—for Harris, cities, and for Spaatz, German synthetic oil. In contrast, General Eisenhower, Supreme Allied Commander, and his deputy, Air Marshal Sir Arthur Tedder, wanted strategic bombers to help isolate the areas around the future Allied beachhead by bombing transportation and communications targets in France (mainly railways and bridges). This, they believed, would make it difficult for the Germans to bring men and supplies to the front and to fight a war of maneuver in the area. Allied High Command carried the day, forcing the air commanders to go to work aiding the ground war, just when they felt they had finally made a breakthrough in the strategic campaign.[29]

Through much of 1944 the Anglo-American bombers were absorbed by all manner of demands, especially the Normandy campaign and breakout, preparation for Dragoon, the invasion of southern France, and the attacks on V-1 and V-2 sites. Eisenhower gave Spaatz permission to go ahead, whenever possible, with raids against German synthetic oil sources, a target set

Spaatz had come to privilege over others, and justifiably so according to intelligence derived from ULTRA. In the autumn, when the bombers were relieved from their focus on the ground war, another debate ensued over targeting. Harris wanted to continue his focus on cities, Spaatz wanted to press hard against German oil supplies, and Tedder advocated extending the campaign against transportation targets into Germany proper. In the end, there was another functional compromise.

Spaatz, with help from Harris, had lowered German oil stocks considerably by September 1944. But by October bad weather closed in and greatly reduced the effectiveness of his forces against such targets. Harris, seeking to press home his city campaign, asserted the correctness of his course against Sir Charles Portal's increasing insistence that bomber command devote as much attention as possible to oil targets. Portal, ironically, had come full circle. After having done so much, early on, to move bomber command to area attacks, he had become frustrated with the slow progress of the approach. Harris, for his part, insisted that city bombing would work if he were only given the freedom to carry it out without distraction or diversion to other targets. The views of the two men were never reconciled, despite a long and detailed correspondence between them over the winter of 1944–45. In the meantime, the combined bomber offensive went forward, striking German transport, oil, and cities—with ever-increasing effectiveness and impunity.

By the late winter of 1945, Germany was in dire shape. The economy was barely functioning, and both Luftwaffe pilots and army units had to ration fuel. Harris was in the midst of smashing what remained of Germany's cities. Desperate to put a final nail in the coffin of the Third Reich (and running out of major targets), both the Americans and the British both turned to attacking Germany's smaller cities and towns. Indeed, the American target list for February 1945 was a grab bag, revealing a willingness to try just about any means to bring a quick end to the war. Still, the Red Army had to fight its way into Berlin. Even massive bombing had not led directly to German collapse, as some bombing advocates had hoped it would. The Japanese endured the firebombing of sixty-six of their cities, including one attack (of March 9–10, 1945) that killed 80,000 residents of Tokyo in one night.[30]

The central assumption underpinning prewar strategic bombing theory —namely that both modern societies and modern economies are fragile— proved flawed. Human beings are remarkably adaptable creatures, and they did not break under pressure—even under the constant, severe pressure that strategic air forces placed on Germany in late 1944 and into 1945. Economies also proved to be rather robust. The problems the Americans faced in

F-15s from the 44th Fighter Squadron, Kadena Air Base, Japan, and F-16 Fighting Falcons from the 51st Fighter Wing, Osan Air Base, South Korea, on a refueling training mission with a KC-135R Stratotanker over the Pacific Ocean near Okinawa, Japan. (USAF photo by M. Sgt. Marvin Krause)

prosecuting the industrial fabric theory were certainly exacerbated by an underappreciation of the difficulties of achieving accuracy and the high demand for excellent intelligence. But at the root of those problems was simply the fact that economies, too, can adjust—through substitution, reallocation of resources, and physical movement. When we consider oil as a target we must keep in mind that it became vulnerable not just because of Allied air attacks but because, at the same time, the Red Army was overrunning many of the main Axis oil sources in the East.

In Europe, strategic bombing did not prove the case of its most outspoken advocates. Harris would claim that it might have done so had he been left alone to prosecute the air war as he saw fit. Others, albeit in more moderate tones, would sometimes appropriate his argument. But the argument had clear problems. In the ensuing struggle over what bombing had or had not achieved in Europe, some of the most valid claims and most sensible perspectives tended to be overshadowed or shunted aside. It was unfortunate, since this information had much merit and would have helped put postwar thinking and planning on a sounder foundation. The problem was exacerbated by the unwieldy and often internally inconsistent nature of the United States Strategic Bombing Survey. In the Far East, the use of atomic bombs altered—and greatly obscured—not only the debate about the effect of strategic bombing (including atomic bombing), but also the lessons that might have been drawn from it. The new weapon looked so different and so revolutionary as to make discussions about mere "conventional" bombing seem both quaint and obsolete.

For all its problems and shortcomings, however, bombing did make a discernible and significant contribution to the war effort. As Richard Overy has pointed out, bombing, as inefficient at it was, placed a ceiling on what Hitler was able to do. Guns used for antiaircraft fire could not be used simultaneously against tanks. The pounding of the German home front required that extensive resources be devoted to defense resources that in most cases would otherwise have been dedicated to the eastern front.[31] And bombing forced the dispersal of industry, making the German war economy—which depended on coal that was delivered by rail lines—ultimately vulnerable to Tedder's transportation plan.[32]

The war required Anglo-American air planners to make constant adjustments and real-time modifications in order to cope with the many problems and frustrations they faced, not only as a result of the gaps they discovered between expectations and realities, but also as a result of the shifting demands placed upon them by the course of the war itself. Despite the eventual resistance of his own air staff, Harris continued to insist that city bombing was the right approach and that the long and costly efforts of his crews were about to pay off handsomely in a German collapse. The Americans, for their part, held tenaciously to their search for the most important bottleneck targets, hoping that the destruction of them might commence the unraveling of the German war economy. Gen. Carl Spaatz's determination to erode the German oil supply (despite difficulties and distractions) reflected his commitment to this American concept of a military-industrial holy grail. Mounting frustration with the length of the war ultimately provoked the Americans to try other approaches as well. By February 1945 the Americans targeted just about everything they could think of, hoping to hit upon some means of affecting enemy behavior, either directly or indirectly. Their actions were implicitly approved by a government increasingly anxious about war weariness, the prospect of a long battle with Japan, and mounting American casualties.

Since 1945, debates over the role and effectiveness of aerial bombing have continued unabated. The World War II experience taught us many things that have remained relevant to these subsequent debates about strategic air warfare, but two are particularly worthy of note. First, bombing is hard to do well and is subject to forces that cannot always be controlled, especially weather. Second, determined populations can hold out for a long time against aerial onslaughts. Even when heavy pressure can be brought to bear on enemy economies and societies, they can prove resilient and robust. In the economic realm, industry can be dispersed, repairs can be made, and resources can be obtained externally. In the social realm, civilians can move out of the way of bombs, they can become acclimated to their effects over time, or they can choose to accept high levels of discomfort and sacrifice. Motivated and mobilized civilians backed by determined governments can sustain very high pain thresholds.[33] Alternatively, discontented civilians may simply lack the mechanisms to convey their discontent into political leverage against the national leadership. North Korea's cities and industry were fire-bombed, and there were few important targets in Vietnam that were not hit hard by bombers (often multiple times). There is an observable pattern in many wars involving strategic bombing, namely: The nation prosecuting strategic air warfare tends to engage in increasingly indiscriminate targeting as frustration over the limited effects of the enterprise grows and political pressure for discernible progress mounts.

Proponents of air power have argued that immediate, all-out aerial assaults are the most effective in producing both material and psychological effects. This position, which was articulated repeatedly in the interwar years, was restated with respect to the war in Vietnam. Quick, powerful strikes against a broad target set, some argued, might have produced the "sharp knock" needed to collapse the enemy's will to wage war. Whatever its merits may be, the argument nonetheless overlooks an important—indeed, central—fact of warfare, which is simply that wars are fought within political constraints and parameters, not to mention legal and ethical ones. For political reasons it will rarely, if ever, be possible to implement what air forces have traditionally believed to be the most effective campaign: an immediate, all-out strike on the enemy's most valuable assets. Even in World War II (the most "total" of modern wars), limits on Anglo-American bombers were lifted only slowly, over a period of years.

In the thirty years since the Vietnam War, bombers have continued to play key roles in wars involving industrialized nations. For America alone, wars in the Persian Gulf in 1991, Bosnia in 1995, Kosovo in 1999, and Afghanistan in 2001 have seen air attacks against opponents' resources and assets in an effort to coerce concessions from hostile regimes. Pinpoint precision attack of leadership, political control, and civilian economic targets is now said to offer coercive leverage unattainable by older methods; some even

The Northrop Grumman B-2 stealth bomber in flight. (USAF photo)

see a "revolution in military affairs" in this apparent combination of new technology and new ideas.[34] But the effectiveness of strategic air warfare remains hotly contested.

If the tools of air warfare have changed dramatically since the canvas and plywood of World War I, the basic ideas behind the *use* of those tools have changed little. Yet these ideas are neither self-evident nor the inevitable consequences of aircraft or the technology of flight. Rather, they are the product of a complex social, political, and cognitive context in which air forces have tried to understand their world and their missions. This context has been formed from overlapping and interactive influences, including the perceived social consequences of industrialization, urbanization, and increasing economic interdependence; popular concerns and expectations about the development of new technologies for war fighting; the intellectual demands of envisioning uses for a new tool of war (and the constraints upon it); and the cognitive errors with which that latter process can be fraught.

NOTES

1. Paul Addison, "Winston Churchill and the Working Class, 1900–1914," in Jay Winter, ed., *The Working Class in Modern British History* (New York: Cambridge University

Press, 1983), 43–64; David Lloyd George, *War Memoirs* (London: Ivor Nicholson and Watson, 1934), 4:1926.

2. Samuel Hynes, *The Edwardian Turn of Mind* (Princeton, N.J.: Princeton University Press, 1968), 42–46.

3. Stewart L. Murray, "Internal Condition of Britain during a War," *Journal of the Royal United Services Institution* 57, no. 430 (Dec. 1913): 1564, 1566, 1587–88.

4. See, generally, Peter Fritzsche, *A Nation of Fliers: German Aviation and the Popular Imagination* (Cambridge, Mass.: Harvard University Press, 1992); Douglas Robinson, *The Zeppelin in Combat, 1912 to 1918* (London: G. T. Foulis, 1962); Barry Powers, *Strategy without Slide Rule* (London: Croom Helm, 1976); Raymond Fredette, *The Sky on Fire* (New York: Holt, Rinehart and Winston, 1966); and S. F. Wise, *Canadian Airmen and the First World War* (Toronto, Ontario: University of Toronto Press, 1980), the first volume of the official history of the Royal Canadian Air Force.

5. Brock Millman, "British Home Defence Planning and Civil Dissent, 1917–1918," *War in History* 5 (Apr. 1998): 204–9, 212, 216–25, 231–32.

6. On the establishment of the RAF, see Malcolm Cooper, *The Birth of Independent Air Power* (London: Allen and Unwin, 1986).

7. B. H. Liddell Hart, *Paris; or, The Future of War* (New York: Garland, 1972 [1925]), 28.

8. See Giulio Douhet, *The Command of the Air* (London: Faber and Faber, 1942).

9. Air Vice Marshall Sir H. R. M. Brooke-Popham, "Air Warfare," Commandant's Lecture, 1924, RAF Staff College, in AIR 1/2385/228/10, Public Record Office, London.

10. See, for instance, the 1928 RAF Manual, Part 1, Operations, Ministry of Defence, London (copies available at the RAF Museum, Hendon, United Kingdom).

11. On Mitchell, see Walter Millis, *Arms and Men: A Study in American Military History* (New Brunswick, N.J.: Rutgers University Press, 1981 [1957]).

12. The early roots of this theory can be attributed to members of the British air staff, whose work impressed the Americans observing it so much so that they appropriated it for themselves. See George K. Williams, "The Shank of the Drill: Americans and Strategical Aviation in the Great War," *Journal of Strategic Studies* 19 (Sept. 1996): 381–431.

13. William C. Sherman, *Air Warfare* (New York: Ronald Press, 1926), 218.

14. See I. F. Clarke, *Voices Prophesying War* (New York: Oxford University Press, 1966).

15. Richard Titmuss, *Problems of Social Policy* (London: Her Majesty's Stationery Office, 1950), 12–13, and n. 1.

16. Report by the Advisory Committee on London Casualty Organization, July 20, 1938, MH 76/128, Public Record Office, London.

17. Titmuss, *Problems of Social Policy*, 19–21.

18. "War, Drunkenness, and Suicide," *Nature* 46 (July 20, 1940): 90; Felix Brown, "Civilian Psychiatric Air Raid Casualties," *Lancet* (May 31, 1941): 691.

19. For an elaboration of the problems the RAF faced in these years, see Tami Davis Biddle, *Rhetoric and Reality in Air Warfare: The Evolution of British and American Ideas about Strategic Bombing, 1914–1945* (Princeton, N.J.: Princeton University Press, 2002), 69–127.

20. For the figure, see John Terraine, *A Time for Courage* (New York: Macmillan, 1985), 85.

21. On the Butt Report, see Charles Webster and Noble Frankland, *The Strategic Air Offensive against Germany, 1939–1945*, 1:178–80, 4:205–13.

22. Webster and Frankland, *Strategic Air Offensive*, 4:143–48.

23. On the ideas developed by and problems faced by the Air Corps, see Biddle, *Rhetoric and Reality*, 128–75. See also Stephen MacFarland, *America's Pursuit of Precision Bombing, 1910–1945* (Washington, D.C.: Smithsonian Institution Press), 26–149.

24. Tami Davis Biddle, "British and American Approaches to Strategic Bombing: Their Origins and Implementation in the World War II Combined Bomber Offensive," *Journal of Strategic Studies* 18 (Mar. 1995): 120–21.

25. On Harris's thinking, see Tami Davis Biddle, "Bombing by the Square Yard: Sir Arthur Harris at War, 1942–1945," *International History Review* 21 (Sept. 1999): 626–64.

26. Biddle, "British and American Approaches," 121–22.

27. Biddle, "Bombing by the Square Yard," 646.

28. Richard G. Davis, *Carl A. Spaatz and the Air War in Europe* (Washington, D.C.: Smithsonian Institution Press, 1993), 508.

29. On the decisions regarding strategic air power prior to Overlord, see W. W. Rostow, *Pre-Invasion Bombing Strategy: General Eisenhower's Decision of March 25, 1944* (Austin: University of Texas Press, 1981).

30. On the air campaign in Japan, see Michael S. Sherry, *The Rise of American Air Power: The Creation of Armageddon* (New Haven, Conn.: Yale University Press, 1987), and Kenneth P. Werrell, *Blankets of Fire: U.S. Bombers over Japan during World War II* (Washington, D.C.: Smithsonian Institution Press, 1996).

31. Richard Overy, *Why the Allies Won* (London: Jonathan Cape, 1995), 127–33.

32. For recent analyses of strategic bombing in World War II, see Biddle, *Rhetoric and Reality*, 270–88, and Sebastian Cox's analytical introduction ("The Overall Report in Retrospect") to *The Strategic Air War against Germany, 1939–1945: The Official Report of the British Bombing Survey Unit* (London: Frank Cass, 1998), xxiii–xli.

33. Robert Pape, *Bombing to Win* (Ithaca, N.Y.: Cornell University Press, 1996), 209; Thomas Griffith, "Air Pressure: Strategy for the New World Order?" at <www.airpower. maxwell.af.mil/airchronicles/apj/apj94/griffith.html>.

34. See, for instance, William J. Perry, "Desert Storm and Deterrence," *Foreign Affairs* 70 (Fall 1991): 66–82; John F. Jones Jr., "Giulio Douhet Vindicated: Desert Storm 1991," *Naval War College Review* 45 (Autumn 1992): 97–101; Thomas Keaney and Eliot Cohen, *Gulf War Air Power Survey Summary Report* (Washington, D.C.: U.S. Government Printing Office, 1993), 235–25; Stephen D. Biddle, "Victory Misunderstood: What the Gulf War Tells Us about the Future of Conflict," *International Security* 21 (Fall 1997): 139–79; David S. Fadok, "John Boyd and John Warden: Airpower's Quest for Strategic Paralysis," in Phillip Meilinger, ed., *The Paths of Heaven: The Evolution of Airpower Theory* (Maxwell Air Force Base, Ala.: Air University Press, 1997), 357–98; Richard G. Davis, "Strategic Bombardment in the Gulf War," in R. Cargill Hall, ed., *Case Studies in Strategic Bombardment* (Washington, D.C.: Air Force History and Museum Program, 1995), 527–621.

Part 4

AVIATION IN THE AMERICAN IMAGINATION

CHAPTER 10
THE ROUTINE STUFF
HOW FLYING BECAME A FORM
OF MASS TRANSPORTATION

Writers have long embraced the frontier—more specifically, the frontier of the nineteenth-century West—as a metaphor for American aviation. They have dubbed airmail pilots as the Pony Express of the skies. They have described the sky as the high frontier, the next frontier, and the new frontier, and have spoken unironically of its conquest. Not to be outdone, the airlines have twice appropriated the metaphor for themselves. There was the original Frontier Airlines, the one that disappeared into the Frank Lorenzo maelstrom, and the second one that began operations out of Denver in 1994. And space, as we all know, is the final frontier.

What's striking, though, is just how superficial, how unthinking, and even how gimmicky so much of the "frontier" talk about aviation has been. The frontier metaphor has appeared most often as a casual evocation of an exciting past, a rhetorical cliché. Few who have used the image have bothered to think through the process of frontier settlement, or the ways in which the western frontier experience actually fit—or didn't fit—the notion of an aviation frontier.

I hope to change that by taking seriously the idea that the sky was, or in some respects still is, a frontier domain. My theme is that the sky, considered as a zone of human activity, underwent a settlement process analogous to that of the nineteenth-century mining and ranching frontiers. Both began with a sprinkling of mostly male adventurers and ended with a much larger and more or less demographically normal slice of the general population. By the last two decades of the twentieth century, if not before, ordinary people could fly in relative comfort and safety and at low cost, but without much excitement or interest. For most Americans the sky had become domesticated, familiar, undifferentiated: Kansas without the Indians.

The routinization of flight has even become the theme of a popular novel. Ryan Bingham, the hero of Walter Kirn's *Up in the Air*, is an airborne Willy Loman, a man who passes his life in airplanes and airports. ("There's always a change in Denver. It's unavoidable. A trip to the bathroom out west means changing in Denver.")[1] Bingham's dream, achieved just before he goes around the bend, is to accumulate a million frequent-flier miles. Though Kirn's novel is satire, there is no mistaking the historical moral. Commercial aviation's frontier days ended long ago. Ryan Bingham is no pioneer of the sky. He's a resident.

Lacking Frederick Jackson Turner's facility for telescoping a grand thesis into a single essay, I'll have to settle for developing one strand of the argument. The question I want to explore is this: How did scheduled passenger flying become a routine experience, a commonplace form of mass transportation, during the mid-twentieth century? Historians have offered five types of complementary answers. They are, in rough order of frequency, the technological, institutional, military, socioeconomic, and social-constructionist explanations.

The technological explanation is the easiest to understand. A large number of people refused to venture aloft until planes became safer, faster, larger, and more comfortable. The first generation of aircraft—dangerous, poky, fragile, open to the elements—appealed mainly to exhibition fliers and the crowds who gathered to watch them risk their lives. World War I brought dramatic gains in performance, though not in commercial utility. Try imagining a carload of Pullman passengers, as one writer put it, going aloft in a dogfighting machine.[2] Only in the late 1920s and early 1930s, when airlines began offering regular service in Fokker and Ford trimotors, did airplanes begin to become a practical means of moving people from city to city. Even then, passengers had to put up with slow speeds, rough and noisy rides, and a questionable safety record, dramatized by the 1931 Fokker crash that killed Knute Rockne and seven other men.

Over the next few years, however, the situation changed completely. By the mid-1930s American carriers could boast advanced, all-metal aircraft with unprecedented safety records. The *annus mirabilis* was 1934, the year in which the Lockheed L-10 Electra, Boeing 247-D, Douglas DC-2, Sikorsky S-42, and Martin M-130 all debuted. That year thirty-one persons—ten pilots and twenty-one passengers—died in scheduled flights. Five years before, in 1929, thirty-nine persons had perished, even though American airlines then accounted for only one-ninth as many passenger miles as in 1934. International comparisons underline the same point. The chance of a British passenger dying in 1934 was five times that of an American, notwithstanding British airlines' international reputation for safety.[3]

American equipment was by then the fastest in the world, as well as the safest. In 1934 more than half of American passengers flew at cruising speeds over 160 miles per hour. By contrast, just 33 of the 616 planes owned by European airlines cruised at more than 125 miles per hour. The best European long-distance *racing* planes could barely outperform the new generation of American airliners.[4] And the American designs kept getting better. During the 1940s the airlines acquired large four-engine pressurized planes capable of flying coast to coast above the worst weather. A plane like the Constellation offered comfort, safety, and a 280-mile-per-hour cruising speed, shaving days off a long-distance trip.

Speed was by far the most important selling point of commercial aviation, an industry whose main competition was the Pullman car on land and the ocean liner at sea. The advent of practical jet airliners in the late 1950s marked the airlines' final triumph over long-distance passenger trains and ships, which were relegated to scenic and leisure-vacation roles. In the technological scheme of things, the rising curve of U.S. airline passengers—from 576,000 in 1933 to 53,070,000 in 1958—tracks the rising curve of aircraft performance.[5] Build them better, bigger, and faster and more people will fly in them.

An interesting subplot of the technological story is why the breakthrough airliners of the 1930s and 1940s were all of American design. Why didn't British or German engineers, no slouches in the design department, come up with something like the DC-3 or the Constellation? One answer, suggested by the historian Richard K. Smith, is that American success was "driven by the unique demands of a continental nation for a profitable high-speed transportation system." That is, the sheer *size* of the United States, the need to carry transcontinental cargos long distances over three mountain ranges, through unpredictable and often violent weather, produced, in Darwinian fashion, superior long-range transport aircraft that combined optimal speed with an optimal weight envelope.[6]

America's aviation success was due to more than its land mass. Without a large industrial base and a growing pool of aeronautical engineers, the country could not have progressed so far or so fast. Still, Smith's ingenious argument prompts a bit of counterfactual speculation. If there had been no nineteenth-century frontier expansion, if the United States had remained a littoral nation of thirteen former colonies strung out along the coast, the country would not have become the aviation hothouse of the twentieth century. Aeronautically, it would have been another Europe, with smaller, less powerful planes hopping from one coastal city to another. America's land expansion had potentiated its aviation expansion.[7]

Government and private institutions had fostered nineteenth-century frontier development through land grants, military protection, and railroad financing. Though it would take a different form, aid from public and private institutions—the second line of explanation—played a similar role in

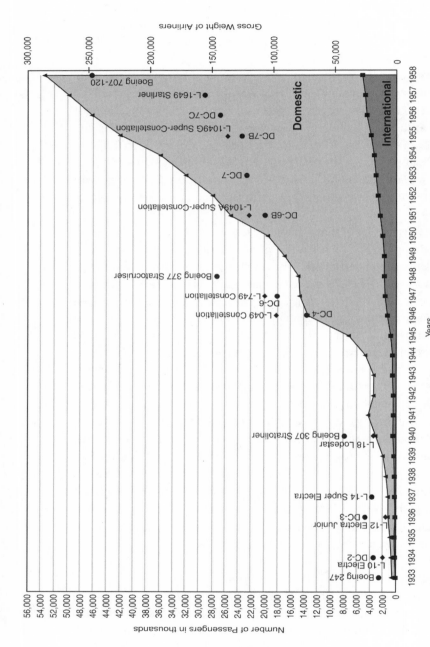

Revenue Passengers Enplaned on U.S. Scheduled Airlines, 1933–1958, with Gross Weights of Airliners by Year of First Commercial Use

The number of passengers on U.S. carriers rose steadily during the mid–twentieth century. Technological determinists attribute the rise to bigger, faster, and safer equipment: two-engine prop planes gave way to four-engine pressurized prop planes, which gave way to jet transports. This position is not so much wrong as it is incomplete. Political decisions like airport subsidies and industry tactics like advertising also played a role in air transport expansion. *Sources:* Air Transport Association historical data and R. E. G. Davies, *Airlines of the United States since 1914* (London: Putnam, 1972), 658–61.

the expansion of commercial aviation. That was especially true during the critical, formative decades of the 1920s and 1930s. The 1925 Kelly Bill, which established government air-mail contracts, laid the financial foundation for the industry, which could not have survived in those years by flying passengers alone. The 1926 Air Commerce Act promoted aviation through such means as licensing pilots, certifying aircraft, and building navigational aids. Funding for the government's principal research arm, the National Advisory Committee for Aeronautics (NACA), rose more than twentyfold, from $200,000 in 1921 to $1.5 million in 1930 to $4.4 million in 1940.[8] NACA research on cowlings, retractable landing gear, engine design and placement, and deicing equipment, among other innovations, improved aircraft safety and efficiency, thus reducing fear and fares, the two great obstacles to expansion.

These were also the aims of the Daniel Guggenheim Fund for the Promotion of Aeronautics, which encouraged "air-mindedness" by such means as underwriting Lindbergh's triumphal 1927 cross-country tour. Of more lasting significance were the fund's grants to create or improve aeronautical engineering programs at seven universities around the country. Though the fund operated for just four years, from January 1926 to February 1930, it left a remarkable legacy. By 1942 the majority of senior aeronautical engineers in the United States had graduated from Guggenheim schools. NACA's Langley Laboratory alone employed fifteen Guggenheim-school alumni.[9]

Funding for all aspects of aviation—research and development, airframe and engine production, air- and ground-crew training, airport and navigational facilities—was especially generous during wartime. Many writers—a sufficient number, at any rate, to label their arguments a third type of explanation—have pointed out that the money lavished on military aviation ultimately paid commercial dividends. "Ultimately," for military aviation had its disruptive side, too. World War I, as Eddie Rickenbacker wrote in 1925, "forced a rapid development of aviation, but at the sacrifice of economy, durability, comfort, and carrying capacity for the sake of speed, altitude, and maneuverability." Despite production and design progress, the war "left the industry on a false foundation, with its efforts wrongly directed with a surplus of war materials on hand that had to be used up before any concrete progress could be made."[10]

Rickenbacker doubtless had in mind the barnstormers, the mostly military-trained pilots flying military-surplus Jennies, powered by military-surplus engines. The economics were simple. Pilots could acquire a plane and engine for as little as $250. Fifty cents worth of gas sufficed to take a first-time passenger on a five-minute flight, for which he might pay, depending on local demand, as much as $5. (In 1925 one man even hired Charles Lindbergh to take him aloft so that he might urinate over his home town.)[11] Rickenbacker envisioned something grander. He looked forward to the day when Americans regarded travel by plane as routine and acceptable as travel by car.

World War II, like World War I, created hothouse conditions for aviation, and the ensuing Cold War kept the technological innovations coming. Though the outbreak of war in late 1941 created short-term disruptions—the government imposed civilian travel restrictions, converted some commercial planes to military use, and delayed delivery of new equipment the airlines had ordered—the long-term consequences were undoubtedly beneficial. The war familiarized GIs with flying and regularized transoceanic air transport, developing routes and traffic volume undreamed of in 1939. Twenty-six scheduled Air Transport Command flights crossed the Atlantic daily in mid-1945, another thirty-eight the still-contested Pacific.[12] Military innovations, such as radar, jets, and transponders, led to major gains in safety, comfort, and efficiency during the quarter-century after the war. Large American manufacturers like Boeing took the lessons they learned in designing and building military aircraft and applied them in their commercial divisions. The B-47, with jet nacelles slung forward and below flexible, swept wings, was the ancestor of the Boeing 707. The 707 in turn provided the basic design for the wide-body Boeing 747, which went into service in the early 1970s and quickly dominated transoceanic routes. By 1990 as many as 143,000 people, the population of a small city, might be aloft at any given time in a 747.[13]

That so many people should want to fly abroad, or could afford to do so, was itself a product of the long-term economic and social trends, the fourth type of explanation mentioned at the outset. The single most important trend was the growth in discretionary income. During the 1920s and 1930s commercial flying was pretty much confined to the wealthy and business travelers with expense accounts. That began to change during the long postwar boom. Adjusted for inflation, Americans' disposable incomes increased an average of 50 percent between 1945 and 1970,[14] years when improvements in engines and airframe design and experiments with economy fares were driving ticket prices down. Vacation and leisure travelers, who accounted for less than 10 percent of air travel in the 1930s, made up about half of it by the 1970s.[15]

Postwar education levels rose even faster than income. In 1940, among Americans who had reached age twenty-five, just one in four had graduated from high school and one in twenty had graduated from college. By 1970 slightly more than one in two had graduated from high school, one in ten from college. By 1990 the rates were three in four from high school, one in five from college.[16] As incomes and education levels increased, so did the demand for air travel to overseas destinations. The marketing logic was elementary. Other things being equal, someone who had taken an art history course was keener on seeing the Louvre than someone who hadn't.

Students traveling or studying abroad—the number of the latter increased eightfold worldwide between 1950 and 1978—formed another obvious market.[17] Charter operators responded by scheduling transatlantic charter flights with fares as low as $120. In the early 1970s Pan Am promoted a special Tuesday evening flight to Amsterdam featuring rock music, health foods, beer, sangria, and guitar-strumming troubadours.[18] (The ensuing jet lag scarcely bears contemplation.) The airline even came up with a slogan calculated to appeal to countercultural youth: "You can't improve a world you haven't seen."[19]

As this anecdote suggests, airline executives (and also aircraft manufacturers, whose commercial sales were tied to passenger volume) used advertising and public relations techniques to expand existing markets and to develop new ones. Though such tactics were hardly unique to the aviation industry, they did serve to counter a unique problem, fear-based sales resistance. That is, the *idea* that flying was something one could do in safety and comfort had to be socially, as well as technologically, constructed. The conscious manipulation of public perception is the fifth line of explanation for commercial aviation's midcentury expansion.

A nine-city survey of business men and women conducted by the J. Walter Thompson Company in late 1945 showed the problem the industry faced. Sixty percent of the sample had traveled by air, 40 percent had not. Though the interviewees acknowledged the timesaving value of flying for business purposes and praised the services of stewardesses (a much-copied innovation to calm and assist passengers), many still expressed reservations about flying. Among the comments transcribed by surveyors:

- Dislike because the air pressure is hard on my ears.
- Have no stomach for airplanes—can't use them.
- Afraid of flying—makes all business trips by train.
- Did not enjoy trip. Was nervous and tense the entire time of trip. No fault of the airline—seems to have phobia—would never take air trip again unless she had to make time.
- If an emergency arose, I might fly. Otherwise, I'm perfectly satisfied to remain on the ground.

- I've paid so little attention, I don't even know [the airlines] by name. You couldn't drag me into a plane.
- I don't want to go up. That's all.[20]

The deepest resistance to going up was predicated on the fear of coming too abruptly down. Passengers didn't have any control in an emergency situation or much of a chance of survival if things went badly wrong. When you covered a flood, as one journalist put it, you saw bodies; when you visited a crash site, you just saw faces.[21] And fire was a worse fate than dismemberment. "I resolved that, if the plane started burning, I would open my mouth and suck in the flames." Eddie Rickenbacker wrote of the 1941 crash that almost took his life. "It's quicker that way."[22]

Those who responded to the 1945 survey had all lived through the 1920s and early 1930s, aviation's carnivalesque era of wing-walkers and barnstormers and racing planes. The sensational publicity accorded such feats magnified their sense of the danger of flying. Popular movies exaggerated the risks, as did aviation pulp stories and comics like *Scorchy Smith*: "His motor cut, Scorchy is attempting to ride the sound beam of the plane below to a glide landing through the dense fog . . . LOOK OUT FOR THE HANGAR!! . . . There goes a wheel! . . . SHE'S OUT OF CONTROL!!"[23]

Though any plane might escape control, high-performance military aircraft were particularly suspect. In June 1942 J. Walter Thompson undertook a survey for the U.S. Navy's Bureau of Aeronautics. The researchers discovered, when they questioned the parents of boys of enlistment age, that fully a third would "emotionally object" to their sons joining the air services. "The principal reason for such objection is the danger that is involved," they concluded, adding that a "carefully planned campaign to point out the Navy's advantages in training and equipment can reduce this resistance."[24]

The commercial airlines planned campaigns of their own, though the object was the more modest one of promoting ticket sales rather than enlistments. One approach was to emphasize the airlines' accumulated experience. For years Pan American advertised itself as the world's most experienced airline, the word *experience* serving as a euphemism for safety. TWA tried a similar tack. A 1945 magazine ad showed a TWA Constellation crossing the Atlantic above the caption "routine stuff." "Flying the oceans isn't a new experience to TWA planes and crews," the copy explained. TWA already had more than 8,700 overseas flights to its credit, much of it gained through wartime flying for the Army Air Transport Command.[25]

One of advertising's most durable maxims is that a pitch like "experience" will always appeal more strongly if it can be personified. People like to

ROUTINE STUFF

Flying the oceans isn't a new experience to TWA planes and crews. TWA has more than
8,700 overocean flights to its credit, starting in February 1942, with service to Cairo, Egypt.
That was the pioneer overocean flight of any domestic airline for the Army Air
Transport Command. Later TWA became the first U.S. airline to operate year-round
schedules across the North Atlantic. Pictured above is TWA's new trans world system, the
foreign portion of which is just as familiar to our crews as are the airways of our
domestic routes. Wherever you live along TWA's transcontinental route, you will soon
have direct, one-carrier service to key foreign centers halfway around the world to India and
Ceylon. If you're going to Europe, Africa or Asia, see TWA or your travel agent now.

NEWFOUNDLAND
IRELAND
FRANCE
SWITZERLAND
ITALY
GREECE
EGYPT
PALESTINE
TRANS-JORDAN
IRAQ
SAUDI ARABIA
YEMEN
OMAN
INDIA
CEYLON
PORTUGAL
SPAIN
ALGERIA
TUNISIA
LIBYA

TWA

TRANS WORLD AIRLINE

This TWA advertisement appeared in the *Saturday Evening Post* and other magazines in 1945. Notice the polar projection. Such maps became increasingly common in the 1940s and 1950s, as long-distance flights linked Northern Hemisphere cities. The Arctic, not Rome, became the *centrum mundi* of the new world of long-distance aviation. (Ad*Access database, <http://scriptorium .lib.duke.edu/adaccess/T/T23/T2317-72dpi.jpeg>, December 20, 2001) read about people, not abstractions. By the 1940s, the airlines could boast of pilots who had acquired unprecedented amounts of flying time. Pan American had a veritable stable of star captains. Jack Knight and "Ham" Lee, two fabled airmail veterans, flew for United. Before his retirement in 1949, Lee managed to chalk up 27,811 hours, more than three years in the air. United's news bureau dubbed him the "flyingest" man in the world, a pioneer who had made the transition from the days of open-cockpit planes to "the present type of precise, scientific airline operations."[26]

Press releases were one thing; passengers' firsthand experiences another. In Tom Wolfe's version of aviation history, the "right stuff" voice emanating from the cockpit was an osmotic subcultural emulation of legendary test pilot Chuck Yeager's no-sweat, top-of-the-status-sphere drawl.[27] In fact, airlines gave pilots speaking lessons, training them to exude confidence. Stewardesses took lessons too. Bernard De Voto, who made notes during a 1952 cross-country flight, found their elocution superior to the pilots'. Part of the stewardesses' training involved taping and playing back their voices, checking to make sure they came across as suitably soft and calming. They rehearsed specific scenarios, such as reassuring passengers anxious about engine sparks. And they were taught to ask for each passenger's name at the beginning of the flight. More an exercise in confidence-building than a roll call, the idea was to expose passengers to the stewardesses' calm, professional voices.[28]

The thinking that went into these campaigns could be quite subtle. Anyone who studies midcentury publicity photos and advertisements will notice that commercial airliners were generally shown flying level or, at most, gradually ascending or descending. This was not coincidental. Advertising experts knew that dramatic angles could prompt conscious or unconscious fears. "Look, let's agree right now never to show airborne planes at a cockeyed angle," said one during a confidential meeting. "It looks either like a model at best or, at worst, like a giant plane out of control."[29]

Postwar advertising executives understood that they had to do more than placate the fears of their mostly male business travelers. They also had to placate their spouses, many of whom begged their husbands not to fly. So advertisers targeted wives, emphasizing that husbands who flew returned home sooner. "She'll be glad you took the fast one," as one advertising executive put it.[30]

Advertisers tried to persuade women to venture aloft themselves, knowing that the actual experience of flying was the surest means of dispelling fear. "Trips with children easy when you travel by TWA," promised one 1948 newspaper ad, which featured an obsequious stewardess serving a young mother with a happy child on her lap. Special food was available for babies, the ad promised, and full-sized lavatories offered female passengers a "powder room in the sky."[31] In fact, few passengers found traveling with infants easy, or confused airplane lavatories with the roomier facilities of a Pullman car. But airline advertising projected the ideal experience, just as nineteenth-century railroads promised immigrants frontier farmland that turned out to be rather more arid than advertised.

Still, it cannot be said that the airlines' promises of comfort and safety were wholly false. To be effective, James Webb Young once observed, an advertising proposition needs to be three things. It needs to be interesting, persuasive, and confidence-building.[32] While it is possible to create effective ads for products that are inherently undeserving of confidence, such as cigarettes, midcentury airline advertisers had a far easier task. The rapid progress in aircraft design and performance, known to any American who had access to a newspaper or radio, lent credibility to their copy. Planes were in fact becoming faster and more efficient, engines more powerful and reliable. Passengers could ride in pressurized cabins above the worst weather. Better than the lurching train, the writer Bernard De Voto noted, as he passed his trip correcting galleys in the soundproofed cabin of his DC-6. The only thing he regretted about air travel, he told his readers, was its comparative dullness.[33]

This was, after all, the point. For commercial flying to become commonplace it had to become safe, predictable, uneventful—dull. Speed was the key, speed without adventure. No more lion-cub mascots outfitted with their own parachutes, à la Roscoe Turner. Just pull down the shades and enjoy the movie. "My only problem was to pass the time," Daniel Boorstin complained of a routine New York-to-Amsterdam flight at 23,000 feet. The experience of high-altitude flying, he added, had ceased to be one of traveling through space and had become one of simply traveling through time.[34] The accuracy of this insight will be apparent to anyone who has suffered jet lag or worried about making a connecting flight.

Boorstin and De Voto were both prominent writers and intellectuals. In social-class terms, they typified the affluent and educated air travelers of the 1950s and early 1960s. Eventually, when high-bypass-ratio turbofan jet engines and wide-body airframes made it possible for each passenger to travel sixty miles for each gallon of fuel and at ten times the speed of a car, and when declining fares and deregulation made it affordable to do so, the vast majority of Americans made at least the occasional trip by air.[35] In 1998 more than four in five adults told pollsters they had flown at some point in their lives, and half of these had made at least one trip during the previous year.[36] Having become a routinized form of long-distance transportation in the mid-twentieth century, flying became democratized in the late twentieth.

I have begun to stray beyond my chronological territory. Let me conclude by saying something provocative. The emergence of commercial aviation as a reliable form of mass transportation for people and cargo was the single most important development in the history of flight in America. Stretching the point, it was one of the most important developments in the history of modern America. The midcentury rise of commercial aviation, together with its supporting manufacturing and service industries, had multiple and profound effects. It increased productivity, and therefore real per capita income, by speeding critical parts, documents, and personnel to their destinations. It bolstered American military power and enhanced the government's ability to project that power into remote corners of the world. It exerted centrifugal forces on families and cities. It consumed vast amounts of land for airport construction. It made possible bicoastal sports leagues and national universities. It hastened the spread of infectious diseases and the degradation of the environment. It helped make the English language and middle-class American consumer standards into international norms, aviation-linked globalization antedating the Internet version by forty years or more.[37]

If I have stressed the socially constructed dimension of this revolution, it is only because other writers have so emphasized its institutional, military, socioeconomic, and, above all, technological origins. The emphasis on technological progress is certainly justified. Better planes were the necessary, if not quite the sufficient, cause of routine commercial flight across (and from) a continental nation. But there was more to the story than build-them-better-and-they-will come. The designers at Douglas and Boeing and Lockheed had a powerful, if unacknowledged, tailwind: the efforts of advertising executives and public-relations experts to persuade wary Americans to trust their equipment and take advantage of the new aerial frontier.

NOTES

1. Walter Kirn, *Up in the Air* (Garden City, N.Y.: Doubleday, 2001), 278.

2. Laurence La Tourette Driggs, "The Future of Aviation," *Outlook* 121 (Apr. 9, 1919): 608.

3. "Air Safety Record Set in Nation Last Year, with One Accident to Every 654,610 Miles," *New York Times*, Mar. 28, 1935, 14. The following year, 1935, the Massachusetts Indemnity Insurance Company informed C. R. Smith, president of American Airlines, that henceforth executives or salesmen who traveled by air would be rated as better risks

than those who traveled by car. See *Aircraft Year Book for 1936* (New York: Macmillan, 1937), 156.

4. Richard K. Smith, "The Weight Envelope: An Airplane's Fourth Dimension . . . Aviation's Bottom Line," *Aerospace Historian* 33 (Spring 1986): 41.

5. Air Transport Association, "Annual Traffic and Capacity," <http://www.airlines.org/public/industry/display1.asp?nid=1032>, accessed Aug. 28, 2002.

6. Smith, "Weight Envelope," 42.

7. The Russian case offers another example. Had it not been for the disruptions of war and revolution (which, among other things, sent the great aircraft designer Igor Sikorsky to America), Russia might well have challenged America's midcentury dominance in the skies. As it was, by 1939 the Soviet Union had almost caught up to Germany as the national air transport leader outside the United States. After the Soviet Union recovered from the devastation of World War II, Aeroflot grew rapidly, jumped to twin-engine jets in 1956, and served a nation whose principal east–west cities, Moscow and Vladivostok, were nearly twice as far apart as New York and San Francisco. In so vast a country, air transport proved the ideal solution to the problem of physical communication. See R. E. G. Davies, *A History of the World's Airlines* (New York: Oxford University Press, 1964), 116–18, 295–96.

8. G. Lloyd Wilson and Leslie A. Bryan, *Air Transportation* (New York: Prentice-Hall, 1949), 576.

9. Richard P. Hallion, "Philanthropy and Flight: Guggenheim Support of Aeronautics, 1925–1930," *Aerospace Historian* 28 (Spring 1981): 13.

10. E. V. Rickenbacker, "Is Aviation Starting to Repeat Auto History?" *Sales Management* 8 (June 13, 1925): 859.

11. A. Scott Berg, *Lindbergh* (New York: G. P. Putnam's Sons, 1998), 82.

12. Oliver La Farge, *The Eagle in the Egg* (reprint, New York: Arno Press, 1972), 4.

13. Clive Irving, *Wide-Body: The Triumph of the 747* (New York: Morrow, 1993), 355, 370.

14. Alfred W. Niemi Jr., *U.S. Economic History*, 2d ed. (Chicago: Rand McNally, 1980), 400.

15. Roger E. Bilstein, "Travel by Air: The American Context," *Archiv für Sozialgeschichte* 33 (1993): 288. I am indebted to Professor Bilstein for calling this article to my attention.

16. U.S. Bureau of the Census, *Census Questionnaire Content, 1990* CQC-13, <http://www.census.gov/apsd/cqc/cqc13.pdf>, accessed Dec. 23, 2001.

17. *Statistics of Students Abroad, 1962–1968* (Paris: UNESCO, 1972), 19; *Statistics of Students Abroad, 1974–1978* (Paris: UNESCO, 1982), 17.

18. Paul Critchlow, "Cut-Rate Air Fares Lure Young," *Philadelphia Inquirer*, July 31, 1972, clipping in Box 259, Pan American World Airways, Inc., Records, Archives and Special Collections Department, University of Miami [hereafter Pan Am].

19. Najeeb E. Halaby, "Serving Youth in the Seventies," typescript, 1971, 8, Box 462, Pan Am.

20. "Pan American World Airways: Survey to Determine Airline Preferences in Nine Cities," typescript, 1945, 33, Box 465, Pan Am.

21. Laurence Gonzales, "Airline Safety: A Special Report," *Playboy* clipping, June 1980, Box 402, Pan Am.

22. Eddie Rickenbacker, *Rickenbacker* (Englewood Cliffs, N.J.: Prentice-Hall, 1967), 237.

23. Frank Robbins, *Scorchy Smith*, "No Sooner Said Than Seen," "One-Legged Landing," and "Whatever Goes Up—!" Aug. 29, 30, 31, 1939.

24. "Nationwide Public Attitude Survey—June 1942—Among Boys of Enlistment Age, Parents and Girls," microfilm reel 338, J. Walter Thompson Company Archives, Special Collections Library, Duke University [hereafter JWT Archives].

25. Ad*Access database, <http://scriptorium.lib.duke.edu/adaccess/T/T23/T2317-72dpi.jpeg>, accessed Dec. 20, 2001.

26. "Biography of Captain E. Hamilton Lee," typescript press release, n.d., Box 2, Records of the Post Office Department, RG 28, National Archives and Records Administration, Washington, D.C.

27. Tom Wolfe, *The Right Stuff* (New York: Farrar, Straus, and Giroux, 1979), chap. 3.

28. Vance Packard, *The Hidden Persuaders* (New York: David McKay, 1957), 64–77; Bernard De Voto, "Transcontinental Flight," *Harper's*, July 1952, 47.

29. Anonymous handwritten notes, Nov. 23, 1959, review board records, Box 22 (Northeast Airlines), JWT Archives.

30. Packard, *Hidden Persuaders*, 65; quote is from Douglas Commercial Aircraft review board meeting, May 10, 1956, review board records, Box 9, JWT Archives.

31. Ad*Access database, <http://scriptorium.lib.duke.edu/adaccess/T/T20/T2031-72dpi.jpeg>, accessed Dec. 20, 2001.

32. James Webb Young, "The Nature of Advertising Knowledge," typescript speech, Feb. 2, 1954, 4, Box 39, speeches and writings collection, JWT Archives.

33. De Voto, "Transcontinental Flight," 48.

34. Daniel J. Boorstin, *The Image: A Guide to Pseudo-Events in America* (New York: Atheneum, 1971), 94.

35. Irving, *Wide-Body*, 371.

36. Gallup Organization, *Air Travel Survey, 1998* (Washington, D.C.: Air Transport Association, 1998), III-2, III-8.

37. Another trend spotted by Boorstin in *Image*, 97–99.

CHAPTER 11
THE EFFECT OF FLIGHT ON ART
IN THE TWENTIETH CENTURY

In the spring of 1912, Pablo Picasso produced three cubist paintings provocatively featuring the cover of a contemporary pamphlet advocating military aviation.[1] "Our future is in the air," the works declared, incorporating the language of the booklet. Significantly, while executed in oil on canvas, the paintings were produced at the same time that Picasso made his first collage, a step that would have a formative effect on the future of twentieth century art. Picasso's images, then, suggest a parallel between the extraordinary accomplishment of heavier-than-air flight and a revolution in the treatment of form occurring simultaneously in the fine arts. By an intriguing coincidence, Louis Vauxcelles's 1908 review of the work of Picasso's collaborator, Georges Braque, which coined the term *cubism*, appeared directly above an article chronicling yet another triumphant flight by Wilbur Wright.[2] The suggestive pairing was preserved by a clipping made by the painters' friend and dealer Daniel-Henri Kahnweiler. As William Rubin notes, "Wilbur Wright's name appears in juxtaposition to Braque's in the clipping, in Kahnweiler's album, of the celebrated review by Louis Vauxcelles of Braque's first Cubist paintings (the exhibition of November 1908), in which he calls Braque 'un jeune homme fort audacieux' (an extremely audacious young man)."[3] The clipping may well have helped inspire Picasso's nickname for Braque, "Wilbourg," a name with which Braque also referred to himself—and which identified the painters' collaboration still further with that of the famous brothers.[4]

Numerous commentators, including the cubists' contemporary, Gertrude Stein, have drawn attention to the fascinating relationship between the invention of the airplane and that of cubism.[5] Indeed, Picasso's friend, André Salmon, noted that the artist painted in "his aviator's outfit" while he was pioneering cubism.[6] Mark Tansey gave visual form to the connection in his 1992 painting, *Picasso and Braque*.[7] The painting, which depicts Picasso rising on a cubist violin—cast as biplane—clearly conflates the monumental technological and artistic breakthroughs. Tansey expands on the structural logic of Picasso's collage until it becomes a three-dimensional form, complete with wings. According to Braque, Picasso himself noted such architec-

Our Future Is in the Air, Pablo Picasso (1912). (Private collection; © 2003 Estate of Pablo Picasso/ Artists Rights Society [ARS], New York)

tonic affinities between Braque's early cubist sculpture and the biplanes of Wilbur Wright.[8]

The origins of the collaborations that lie behind flight and cubism are so well known that one instinctively associates Picasso and Braque with the famous photograph of the first flight at Kitty Hawk, December 17, 1903, captured by John Daniels. Following this analogy, and tying it to history, Picasso and Braque are appropriately cast, respectively, as Orville—who piloted the first flight—and Wilbur, who ran alongside. However, the figural arrangement of the artist-partners is actually based on another image— a picture of A. M. Herring piloting an oscillating-wing glider designed by Octave Chanute.[9] Herring, who flew a craft designed by another, would ultimately fail in his aspiration to be the first to achieve heavier-than-air flight. Yet Tansey has not altogether overlooked the John Daniels photograph of Orville and Wilbur. As Tom Crouch has observed, "the shovel and detritus in front, the grease can and the tools [converted in Tansey's painting to cans of paint and brushes] belong with the first flight photo." Present as well in Tansey's painting is the small bench against which one of the wings of the Wright Flyer rested and a shovel for the sand anchor that held the craft in place. Tansey has also included the footprints visible in the photograph of the first flight.[10]

Picasso and Braque, Mark Tansey (1992). (Los Angeles County Museum of Art, Modern and Contemporary Art Council Fund; photo © 2003 Museum Associates/LACMA; photo courtesy of Gagosian Gallery)

The relationship of Tansey's painting to the early history of flight and the development of cubism is clear. However, even more important is Tansey's implicit suggestion that the "takeoff" of modernism, intimately linked to the success of cubism, has roots in the development of flight. This essay examines the effect of flight on art of the twentieth century. I focus on five diverse artists whose work has been marked indelibly by the legacy of heavier-than-air flight. The selection is by no means comprehensive, but, rather, designed to indicate the breadth of responses to the airplane. Previous scholarship, particularly that of Robert Wohl, Gerald Silk, and Julie Wosk has primarily concentrated on airplane imagery.[11] However, this represents only one aspect of flight's influence on the art of the twentieth century. As I argue, the legacy of flight is even more pervasive, shaping attitudes toward materials, metaphor, and even artistic identity.

Twentieth-century artists have been moved not only by the appearance of the airplane, but also by the principles it represents—a mastery of mass and weight, a victory over the force of gravity through motion. My discussion opens with a look at two sculptors for whom these dimensions of flight have been particularly influential, Alexander Calder and Richard Serra.

The Wright Brothers' first flight, December 17, 1903. (Photo by John Daniels)

Alexander Calder was a boy of five when the Wright Brothers first flew at Kitty Hawk. Almost twenty-four years later, as an American artist living in Paris, he was present at Charles Lindbergh's triumphal landing at Le Bourget airfield. Describing the festive atmosphere following Lindbergh's touchdown, Calder recalled that "People wanted Lindbergh, so they cried, 'L'aviateur!' I tried to cry 'L'aviateur!' too, and drew quite a laugh. Finally Lindbergh appeared."[12] Even before participating in this historic event, Calder's early wire sculpture showed an affinity for motion and airborne movement. His *Circus*, developed between 1926 and 1930, featured acrobats on a flying trapeze, seals tossing a ball, and a high-wire act. In 1929 the artist created an even more direct homage to flight with his wire sculpture *Spirit of St. Louis*, commemorating Charles Lindbergh's daring 1927 solo flight from New York to Paris—a trip Calder himself had made by ship in 1926.[13] Calder's playful representation of the event features Lindbergh's plane literally held aloft by the "spirit" of St. Louis, a muse who seems to emerge from the waves of the Atlantic Ocean.[14] However, if Calder's representation can be considered traditional in many respects, even reminiscent of Peter Paul Rubens's early-seventeenth-century portrayal of the arrival of Marie de Medici by ship at the port of Marseilles, his response to Lindbergh's accomplishment, and to

Spirit of St. Louis, Alexander Calder (1929). (Private collection; photo courtesy of Vance Jordan Fine Art, Inc.)

the significance of flight in general, would soon become more subtle. In approximately 1931, Calder developed the first of the kinetic works for which he would become famous, and which Marcel Duchamp would term "mobiles."[15] According to Calder the very earliest of these pieces "swayed in the breeze"—exploiting moving air just as a plane does.[16]

Calder's mobiles have generally depicted entities that themselves seem to float or glide in a manner analogous to the structures themselves: sea creatures, constellations, the universe. Calder's early mobile, *Starfish*, of 1936–40, collapses these themes. The title *Starfish*, suggests both a sea creature and a constellation seen in the sky, while wire loops and three dangling masses evoke the orbits of celestial bodies. At the same time, the very shape and upward inclination of the sculpture suggests an airplane. Calder's 1949 mobile, *Blériot (I)*, honoring the achievement of Louis Blériot, the first aviator to cross the English Channel, comments even more directly on the inspiration provided by flight.[17] According to Calder, "The important thing is for the mobile to catch the wind. It has to move. . . . [T]he mobile catches any wind whatever."[18]

The tendency of Calder's work to embody principles of flight is reflected in his friend Robert Osborn's remark that "Calder, a generation before our time, was in full orbit. Long before atoms were ruptured into flying particles or rockets ricocheted from planet to planet, this weighty man had blasted off."[19] In 1973 Calder received an opportunity to literalize the metaphoric relationship of his sculpture arrangements of color and form to flight when he was commissioned by Braniff International to decorate a McDonnell-Douglas DC-9 jet airliner, 157 feet long. The resulting work, *Flying Colors*, was followed two years later by a second commission from Braniff to paint a second plane in honor of the U.S. bicentennial. The red, white, and blue result, *Flying Colors of the United States*, evoked the sensation of a flag streaming in the wind.[20]

If fascination with flight marked the starting point of Calder's development of the mobile, aeronautical engineering also played a critic role in the development of a mobile at the end of his career. In 1972 the National Gallery of Art in Washington, D.C., invited Calder to create one of the floating sculptures for its East Building, then under construction. Following the requirements of the commission, Calder created a small model, or maquette, of the piece. It balanced beautifully and promised to circulate freely in the air currents generated by the new building's ventilation system. However, as Calder embarked on enlarging the mobile, it became evident that the work, to be executed in steel, would be far too heavy to hang freely. Calder then entered into discussions with the engineer Paul Matisse, who, together with Calder, struck upon the solution of fabricating the piece with the honeycombed aluminum used in aircraft design. The work's final weight was a mere 920 pounds.[21]

Like Calder, though employing different techniques, the work of Richard Serra reveals a deep affinity for problems of balance and gravity. However, while Calder's mobiles seem to float in the air, Serra's work is brutally heavy, making us aware, through opposite means, of the mass and force associated with our own bodies and the action of gravity upon us. As Serra has written, "Weight is a value for me, not that it is any more compelling than lightness, but I simply know more about weight than lightness and therefore I have more to say about it. . . . I have more to say about the perpetual and meticulous adjustments of weight, more to say about the pleasure derived from the exactitude of the laws of gravity. . . . [T]here is an imponderable vastness to weight."[22]

Serra's interest in problems of the heaviness and lightness of mass dates back to an early childhood experience with buoyancy. On his fourth birthday, in the fall of 1943, Serra accompanied his father to witness the launch of an enormous steel tanker. He was awestruck by the spectacle of the massive vessel plunging into the sea and bobbing back to the surface. Describing the experience, Serra recalled, "It was a moment of tremendous anxiety as the oiler en route rattled, swayed, tipped, and bounced into the sea, half submerged, to then raise itself and find its balance. Not only had the tanker collected itself, but the witnessing crowd collected itself as the ship went through a transformation from an enormous obdurate weight to a buoyant structure, free, aloft, and adrift. My awe and wonder of that moment remained. All the raw material that I needed is contained in the reserve of this memory which has become a recurring dream."[23]

Serra's account of his amazement at witnessing something that seemed to defy physical laws has strong resonance with Amos Root's description of the awe-inspiring sight of Orville Wright completing the first loop ever in an airplane almost exactly forty years earlier, in September 1904: "When it turned that circle, and came near the starting point, I was right in front of it; and I said then, and I believe it still, it was . . . the grandest sight of my life. Imagine a locomotive that has left its track, and is climbing up in the air right toward you—a locomotive without any wheels . . . but with white wings instead . . . a locomotive made of aluminum. . . . I tell you friends, the sensation that one feels in such a crisis is something hard to describe."[24]

In 1983, with his sculpture *Kitty Hawk*, Serra produced his own response to the Wright Brothers' remarkable accomplishment of lifting a metallic structure into the air. The work grew out of a decade and a half of experimentation with "prop pieces," sculptures that challenge our intuitive understanding of weight and gravity. Serra's 1969 piece, *One Ton Prop (House of Cards)*, for example, reverses our traditional understanding of what is meant by a house of cards, typically something unstable and transitory, as in Jean Baptiste Siméon Chardin's moralizing genre painting, *House of Cards*, of 1735. In Serra's work, by contrast, the very heaviness of these typically light elements is precisely what ensures the structure will endure. As Serra later put it, "The four plates' tendency to implode, to collapse inward and downward was . . . gravity-defining."[25]

In an interview conducted approximately five years after completing *House of Cards*, Serra made reference to a host of his recent works that followed similar principles—pieces such as *Stacked Steel Slabs*; the Skullcracker Series, 1969; and *Sight Point*, 1971–75, a tower in which a viewer could stand, with attention directed upward toward the light from the sky. Each of these works Serra tied to the influence of Constantin Brancusi, an early-twentieth-century sculptor with an abiding interest in flight, worked out, perhaps most graphically, through multiple *Bird in Space* sculptures.[26] Regarding this work, Serra commented, "[Brancusi's] better pieces achieve

Kitty Hawk, Richard Serra (1983). Cor-ten steel, unframed, 48 × 72 × 4", two pieces, Albright-Knox Art Gallery, Buffalo, New York, Mildred Bork Connors, Edmund Hayes, George B. and Jenny R. Matthews, and General Purchase Funds, 1991. (© 2002 Richard Serra/Artists Rights Society [ARS], New York)

weightlessness. Although a lot of my earlier work in this regard tends to be somewhat heavy-handed in this respect, since then I've found at certain balance points the weight would actually negate itself. One does not sense force or weight acting as a fulcrum, a lever or a counter balance. If the pieces are equally balanced, the weight is cancelled out, you have no thought of tension nor of gravity. The result is a hovering of two discrete elements touching in a suspended state. . . . But that is something that one arrives at by doing. That has to come out of one's relationship to the material, or one's understanding of one's own body in relation to the ground."[27]

Serra accomplished just such a sense of weightlessness in *Kitty Hawk*. Although his interpretation of the Wright Flyer defies expectations, it succeeds in evoking the extraordinary nature of their accomplishment. Executed in cor-ten steel, made by the U.S. Steel Corporation where Serra had worked during college, the piece consists of two plates that together total five tons, a weight over fourteen times that of the 700-pound Wright Flyer.[28]

Visually, the work resembles an abstracted airplane with its upper, horizontally oriented member suggesting wings and its lower, vertical piece evoking a fuselage.[29] Perhaps even more important, however, the work itself seems to simulate the Wright Brothers' accomplishment with its own apparent achievement of weightlessness. One piece of steel, fourteen by four feet, appears to float above the lower plank, a slightly thicker piece of steel, measuring six by four feet. The work's poignancy is heightened by the fact that the upper piece of steel is thrust all the way forward, so far forward that its two-inch thickness actually extends beyond the edge of the lower piece. And yet the large plank of steel, resting only against the narrow beam below it and against two points on the wall, balances. It almost defies the imagination—just as flight itself did. Under the right conditions, one can imagine steel floating on air as effortlessly as it glides through water. According to Serra, "It is the distinction between the prefabricated weight of history and direct experience which evokes in me the need to make things that have not been made before. I continually attempt to confront the contradictions of memory and to wipe the slate clean, to rely on my own experience and my own materials even if faced with a situation which is beyond hope of achievement. To invent methods about which I know nothing, to utilize the content of that experience so that it becomes known to me, to then challenge the authority of that experience and thereby challenge myself."[30]

The challenge of invention articulated by Serra, and taken up mechanically by the Wright Brothers, was adopted quite literally by the Belgian pop sculptor, Panamarenko. Like Serra, Panamarenko was sensitive to his use of materials and sought to take advantage of new products that were becoming available in the early 1960s, asserting that "I don't think their special qualities have been sufficiently exploited yet."[31] But while deeply interested in the physicality of art, and of flight, Panamarenko also sought to associate himself with the mythology of flight, casting himself into the role of inventor of flying machines, like Leonardo da Vinci during the Renaissance.

In adopting the persona of inventor Panamarenko emulated a strategy of his friend and mentor, the German conceptual artist Joseph Beuys. Beuys presented himself as a latter-day Icarus. A member of the Nazi Air Force during World War II, Beuys claimed to have crashed over Crimea during a snowstorm. He said native Tatars nursed him back to health, wrapping him in fat and felt, materials that became central components of his artwork.

As with Beuys, an association with flight became central to Panamarenko's very identity. The artist's unusual name, the only one by which he is known to the public, was inspired by the popular international airline Pan American. Panamarenko emblazoned his name on a hat resembling that worn by an airline captain and made it part of his attire. The artist rechristened himself and took up flight-oriented art in 1962, the year after Russian cosmonaut Yuri Gagarin became the first human to fly in space—to be followed shortly by Alan Shepard, the first American astronaut in space.

Initially, Panamarenko's work focused on the allure of space travel. He conducted a series of performances known as the "Milkyway Happenings" in 1966 and in the same year publicly demonstrated the efficacy of his *Magnetic Shoes*. This unique artwork, a pair of boots equipped with electromagnets, enabled the wearer to defy gravity by walking upside down on metallic surfaces. The piece included a padded cap, lest the power fail.[32] Panamarenko's work did not merely picture the achievements of space travel but enacted them, actually permitting an earthbound person to escape the force of gravity. The success of spaceflight led Panamarenko to declare that the "highest purpose one can have is to devise a way to leave the earth."[33]

If Panamarenko's magnetic shoes simulated the effects of weightlessness, subsequent projects focused more literally on this goal. In 1967 Panamarenko constructed the first of several flying machines, *Das Flugzeug*. Resembling a cross between a whirligig and a bicycle, the contraption is something that one could almost imagine having been created by the Wright Brothers, whose own aeronautical inspiration came largely from bicycle design.[34] The work also suggests the influence of Leonardo, whose *Standing Ornithopters* of 1486–1490 involved the use of similar systems of treadles and drums.[35]

The failure of Panamarenko's machine to fly did not discourage the artist. Inspired in 1969 by the moon landing and the first Concorde flight, Panamarenko continued to devise more flight-oriented projects and devices, such as his *"One Seat" Rocket*, complete with remote power units. Perhaps most significant, Panamarenko dedicated himself to yet another inhabitable flying machine, the *Aeromodeller*. Indeed, he marked the first moon landing by staging a lecture on July 21, 1969, about his new project, which resembled a dirigible, consisting of a motor-driven balloon with a wicker basket attached.[36] Panamarenko intended to live aboard the airship upon its completion in 1971, but plans for its maiden voyage were abruptly aborted because of "official resistance and poor weather conditions."[37] Had cancellation not been forced upon the artist, it is unclear that the ship could have taken successfully to the skies.

While such repeated failure would be devastating to the engineer, for the artist it did not represent a setback. "But then is it flying itself that really intrigues you?" an interviewer asked the artist in 1970. "If I start making something it is, of course, flying in itself," responded Panamarenko, "but once I'm busy *the most important thing really is the process of making it* and when it's finished I feel I have to try it and so on, but there's no hurry then, because *there are several aspects of propellers and motors that are actually*

The Aeromodeller, Panamarenko (1969–71). SMAK, Stedelijk Museum voor Actuele Kunst, Ghent.

more interesting, that are actually more like flying than flying itself."[38] Perhaps even more important for Panamarenko, to work in a well-defined, predictable arena represents no challenge. As he asserted, "When you know beforehand that a thing is going to work, there is no more emotion and all you do is carry out an existing plan."[39]

Panamarenko's failed flying machines form an interesting counterpart to the grounded airplanes of Anselm Kiefer, who also began as a protégé of Joseph Beuys. Throughout his career, Kiefer has explored the political and cultural myths of the German state, seeking to come to terms with the devastating legacy of World War II by examining themes of destruction and regeneration on many levels—political, cultural, physical, and metaphysical. Since the late 1980s, the airplane has assumed a vital role Kiefer's iconography, both for its political and for its spiritual symbolism, as a close study of Kiefer's poignant 1989 sculpture, *Angel of History*, reveals.[40]

The casting of an airplane—in this case a version of what appears to be a B-52 bomber—into the role of the "angel of history" is provocative in itself.[41] But Kiefer complicates this metaphor by constructing an object that is utterly

Angel of History, Anselm Kiefer (1989). (Eugene L. and Marie-Louise Garbáty Fund; photo © 2002 Board of Trustees, National Gallery of Art, Washington, D.C.)

incapable of flight. Executed in lead, the work weighs approximately 2,000 pounds. Its structure is sagging and dilapidated. A white patina on its surface indicates oxidation, and red rust surrounds iron rivets in its tail. The passage of time—or history—has clearly acted on this decaying aircraft. Books—nine lead tomes in all—layered on its wings suggest yet another manifestation of "the weight of history" both carried and represented by this vehicle. Kiefer's title, *Angel of History*, calls attention to the books. The phrase refers to a passage from Walter Benjamin's "Theses on the Philosophy of History."[42] Benjamin, a prominent German historian and literary critic, who was Jewish, became a victim of the Holocaust when he took his own life during an attempt to flee the Nazis. Dried poppies, which extrude from the books, the engine, and even the tailpiece of the plane, function as a symbol for the fallen scholar and other victims of persecution.[43]

While this work clearly references the horrors of World War II, it speaks simultaneously to another more positive chapter in Germany history. Made in 1989, it marks the fall of the Berlin Wall and the beginning of the end of the iron curtain. Indeed, a version of this work, along with similar airplane sculptures, *Melancholia* and *Jason*, were exhibited in Cologne, Germany, immediately in the wake of this momentous event.[44]

The prospect of the sudden reunification of Germany may have suggested yet a further association for Kiefer with the imagery of the grounded bomber. As an artist who has always found an "inextricable rapport . . . between historical events and celestial metaphysics," the image of the bomber may well have called to mind images of the nuclear bombing of Japan that concluded World War II.[45] The atomic bomb, of course, relied on a process of nuclear fission, or the splitting of atoms. Nuclear fusion would release a greater, and less dangerous, energy. According to Kiefer, "I see the universals wherever I go in the world. I see the myths living, like the parts of Osiris that were gathered up to make new energy. That same process lives in nuclear fusion. You see, I want to show something that was lost but is still here, transformed, today."[46] For Kiefer, the reunification of Germany may well have signaled such a potential energy explosion.

The reversals at work in such an association—fusion rather than fission and rebuilding rather than destruction—are present in all aspects of this work. This is a bomber that seems to sprout flowers, and which carries a cargo of poppy seeds, rather than bombs. And it is an earthbound flying machine in a state of decay, perhaps, it seems, in the process of being consumed by nature. Indeed, the oxidation and rust alluded to earlier in my talk have strong metaphysical associations for the artist, suggesting the transformation of base metals, lead and iron, into their "purer" forms, silver and gold.[47]

At the same time, it is possible to see the plane in terms of an aborted flight, a failed flight, bringing to mind images of Icarus, with whom the artist has indeed identified himself.[48] According to Kiefer, "Art for me is simply the possibility of creating relationships between disparate things and in that way giving them a meaning."[49] Here, the airplane stands for exactly such union of opposites—heaven and earth, the political and the metaphysical. The flying machine becomes a contemporary symbol for the creativity of the artist.

A similar dynamic animates the work of the British pop artist Malcolm Morley, whose 1995 sculpture *Flight of Icarus* pictures a German Fokker Drɪ *Dreidecker* flush against the wall. The large red circle at the nose of the plane evokes simultaneously the whirl of the plane's propeller and the image of the sun whose heat melted Icarus's waxen wings. Morley's plane embodies a crisis of representation both personal and pictorial.[50] The three-dimensional structure pushes outward from the wall, breaking free of the two-dimensional plane of painting, but it is at the same time inextricably bound to it. Juxtaposed with a barrier, the plane seems doomed to crash. The crumpling of surfaces is one device employed by Morley to explore the relationship of two- and three-dimensional form.[51] The sculptural rendition

Flight of Icarus, Malcolm Morley (1995). (Collection Timothy Eggert, Washington, D.C.; courtesy Sperone Westwater, New York)

of the German military aircraft also evokes the model-building that Morley engaged in as a child.[52]

The sense of tension implicit in the work is heightened by the fact that this beautiful object represents a military aircraft. It is a haunting precursor to the German air power that destroyed Morley's own childhood home during World War II, a well-placed bomb taking with it the young boy's model ship on the verge of completion. This personal tragedy, the memory of which Morley recovered during therapy, may have helped spur a number of paintings made by the artist during the 1970s depicting airplane disasters.[53] These include three 1976 paintings depicting planes nosediving into battleships: *Age of Catastrophe*, *Little Corner of Plane-Ship Catastrophe and Central Park*, and *Burial of Catastrophe*.

Indeed, Morley's work responds on a number of levels to the implications of the airplane for him as a contemporary artist. Not only does the plane serve as a type of pun to get at problems of pictorial representation and as an agent of childhood trauma, but Morley has also created works that allude to the role of the airplane in popular culture. More recent works, such as *Air Circus over Maine* (1990) and *André Malraux Flying the Spad-Herbement*

S 20 bis over Gustavia (1991), deliberately invoke the cover of the Captain Biggle adventure stories Morley read growing up.[54]

At the same time, Morley's artwork addresses the problem of the shifting nature of sight and perception, an issue intrinsic to the experience of flight itself. According to Morley, "the goal of any painting is to bring together the perceptual problems of seeing—its structure—with the actual experience of seeing."[55] The most recent installation of Morley's *Flight of Icarus* in the main stairwell of the Corcoran Gallery of Art manifested this conflict by enabling the viewer to approach the sculpture from different heights and views—in a manner analogous to moving up and down in a plane.

This essay provides only a brief sketch of the effect of flight on the development of twentieth-century art. I have deliberately incorporated artists whose work has not been widely discussed in this regard, but for whom the substance and symbolism of flight represent an essential point of departure. As I have attempted to show, flight has influenced artists not simply by providing new imagery, but on a far deeper level. It has inspired a new understanding of materials and mythologies—an appreciation for the meaning of weight and gravity and for the mystical and political dimensions of human flight that became a reality nearly a century ago. In the words of the French kinetic sculptor Jean Tinguely, who claimed rocket scientist Wernher von Braun as one of the artists he most admired, "We're living in an age when the wildest fantasies become daily truths. Anything is possible."[56]

NOTES

I benefited from the opportunity to present a version of this paper in October 2001 at "They Taught the World to Fly," First Flight Centennial Commission International Symposium, in Raleigh, North Carolina. I thank Tom D. Crouch of the National Air and Space Museum, Smithsonian Institution, for the invitation to speak and for his subsequent contributions to my thinking about this topic, as well as Huston Paschal, of the North Carolina Museum of Art, for her thoughtful reading of and response to this essay.

1. On these works, see Gerald Silk, "Our Future Is in the Air," manuscript, to appear in Dominick A. Pisano, ed., *The Airplane in American Culture* (Ann Arbor: University of Michigan Press, forthcoming), 14–15; Robert Wohl, *A Passion for Wings: Aviation and the Western Imagination, 1908–1918* (New Haven, Conn.: Yale University Press, 1994), 272; Francis Frascina, "Realism and Ideology: An Introduction to Semiotics and Cubism," in Charles Harrison, Francis Frascina, and Gil Perry, *Primitivism, Cubism, Abstraction* (New Haven, Conn.: Yale University Press, 1993), 155–57; Jeffrey S. Weiss, "Picasso, Collage, and the Music Hall," in Kirk Varnedoe and Adam Gopnik, eds., *Modern Art and Popular Culture: Readings in High and Low* (New York: Museum of Modern Art, 1990), 96; and William Rubin, *Picasso and Braque: Pioneering Cubism*, exhibit catalog (New York: Museum of Modern Art, 1989), 33–34.

2. Silk, "Our Future," 15.

3. Rubin, *Picasso and Braque*, 33 (translation mine).

4. Ibid.; Silk, "Our Future," 15. See Picasso's letters to Kahnweiler of Aug. 11, 1912, and Aug. 15, 1912, in which he makes reference to "Wilbourg Braque"; and see Braque's letter to Kahnweiler of Aug. 16, 1912, signed "Your Wilburg [*sic*] Braque," Rubin, *Picasso and Braque*, 401–2.

5. See Silk, "Our Future," 14–16; Weiss, "Picasso, Collage, and the Music Hall," 96; Rubin, *Picasso and Braque*, 32–34; Gertrude Stein, *Picasso* (London: B. T. Batsford, 1938), 49–50; on Stein's comparison of cubist imagery to "the earth seen from an airplane," see Stephen Kern, *The Culture of Time and Space, 1880–1918* (Cambridge, Mass.: Harvard University Press, 1983), 245.

6. Rubin, *Picasso and Braque*, 33.

7. Tansey was inspired in part to do this painting by William Rubin's discussion of the links between flight and cubism in the 1989 Museum of Modern Art exhibition, "Picasso and Braque: Pioneering Cubism" (see n. 1 for catalog reference) (Judi Freeman, "Metaphor and Inquiry in Mark Tansey's 'Chain of Solutions,'" in *Mark Tansey*, exhibit catalog, Los Angeles County Museum of Art, June 17–Aug. 29, 1993 [San Francisco: Chronicle Books, 1993], 61).

8. William Rubin points out that in an unpublished manuscript by Jean Paulhan of 1939–45 "Braque tells us specifically that the 'scaffoldings' of his paper sculptures reminded Picasso of Wright's 'biplanes'" (Rubin, *Picasso and Braque*, 33).

9. I am indebted to Tom Crouch for making me aware of this relationship (personal communication, Nov. 20, 2001). The photograph of Herring in Chanute's glider and of the Wright Brothers' first flight in 1903 are reproduced in Marvin W. McFarland, ed., *The Papers of Wilbur and Orville Wright*, vol. 1 (New York: McGraw-Hill, 1953), plates 55 and 70, respectively.

10. I thank Tom Crouch for these observations (personal communication, Nov. 20, 2001).

11. Silk, "Future Is in the Air"; Robert Wohl, *A Passion for Wings*, 157–201; Julie H. Wosk, "The Aeroplane in Art," *Art and Artists*, no. 219 (Dec. 1984): 23–28.

12. According to Calder's autobiography, he did not actually witness the landing, speculating, "I guess he came down behind us." The day turned out to be eventful for Calder for another reason. He met an American banker from Wisconsin who introduced him to the Gould Manufacturing Company, which would later manufacture toys by Calder (Alexander Calder, *An Autobiography with Pictures* [New York: Pantheon Books, 1966], 83–84).

13. In 1929, the same year Calder created his sculpture, *The Spirit of St. Louis*, Filippo Tommaso Marinetti founded a new movement he labeled "Futurist Aeropainting," a movement allied with Benito Mussolini's Fascist government and devoted to the cult of flight (Silk, "Our Future," 10–11).

14. I thank Dodge Thompson for this interpretation (personal communication, summer 2001).

15. Marcel Duchamp suggested this name in late 1931 when he saw a motor-driven sculpture in Calder's studio. The double entendre implicit in the term seems to have

pleased Calder: "In addition to something that moves, in French [mobile] also means motive." (Calder, *Autobiography*, 126–27).

16. Ibid., 118.

17. French painter Robert Delaunay depicted Blériot's 1909 triumph in his *Hommage à Blériot* of 1914. On the painting, see Silk, "Our Future," 12–13, and Wohl, *Passion for Wings*, 192–99.

18. Calder originally quoted by Yvon Taillandier in *XXe Siècle*, Mar. 1959, reprinted in Daniel Marhesseau, *The Intimate World of Alexander Calder*, trans. Eleanor Levieux and Barbara Shuey (Paris: Solange Thierry, 1989), 232.

19. Robert Osborne, Foreword to Calder, *Autobiography*, 7, Quoted in part in Jean Lipman, *Calder's Universe* (New York: Viking Press, 1976), 186.

20. Lipman, *Calder's Universe*, 187.

21. Marla Prather, *Alexander Calder, 1898–1976*, exhibit catalog (Washington, D.C.: National Gallery of Art, Mar. 29–July 12, 1998), 284; Richard B. K. McLanathan, *East Building, National Gallery of Art: A Profile* (Washington, D.C.: National Gallery of Art, 1978), 31.

22. Richard Serra, opening statement in *Richard Serra: Sculpture*, exhibit catalog, New York: Pace Gallery, Sept. 15–Oct. 14, 1989, n.p.

23. Serra, *Richard Serra: Sculpture*, n.p.

24. Quoted in Tom D. Crouch, *The Bishop's Boys: A Life of Wilbur and Orville Wright* (New York: W. W. Norton, 1989), 284–85.

25. "Interview: Richard Serra and Bernard Lamarche-Vadel," New York City, May 1980, in Clara Weyergraf, ed., *Richard Serra: Interviews, etc., 1970–1980* (Yonkers, N.Y.: Hudson River Museum, 1980), 142.

26. For one interpretation of Brancusi's fascination with flight, see Anna Chave, *Constantin Brancusi: Shifting the Bases of Art* (New Haven, Conn.: Yale University Press, 1993), 119–20.

27. "Interview: Richard Serra and Friedrich Teja Bach," Mar. 14, 1975, in Weyergraf, *Richard Serra*, 53–55.

28. Crouch, *Bishop's Boys*, 260.

29. The effect is not unlike Kazimir Malevich's renditions of airplanes, as in *Suprematist Composition: Airplane Flying*, 1915. On Malevich's interest in the airplane, see Wohl, *Passion for Wings*, 157–79.

30. Serra, *Richard Serra: Sculpture*, n.p.

31. Let Geerling, "Exploring Space at Very Close Quarters," in *Metafor och materia: I vetenskapens närhet: Panamarenko, Rollof, Shannon*, exhibit catalog (Stockholm: Moderna Museet, Mar. 23–May 5, 1991), 81.

32. Wim Van Mulders, "On Earth and in the Air: Panamarenko," *Artforum* 25, no. 8 (Apr. 1987): 95; Hans Theys, *Cars and Other Stuff* (Tokyo: Galerie Tokoro, 1993), 48.

33. Jon Thompson, "Panamarenko: Artist and Technologist," in *Panamarenko*, exhibit catalog (London, Hayward Gallery, Feb. 10–Apr. 2, 2000), 38.

34. Bicycles were of particular interest to Panamarenko while he worked out his interest in space and flight in the mid-1960s. His 1965 *Swiss Cycle Collage* announced many of the cosmic themes his work would pursue in years to come, and in 1967, the

same year he built *Das Flugzeug*, Panamarenko created his *Swiss Bicycle* sculpture. See Thompson, "Panamarenko," 20, 22, 24.

35. For a discussion of Leonardo da Vinci's *Standing Ornithopters*, see Charles H. Gibbs-Smith, *Leonardo da Vinci's Aeronautics: A Science Museum Booklet* (London: Her Majesty's Stationery Office, 1967), 14–17.

36. With his choice of such an unconventional material for the passenger compartment, Panamarenko had safety in mind, believing that collisions between such baskets would not be hazardous (Van Mulders, "On Earth and in the Air," 96).

37. Ibid., 98.

38. Quoted in Stroop, "Short Survey of Panamarenko's Life," 26 (emphasis is Panamarenko's).

39. Quoted in Geerling, "Exploring Space at Very Close Quarters," 86.

40. Kiefer's sculpture was originally called *Poppies and Memories* after a volume of poetry by the German-Jewish poet Paul Celan. Celan survived World War II but lost his parents to the Holocaust. He later took his own life. The work was renamed in 1994 during a conversation between Kiefer and Mark Rosenthal (Melissa Geisler, Note for Curatorial File, Mar. 30, 1994; Curatorial File, Anselm Kiefer: *Angel of History*, National Gallery of Art, Washington, D.C.). It should be noted that a second sculpture entitled *Poppies and Memories* exists. While having similar physical attributes, the "airplane" structure of this second version is different from that of the sculpture owned by the National Gallery of Art. One commentator describes this second version of the sculpture as resembling a B-1 bomber. See Ida Panicellis, "Theory of Flight," *Artforum* 28, no. 5 (Jan. 1990): 132.

41. The B-52 bomber has strong historical associations. It played an important role in the Cold War, being used by the United States to contain the spread of communism, a struggle that split Kiefer's native land asunder. See Peter R. March, *Warplanes: A Hundred Years of Military Aviation* (London: Cassell, 2000), 176. Director Stanley Kubrick attests to this aspect of the B-52's strategic importance by giving the aircraft a prominent role in his 1964 black comedy, *Dr. Stangelove; or, How I Learned to Stop Worrying and Love the Bomb*. On Kubrick's use of the B-52, see Charles Maland, "*Dr. Strangelove* (1964): Nightmare Comedy and the Ideology of the Liberal Consensus," in Peter C. Rollins, ed., *Hollywood As Historian: American Film in a Cultural Context* (Lexington: University Press of Kentucky, 1983), 203. Today, of course, the B-52 continues to play an important role in the air war against terrorism.

42. Walter Benjamin, "Theses on the Philosophy of History," in Hannah Arendt, ed., *Illuminations* (New York: Schocken Books, 1968), 257. Benjamin's vision of the "angel of history" is inspired by Paul Klee's 1920 watercolor *Angelus Novus*. Benjamin writes: "A Klee painting named 'Angelus Novus' shows an angel looking as though he is about to move away from something he is fixedly contemplating. His eyes are staring; his mouth is open, his wings are spread. This is how one pictures the angel of history. His face is turned toward the past. Where we perceive a chain of events, he sees one single catastrophe which keeps piling wreckage upon wreckage and hurls it in front of his feet. The angel would like to stay, awaken the dead, and make whole what has been smashed. But a storm is blowing from Paradise; it has got caught in his wings with such violence

that the angel can no longer close them. The storm irresistibly propels him into the future to which his back is turned, while the pile of debris before him grows skyward. This storm is what we call progress." It is worth noting that Klee himself was a pilot during World War I; for a discussion of the effect of this experience on his imagery, see O. K. Werkmeister, "Walter Benjamin, Paul Klee, and the Angel of History," *Oppositions* 25 (Fall 1982): 102–13.

43. The poppies also serve as a direct reference to Paul Celan. See note 40, above.

44. The exhibition "The Angel of History" took place in Cologne at the Galerie Paul Maenz from November 17, 1989, to January 13, 1990. A related exhibition, "Poppies and Memory," took place in Warsaw at the Galeria Foksal in December 1989.

45. Ida Panicelli, "Theory of Flight," *Artforum* 28 (Jan. 1990): 131.

46. Steven Henry Madoff, "Anselm Kiefer: A Call to Memory," *Art News* 86 (Oct. 1987): 130.

47. On Kiefer's interest in alchemy, see Mark Rosenthal, Anselm Kiefer, exhibit catalog, Art Institute of Chicago, Dec. 5, 1987–Jan. 31, 1988 (Philadelphia: Philadelphia Museum of Art, 1987), 115. On Kiefer's use of aging to evoke alchemical associations, see James Hyman, "Anselm Kiefer As Printmaker—II: Alchemy and Woodcut, 1993–1999," *Print Quarterly* 17 (2000): 38.

48. Rosenthal and Kiefer, exhibit catalog, 80.

49. Quoted by Peter Winter, "Whipping Boy with Clipped Wings," *Art International*, Spring 1988, 70.

50. Enrique Juncosa, "The Flight of Icarus (Or a Pilot among the Angels)," in *Malcolm Morely, 1965–1995*, exhibit catalog (Madrid: Sala de Exposiciones de la Fundación "la Caixa," Sept. 20–Nov. 12, 1995).

51. Sarah Whitfield, *Malcolm Morley in Full Colour*, exhibit catalog (London: Hayward Gallery, June 15–Aug. 27, 2001), 100–101.

52. Whitfield, *Malcolm Morley*, 98.

53. Juncosa, "Flight of Icarus," 32; cf. Michael Compton, "Malcolm Morley," *Malcolm Morley: Paintings, 1965–82*, exhibit catalog (London: Whitechapel Gallery, Jan. 22–Feb. 27, 1983), 8.

54. Juncosa, "Flight of Icarus," 34.

55. Michael Klein, "Malcolm Morley: Traveling in Styles," *Artnews* 82 (Mar. 1983): 93, quoted in Juncosa, "Flight of Icarus," 36.

56. The impact of flight on recent art will be the subject of an exhibition at the North Carolina Museum of Art in Raleigh, "Defying Gravity: Contemporary Art and Flight," November 2, 2003–March 7, 2004. A catalog with the same title, authored by Huston Paschal and Linda Dougherty, with contributions by Robert Wohl, Anne Collins Goodyear, and Laura André, will accompany the show. In a response to a questionnaire sent to him by Douglas Davis, Jean Tinguely listed Wernher von Braun as one of the artists whose "work now with machine and technological methods" he "most admire[d]" (Jean Tinguely, completed questionnaire for Douglas Davis, Dec. 1, 1969, reproduced in Douglas Davis, *Art and the Future* [New York: Praeger, 1973], 125). Tinguely's 1966 statement was quoted in Laura Hoptman and Michael Carter, "Useless Science," in *Making Choices*, exhibit brochure (New York: Museum of Modern Art, 2000), n.p.

CHAPTER 12
THE *SPIRIT OF ST. LOUIS*—FACT AND SYMBOL
MISINTERPRETING A HISTORIC CULTURAL ARTIFACT

The New York–Paris flight is past. I suppose that it helped aviation by interesting the people, and probably it had a certain amount of pioneering value. But we need to go ahead in commercial flying, and we should not live in the past. I wish that people would just remember my flight to Paris as something that happened in 1927, and then forget about me.—Charles A. Lindbergh to Donald E. Keyhoe, in Donald E. Keyhoe, Flying with Lindbergh

On entering the Smithsonian Institution's National Air and Space Museum and seeing Charles A. Lindbergh's Ryan NY-P *Spirit of St. Louis*, a visitor will not be fully conscious of its meaning as an object of American culture. One cannot fathom simply by looking at this historic artifact its true significance or the complex and often controversial character of Lindbergh, who is for-ever identified with it.

The exhibit label for the *Spirit* is spare. It notes that on May 21, 1927, the aircraft made the first nonstop solo transatlantic flight. The flight was made from New York's Roosevelt Field to Paris, France, and it was 3,610 miles long. Lindbergh flew for thirty-three hours, thirty minutes. He became the first person to fly nonstop and solo across the Atlantic. In so doing Lindbergh won the $25,000 prize offered by hotel owner Raymond Orteig. After the transatlantic flight, the *Spirit* made a tour of the United States and South and Central America. The aircraft was named for St. Louis, Missouri, the city where Lindbergh's financial backers were based. The label then goes on to describe who designed the aircraft—Donald Hall (with assistance from Lindbergh)—and notes how, because the fuel tanks were located ahead of the cockpit, the pilot's forward vision was blocked; he had to look out a side window or through a periscope. Finally, the label provides specifications of the aircraft—wingspan, length, height, weight gross and empty—and infor-mation about the engine and the aircraft's manufacturer.

While this information is useful, it will not enable one to get at the heart of the meaning of the artifact, the real significance of Lindbergh's flights in it, the contributions he made to aviation in a critical period of its devel-opment, the controversy that surrounded him before America's entry into

World War II, and the puzzling questions of American heroism and celebrity. In many ways, the artifact is misleading: the complexity of Lindbergh and the *Spirit of St. Louis* remain to be discovered.

Steven Lubar, a curator at the Smithsonian Institution's National Museum of American History, has commented that it is not enough to understand merely how machines work, and that this is only the means to an end; the end being "to better understand some bigger issue, whether it be, on the one hand, cultural change, social change, class, ethnicity, gender, race—the big questions of American history, or, on the other hand, the nature of technological knowledge, the relations of science and technology, or the processes of technological change or technological creativity and design—the key issues of the history of technology." Prompted by Lubar's exhortations to historians of technology to find a bigger issue with which to grapple than the artifact in isolation, I will attempt to provide some context for the *Spirit of St. Louis* as an artifact.[1]

It is not my intention here to review the facts of Lindbergh's life before or after he became famous or to delve into the controversy that often surrounded him. There are a number of good biographies that do an admirable job of that. Rather, I would like to discuss Lindbergh and his airplane in factual and symbolic terms and to ask the following questions. What is the factual weight, if you will, of the artifact? What is the symbolic weight of the artifact? How does the artifact's factual weight square with its symbolic weight? To do this, one needs to review some of the facts about Lindbergh and the *Spirit of St. Louis*. Then one must proceed to the flights themselves—the transatlantic flight, the *Spirit*'s tour of the United States for the Daniel Guggenheim Fund for the Promotion of Aeronautics, and, finally, its Latin American tour, undertaken on behalf of the U.S. ambassador to Mexico, Dwight Morrow. Then I would like to offer some speculations about its importance and its interpretation as an artifact of American culture.

The *Spirit of St. Louis* and Its Travels

In September 1926, on a flight from Peoria, Illinois, to Chicago for the Robertson Aircraft Corp., a contract airmail firm in St. Louis, Lindbergh began to think about the possibility of flying across the Atlantic. In his mind the only possibility of success lay in a solo flight: "I'll fly alone. That will cut out the need for any selection of crew, or quarreling. If there's upholstery in the cabin, I'll tear it out for the flight. I'll take only the food I need to eat, and a few concentrated rations. I'll carry a rubber boat for emergency, and a little extra water."[2]

He knew he would need financial backing, and so he arranged to see Earl Thompson, an insurance executive whom he had taught to fly. Lindbergh had it in mind to purchase for the flight an already existing single-engine airplane designed and built by Giuseppe Bellanca, a noted aircraft designer and manufacturer, and powered by a Wright Whirlwind radial engine. His preference for a single-engine craft was motivated by his desire to reduce his exposure to engine failure and his desire to keep weight to a minimum. When he discussed the idea with Thompson, Thompson listened intently but questioned using the Bellanca, a landplane, and suggested a flying boat or trimotor aircraft. Lindbergh explained the disadvantages of such craft, and by the end of the conversation, Thompson had become interested in the idea.[3]

Lindbergh also toyed with the idea of having the Fokker Company build the airplane, and when he spoke with a company representative about it, the man suggested a price of $90,000 to $100,000 for a multiengine air-craft. When Lindbergh persisted in his idea of a single-engine aircraft, he was turned down. The more engines, he thought, the higher the chances of engine failure. Also, additional engines added weight to the aircraft. This turn of events did not deter him, and he made an appointment to see Maj. Albert Bond Lambert, who had commanded a school for balloon pilots during World War I and was owner of Lambert Field in St. Louis. Lambert promised Lindbergh $1,000 for the flight. Next Lindbergh went to his boss, Maj. Bill Robertson, to make arrangements to have someone fly his mail route while he was making the transatlantic flight. Although Robertson could not pledge financial support, he approved of the idea and suggested that the St. Louis *Post-Dispatch* might finance the flight.[4]

An editor at the *Post-Dispatch* told Lindbergh that the paper would never finance such a dangerous flight, but the editor's response only strengthened Lindbergh's resolve. Lindbergh next visited the Wright Aeronautical Corporation in Paterson, New Jersey, where he inquired about the possibility of acquiring the Whirlwind-powered Bellanca for the flight, but he was told that Wright Aeronautical had built the Bellanca merely to demonstrate the capabilities of their engine. The Wright company representative suggested that Lindbergh speak with the aircraft's designer, Giuseppe Bellanca, which he did the following evening at the Waldorf-Astoria hotel in New York. While Bellanca made no commitment, Lindbergh left New York confident that that Bellanca would support him.[5]

He next visited Harry H. Knight of Knight, Dysart and Gamble, a brokerage firm in St. Louis. Knight asked Harold Bixby, a vice president of the State National Bank in St. Louis and president of the St. Louis Chamber of Commerce, to come to his office. Bixby agreed and, in a matter of weeks, Knight and Bixby made the decision to back Lindbergh's attempt.[6]

Not having heard from Bellanca, Lindbergh decided to inquire of the Travel Air Company in Wichita, Kansas, about an aircraft built to his specifications. Travel Air refused, and so in February 1927 he contacted a small company in San Diego, Ryan Airlines, which had built mail planes for the West Coast route: "Can you construct Whirlwind engine plane capable flying nonstop between New York and Paris stop if so please state cost and delivery date." Ryan replied that the company was capable of building an aircraft similar to its model M, but with larger wings, that would meet Lindbergh's needs. Meanwhile, Giuseppe Bellanca wired encouragingly that he would be willing to make Lindbergh an "attractive proposition" concerning a New York–Paris aircraft. In addition, Knight and Bixby had come up with the $15,000 needed to finance the flight. It was at that point that Bixby suggested to Lindbergh that he name the aircraft the *Spirit of St. Louis*. When Lindbergh went to visit Clarence Chamberlin, a pilot for Columbia Aircraft, Bellanca's affiliate and representative, with the money, he was told that the company would close the deal only if it had the right to select the crew who would fly the aircraft. Lindbergh departed, angry that he had wasted his time and money in attempting to obtain the Bellanca. At Harold Bixby's urging, Lindbergh decided to go to San Diego to discuss the manufacture of an aircraft with Ryan Airlines.[7]

Lindbergh was impressed by the sincerity of the assurances of B. F. Mahoney and Donald Hall, Ryan's president and its chief engineer, respectively, that the company could build the *Spirit of St. Louis* in the time necessary for Lindbergh to make the transatlantic attempt. Lindbergh and Hall set about designing the aircraft, designated the Ryan NY-P (New York–Paris) monoplane. After the flight, Hall wrote a technical note for the aircraft that was published by the National Advisory Committee for Aeronautics. Hall stated that "the development of the airplane was begun with the idea of using a standard model Ryan M-2 [a mail plane] and making modifications to suit the special purpose. Upon Colonel Lindbergh's arrival at the factory, it was quickly determined that modification of the M-2 was less practicable than redesign. Colonel Lindbergh laid out the following basic specifications: That the airplane should be a monoplane type, powered with a single Wright J5C engine, have a good power reserve on take-off when carrying more than 400 gallons of gasoline and must have the pilot located in rear of all tanks for safety in a forced landing." Lindbergh later commented on Hall's persistence and attention to detail. "Donald Hall worked for one stretch of thirty-six hours, without sleep. The only good-natured grumbling I've heard in the

Charles A. Lindbergh flew this Ryan NY-P *Spirit of St. Louis* nonstop from New York to Paris in the period May 20–21, 1927. (NASA photo no. L-3094)

shop was when Hall sent down drawings that called for fuselage fairings to an accuracy of one thirty-second of an inch. Then Superintendent Bert Tindale remarked that he'd never before been asked to hold such accuracy. But I saw him working there the rest of the afternoon with a scale on which the inches were divided into thirty-seconds." In a matter of months from Lindbergh's initial contact with Ryan, the *Spirit* was designed and built. Lindbergh test-flew the aircraft twenty-three times in flights varying from five minutes to more than an hour. On May 4, he flew the aircraft nine times with assorted quantities of fuel, ranging from thirty-eight to three hundred gallons.[8]

After waiting for the weather to clear, on May 10 Lindbergh flew the *Spirit* 1,600 miles across the continent, making a single stop in St. Louis. On the way from St. Louis to New York, Lindbergh later recalled the landscape below and his thoughts on the upcoming flight: "Land below is rolling. Shadows of white, fluffy clouds slide across alfalfa, corn, and wood lot. I've caught up with the tail of the storm that delayed my San Diego take-off. Will the Allegheny Mountains be clear? Maybe I started too early. Maybe I can't get through to New York today. But I've got to follow as closely as I can on the area of bad weather. If only it will hover over the Atlantic and hold my competitors on the ground for two or three days more, I'll have the engine checked, the compasses in, and the *Spirit of St. Louis* refueled." He landed at Curtiss Field, near Mineola, on Long Island, setting a transcontinental flying record of twenty-one hours, forty minutes. He was now ready to undertake the most important flight of his life.[9]

In 1902, the year of Lindbergh's birth, a former French-born maitre d' named Raymond Orteig had purchased the Brevoort Hotel on Fifth Avenue between East 8th and 9th streets in New York, renovated it, and made it into a haven for Greenwich Village artists and writers. In the same year he purchased the nearby Lafayette Hotel at University Place and East 9th Street, which became an attraction for international celebrities, who were drawn by Orteig's carefully selected menus and wine lists. During World War I, he had become enthralled by the adventuresome "Knights of the Air" who flew in the war. Kenneth Davis describes Orteig's fascination as follows: "Brave young men placed themselves with gay defiance, between the very wings of death and soared upon them into a clean sweet sky far above the sordid horrors of Verdun, the Aisne, the Somme. In such skies they met and killed their enemies, or were themselves killed, in duels which measured their personal courage and their skill as pilots. To them and their battles adhered many of the emotions and some of the forms of ancient chivalry. Guynemer, Ball, Richthofen, Mannock, McCudden, Lufbery, all great aces, all killed; Fonck, Nungesser, Udet, Bishop, Rickenbacker, all great aces who, miraculously, survived—these became knightly heroes, not unlike Galahad and Roland and Lancelot, to Orteig's mind as to the minds of millions of others. Particularly were his emotions stirred by the young men of the Lafayette Escadrille, Americans flying the French air force, whose brave deeds, fully exploited by the masters of Allied propaganda, were as so many bonds of ardent affection between France and the United States."[10]

Beyond this, Orteig, like Jacques Schneider, who had offered the Schneider Cup in the hope of encouraging over-water flying in 1913, believed that the future of aviation lay in the peaceful pursuit of transoceanic flying. So, in 1919, through the Aero Club of America, Orteig decided to offer a prize of $25,000 "to the first aviator who shall cross the Atlantic in a land or water aircraft (heavier-than-air) from Paris or the shores of France to New York, or from New York to Paris or the shores of France, without stop," with the stipulation that the flight be made within five years. In April 1913, Alfred Harmsworth, owner of the *London Daily Mail* and *London Daily Mirror*, and who would later be known as Lord Northcliffe, offered a 10,000-pound prize for the first person to cross the Atlantic in an airplane from any point in Great Britain or Ireland to any point in Newfoundland, Canada, or the United States.[11]

While no transatlantic flights had been made that conformed to the New York–Paris/Paris–New York requirement laid down by Orteig, it is estimated that 117 people had crossed the Atlantic by air before Lindbergh successfully made the solo crossing in May 1927. The first of them, on June 14, 1919, flown by two British aviators, Captain John Alcock and Lt. Arthur Whitten Brown in a Vickers Vimy, from Trespassy Bay, St. John's, Newfoundland, to Clifden, Ireland, was a difficult 1,936-mile affair that took fifteen hours, fifty-seven minutes nonstop. In completing their flight, Alcock and Brown won the *Daily Mail* prize. About a month earlier, in May 1919, U.S. Navy Curtiss NC flying boats, led by Lt. Cdr. Albert C. Read, made the crossing from St. John's, Newfoundland, to the Azores, to Lisbon, to Plymouth, England, in approximately three weeks. Only one of the aircraft, the NC-4, completed the flight. In July 1919 the British airship R-34 completed the first successful airship crossing of the Atlantic. In 1924, in their around-the-world flight, three Army Air Service Douglas World Cruisers crossed the Atlantic, going from Nova Scotia to Brough, England, by way of Greenland and Iceland.[12]

Meanwhile, in the five years since Orteig had offered it, nobody had been able to do what was required to win the prize, primarily because the technological limitations of aircraft would not allow it, and so Orteig renewed it for an additional five years. In addition, no one had successfully crossed the Atlantic nonstop in an airplane since Alcock and Brown in 1919. By September 1926 it appeared that the French ace René Fonck and a crew of four were ready to compete for the prize. They took off in a three-engine Sikorsky S-35 biplane, but the aircraft, loaded down with fuel and baggage and encumbered with a supplementary landing gear, crashed on takeoff when the second gear failed to release. Fonck's mechanic and wireless operator were killed, but he and his copilot Lawrence Curtin managed to survive. By 1927 there were some strong contenders, including Comdr. Richard Byrd, USN, Noel Davis, and Stanton Wooster, naval aviators with much experience; Clarence Chamberlin, another seasoned pilot, and Charles Levine, his financial backer and owner of their aircraft; and French ace Charles Nungesser and François Coli.[13]

At the time he conceived the idea of flying the Atlantic, Lindbergh had been interested more in the accomplishment than in winning the Orteig Prize. Nevertheless, he was aware that he only had a small window of opportunity to get the jump on his competitors for the New York to Paris flight. Tragically, Davis and Wooster had been killed on April 26 when their Keystone Pathfinder *American Legion* crashed on takeoff on its last test flight. Nungesser and Coli, who had taken off from Le Bourget airport in Paris in their Levasseur PL-8 *Oiseau Blanc* on May 8, had disappeared somewhere over Newfoundland. Richard Byrd and his pilot Floyd Bennett were poised to make the attempt in their Fokker Trimotor *America*, financed by Rodman Wanamaker. Finally, Clarence Chamberlin and Charles Levine, in the Wright-Bellanca W.B.2 that Lindbergh had hoped to acquire for the transatlantic flight, were also viable candidates for the flight, regardless of legal battles and personality conflicts. The press, however, believed Lindbergh to be the front-runner.[14]

After he arrived in New York on May 12, Lindbergh set about making several days of final preparations and waiting for the weather to clear over the transatlantic route. He later recollected his thoughts of the evening of Monday, May 16, four days before the flight, providing an eerie harbinger of what his life would be like after he made the flight. "The Spirit of St. Louis is ready to take off, but my route to Paris is still covered with fog and storm. These have been the most extraordinary days I've ever spent; and I can't call them very pleasant. Life has become too strange and hectic. The attention of the entire country is centered on the flight to Paris, and most of all on me—because I'm going alone, because I'm young, because I'm a 'dark horse.' Papers in every city and village are headlining my name and writing articles about me. Newspaper, radio, and motion-picture publicity has brought people crowding out to Curtiss and Roosevelt Fields until the Nassau Country police are faced with a major traffic problem. Seventy-five hundred came last Saturday, the *New York Times* said. On Sunday, there were thirty thousand!"[15]

The night before the flight, Lindbergh had intended to see a Broadway review to ease some of his tension, but the weather forecast made it appear that the skies would clear sufficiently for him to take off from Roosevelt Field the next morning. He therefore returned to his hotel, but he found that he could not sleep. He rose well before dawn and went to the airfield, which was full of mud and enveloped in a drizzling rain. Nevertheless, he made the decision to take off a few minutes before 8:00 A.M., clearing the telephone wires at the end of the runway by a mere twenty feet. He set his course over New England, Nova Scotia, Newfoundland, Ireland, England, and Paris.[16]

The flight across the Atlantic was an especially grueling one, during which Lindbergh continually struggled to stay awake and encountered a variety of weather—clouds, fog, storms, ice. He later recalled his apprehension during the fourteenth hour of the journey as he flew through clouds. "Great cliffs tower over me, ward me off with icy walls . . . to plunge into these mountains of the heavens would be like stepping into quicksand. They enmesh intruders. They toss you in their inner turbulence, lash you with their hailstones, poison you with freezing mist. It would be a slow death, a death one would have long minutes to struggle against, trying blindly to regain control of an ice-crippled airplane, climbing, stalling, diving, whipping, always downward toward the sea."[17]

During the twenty-second hour, Lindbergh began to imagine apparitions and to hear voices in the cockpit of the *Spirit of St. Louis*, an occurrence that he had omitted from *We*, his account of the transatlantic flight published in 1927. In his Pulitzer Prize–winning book, *The Spirit of St. Louis*, Lindbergh recalled

> While I'm staring at the instruments, during an unearthly age of time, both conscious and asleep, the fuselage behind me becomes filled with ghostly presences—vaguely outlined forms, transparent, moving, riding weightless with me in the plane. I feel no surprise at their coming. There's no suddenness to their appearance. Without turning my head, I see them as clearly as though in my normal field of vision. There's no limit to my sight—my skull is one great eye, seeing everywhere at once. . . . These phantoms speak with human voices—friendly, vapor-like shapes. . . . Now, many are crowded behind me. Now, only a few remain. First one and then another presses forward to my shoulder to speak above the engine's noise, and then draws back among the group behind. At times voices come out of the air itself, clear yet far away, traveling through distances that can't be measured by the scale of human miles; familiar voices, conversing and advising on my flight, discussing problems of my navigation, reassuring me, giving me messages of importance unattainable in ordinary life.[18]

During the twenty-seventh hour, Lindbergh sighted some fishing boats, which led him to believe that the coast of Europe could not be far away. He descended in a circling pattern and called out "WHICH WAY IS IRELAND?" No one answered, and so he straightened the *Spirit of St. Louis* out and headed on in an eastward direction. During the twenty-eighth hour he passed over Ireland: "Ireland, England, France, Paris! The night at Paris! *This* night at Paris—less than six hours from now—*France and Paris*! It's like a fairy tale. Yesterday I walked on Roosevelt Field; today I'll walk on Le Bourget." During the thirty-first hour he spotted the English Channel: "The channel coast of England is gliding range ahead; the coast of France, an hour's flight beyond. Memories of school texts and childhood stories flood my mind. These are the countries of Robin Hood and King Arthur, of Henry the Eighth and the Redcoats, of Joan of Arc and Lafayette and Napoleon."[19]

By the thirty-fourth hour, Lindbergh had reached France, but he was unsure whether the airport he spotted below was actually Le Bourget (the airport was not recorded on his map), so he circled and began his approach. "If I land at the wrong field, it won't be too serious an error—as long as I land safely. I look around once more for other floodlights or a beacon. There are none—nothing even worth flying over to investigate. I spiral lower, left wing down, keeping close to the edge of the field." All the while he was going over in his mind the difficulties he was having in landing the *Spirit of St. Louis*—"I've never landed the Spirit of St. Louis at night before. It would be better to come in straight. But if I don't slideslip, I'll be too high over the boundary to touch my wheels in the area of light. . . . Still too high. I push the stick over to a steeper slip, leaving the nose well down—Below the hangar roofs now. . . . Sod coming up to meet me. . . . The Spirit of St. Louis swings around and stops rolling, resting on the solidness of earth, in the center of Le Bourget. I start to taxi back toward the floodlights and hangars—But the entire field ahead is covered with running figures!"[20]

In *The Spirit of St. Louis* Lindbergh wrote that his

reception by the French people, in 1927, cannot be compressed into a final chapter of this book. After the warnings I had been given in America, I was completely unprepared for the welcome which awaited me on Le Bourget. I had no idea that my plane had been so accurately reported along its route between Ireland and the capital of France—over Dingle Bay, over Plymouth, over Cherbourg. When I circled the aerodrome it did not occur to me that any connection existed between my arrival and the cars stalled in traffic on the roads. When my wheels touched earth, I had no way of knowing that tens of thousands of men and women were breaking down fences and flooding past guards.

Moreover, Lindbergh could not have known when the touched down in Paris the immediate and lingering reaction his flight would provoke.[21]

A crowd estimated at one hundred thousand people swarmed the *Spirit of St. Louis*. They lifted Lindbergh on their shoulders and manhandled him in the air. Two French fliers, Michel Détroyat and George Delage, rescued him and took him to a hangar where he would be safe from the crowd. While in the hangar, Lindbergh met Myron Herrick, the U.S. ambassador to France. Lindbergh requested Herrick's assistance to ensure that the *Spirit of St. Louis* be given a place safe from the crowd and souvenir hunters. After an early-morning supper at the ambassador's residence, Lindbergh finally went to sleep at 4:15 A.M. on the morning of May 22 after having been awake for sixty-three hours. Over the next two months, Lindbergh was overwhelmed by crowds on his visits to Belgium and England, where he was received by the monarchs of those countries. President Calvin Coolidge sent the U.S.

Navy cruiser *Memphis*, flagship of the European fleet, to fetch him and bring him back to the United States.[22]

There is no need to go into the aftermath of the transatlantic flight here, except to note that it had an enormous effect on the public's perception of aviation. Many commentators have given the transatlantic flight more credit than it deserves as a spur to the aviation business. The historiographical debate that surrounds this question must take into account a number of other contributory factors. They include the enactment of the Air Commerce Act in 1926, which enabled the federal regulation of aviation, as well as the passage of the Army Five Year Aviation Program Act and the Navy Five Year Program Act in the same year, which made it possible for both branches of the service to procure large numbers of aircraft. In addition, the aviation industry had been well on its way to rationalization since World War I, when the patent suits initiated by the Wright Brothers were settled, making it possible to carry on business without legal entanglements. Since that time the industry had seen the influence of the automotive industry and the formation of an interest group (the National Aeronautic Association), the growth of the Curtiss company and its relationship to Wall Street financial interests, and the professionalization of aeronautical engineers. All these events had greatly influenced aviation long before Lindbergh's transatlantic flight. Suffice it to say that the transatlantic flight awakened public interest in aviation, and their interest was reflected in several aspects of American culture—popular literature, films, advertising, industrial design—in the 1930s.

Wanting to take advantage of public interest, the Daniel Guggenheim Fund for the Promotion of Aeronautics undertook to promote a nationwide tour in which Lindbergh would pilot the *Spirit of St. Louis* to all forty-eight states to make the public aware of aviation's reliability and to promote the cause of commercial aviation. The fund had previously promoted a nationwide tour involving Comdr. Richard E. Byrd and his pilot Floyd Bennett upon their return from a successful flight over the North Pole in May 1926 in the Fokker Trimotor *Josephine Ford*. As it turned out, Byrd had to tend to his duties as a naval officer, and so Bennett agreed to fly the aircraft across the country.[23]

The Lindbergh tour would have roughly the same goals as that of the Bennett tour, but the goals were more specific: to stimulate public interest in aviation; to encourage use of existing commercial mail and passenger air services; and to promote the development of airports and communications by air. Ironically, Harry Guggenheim, the director of the fund, had not been very impressed with Lindbergh before the transatlantic flight, and he only reluctantly agreed to pose with him for a photographer at Curtiss

Field, Long Island, in the spring of 1927. He quickly changed his mind when he realized what a marvel of calculation and planning the flight had been. What better reasons to have Lindbergh fly around the country than to disabuse the public of the perception that aviation was unsafe and that all pilots were reckless daredevils. Dwight Morrow, a partner in J. P. Morgan and Co. and a friend of President Coolidge—who had chaired the President's Aircraft Board (the Morrow Board), established to study the state of aviation in the United States—was a trustee of the fund. Morrow believed that the fund could make good use of Lindbergh and he did not want Lindbergh to be exploited by interests he felt would be harmful to Lindbergh's image. Harry Guggenheim felt that the tour would be an appropriate use of Lindbergh's time and would benefit from the young pilot's reputation as an American hero. On July 19, 1927, Guggenheim and Lindbergh signed a contract in which it was agreed that Lindbergh would receive $50,000 for his services for flying around the country. An additional $18,721.27 would be disbursed to defray the expenses of the tour.[24]

The tour, which had the backing of the U.S. Department of Commerce's recently organized Aeronautics Branch, departed from Mitchel Field on Long Island on July 20, 1927. With Lindbergh were his flying companion from the Army Air Service and Robertson Aircraft Corporation, Philip R. Love, who flew the lead airplane, supplied by the Aeronautics Branch, accompanied by Lindbergh's aide, Donald E. Keyhoe, from the Department of Commerce and Theodore R. Sorenson, a mechanic from the Wright Aeronautical Corporation, who was later replaced by C. C. Maidment. Milburn Kusterer was the tour's advance man, and Ivy Lee, the Guggenheim's press agent, was responsible for public relations.[25]

Lindbergh insisted that the tour be on time at each of the destinations, and he managed to land promptly at 2:00 P.M. at every stop, with the exception of Portland, Maine, where he was delayed by fog. At each destination, Lindbergh gave a short speech about aviation, highlighting the fact that during 1926, on six airmail routes, more than 34 million letters had been carried with almost 4 million miles flown, with a 95 percent performance rate with only one fatality. From Mitchel Field, the tour proceeded to New England, westward toward the Great Plains and into the Northwest to Seattle, from there along the Pacific Coast to San Diego, then eastward to Texas and the South, and, finally, back to Mitchel Field on October 23, ninety-five days after it had begun. By the time it was over, Lindbergh had visited eighty-two cities and twenty-three state capitals, flown 22,350 miles, and dropped 192 messages. These messages, in the form of an elaborate scroll, contained the signatures of Lindbergh, Harry Guggenheim, and

William P. MacCracken Jr., assistant secretary of commerce for aeronautics over the message "we feel that we will be amply repaid for all of our efforts if each and every citizen in the United States cherishes an interest in flying and gives his earnest support to the airmail service and the establishment of airports and similar facilities."[26]

The fund estimated that 30 million had seen the *Spirit of St. Louis*, as compared to the newspapers, which placed the estimate at 50 million. The tour had logged 260 hours in the air. Lindbergh had made 147 speeches and attended 69 dinners. Some fifteen years later, aviation journalist Reginald M. Cleveland concluded that "the purpose of the tour had been achieved in full measure. Through its agency, millions of American citizens had been brought into personal touch with flying. The awakening of the nation to the possibilities of aviation at last was genuine. For the awakening, the Daniel Guggenheim Fund was in great measure responsible."[27]

More relevant for the study of Lindbergh and the *Spirit of St. Louis* is that the aircraft and its pilot were seen by some 50 million people in every state of the Union. This added to the public perception of Lindbergh as a hero and of the *Spirit* as a national icon even before it was enshrined in the Smithsonian Institution. Moreover, as a result of the tour, Lindbergh became a trustee of the Daniel Guggenheim Fund for the Promotion of Aeronautics, a very important position whereby he was able to rub shoulders with important and influential people. Finally, the tour made Lindbergh not merely the man who had made a solo crossing of the Atlantic by air, but the acknowledged expert on aviation, a role that he played nearly to the end of his life, and one which opened many doors to him in aviation—most notably, for that period of his life, his association with Transcontinental Air Transport and Pan American World Airways.

But the Guggenheim-sponsored tour of the United States was only the first step in showing the *Spirit of St. Louis* and its pilot to large numbers of people. A few short months after its return from the U.S. tour, Lindbergh and the aircraft were called upon to undertake a 9,500-mile, two-month journey from Washington, D.C., to Mexico, Guatemala, British Honduras, El Salvador, Honduras, Nicaragua, Costa Rica, Panama, Colombia, Venezuela, Virgin Islands, Puerto Rico, Dominican Republic, Haiti, and Cuba. Unlike the tour of the United States, however, the Latin American tour was brought about by the confluence of several events, all of which operated together to promote aviation in Latin America.[28]

As in the case of the U.S. tour, Dwight Morrow was the catalyst for Lindbergh's Latin American flight. Morrow had met Lindbergh on the young aviator's trip to Washington after his celebrated transatlantic flight. The two took a liking to each other. Because of his financial acumen and affiliation with the House of Morgan, Morrow soon became Lindbergh's financial adviser. Then, not long after having met Lindbergh, Morrow became ambassador to Mexico. Lindbergh offered his services to Morrow, and Morrow, realizing that he had a difficult situation on his hands in Mexico because relations between it and the United States had been strained, thought that a flight to Mexico by Lindbergh might help to mend the rift between the two countries.[29]

Before anything could happen, however, Morrow had to get approval from the State Department, which had serious reservations about allowing Lindbergh to make the trip. The department had rejected a trip planned by the Houston, Texas, Chamber of Commerce on behalf of aviator Frank Hawks, and a 1926–27 U.S. Army Air Corps Pan American flight had not been well received. The State Department, however, did allow Morrow to make inquiries about the kind of reception Lindbergh might be expected to receive, and he did so with positive results. Lindbergh himself had received an invitation from the American Legion in Cuba to visit Havana during the Sixth Pan American Meeting to be held in February 1928, and the invitation made Lindbergh enthusiastic about the trip. Undersecretary of State Robert E. Olds advised Morrow that there would be no objection to Lindbergh's undertaking the flight, and Olds suggested that the tour could be widened to include other places in Latin America. Olds's motive was to use Lindbergh to improve the delicate U.S. diplomatic position in Central America after its involvement in the domestic affairs of Nicaragua, which had created a great deal of bad feeling toward the United States.[30]

It appears that Lindbergh and the *Spirit*'s trip to Mexico, and, subsequently, to Central America, South America, and the Caribbean, were coincidentally part of a larger strategy to introduce a U.S. airline in those parts of the Western Hemisphere. Historian Wesley P. Newton contends that late in 1927 and early 1928 "several events occurred, not always simultaneously or directly connected, but all bearing toward the same end—implementation of U.S. air domination in Latin America. Many pieces fitted into the imperfectly formed picture: the charm and flying skill of Charles A. Lindbergh; the promotion of commercial aviation sales in the nations to the south; the emergence of Pan American Airways as an instrument of government policy . . . the attempt made at the Sixth Pan American Conference to smooth the way for U.S. aviation domination. But binding all these actions together was an interest in promoting U.S. aviation. Here the role of Lindbergh was singularly important." Similarly, Marylin Bender and Selig Altschul believe that the route of Lindbergh's Latin American tour, which was called the Lindbergh Circle, "was the dramatic spur for an air link between the northern and southern continents of the Western Hemisphere—a route to be woven through Mexico and Central America by a privately owned U.S. government-approved transport company."[31]

At any rate, whether Lindbergh knew that his flight was coincidental with federal interest in making Pan American Airways the so-called "Chosen Instrument" of the U.S. desire to forge airline routes into Latin America is beside the point. In 1970 he told Wesley Newton that at the time he was "tremendously interested in air routes (potential) to South America [as] . . . the second stage of development of world commercial air transport. (First, continental air routes; second, intercontinental air routes; third, transoceanic air routes.)" In that way, the Guggenheim tour of the United States satisfied Lindbergh's wish to plant the seed in the mind of the American public that continental air routes could be developed, and the Latin American tour similarly paved the way for Pan American Airways' incursion into Latin America. The formalization of Lindbergh's involvement in setting up these airways was accomplished when in 1929 he became chairman of the Technical Committee of Transcontinental Air Transport (TAT), whose president, Clement M. Keys, owned the Curtiss Aeroplane and Motor Company, the foremost aviation company in the United States at the time. For his services to TAT, Lindbergh received a salary of $10,000 a year plus a bonus of $250,000, which he could use to purchase twenty-five thousand shares in the company at ten dollars a share. On July 7, 1929, TAT inaugurated an airrailway transcontinental service from New York to Los Angeles, whose route Lindbergh had personally selected. Shortly after he became affiliated with TAT, he became technical adviser to Pan American Airways for an annual salary of $10,000 and the privilege of purchasing approximately one-tenth of the company's stock at half its current value. His role in the airline would continue for more than four decades. Thus, in a very real sense the flights of the *Spirit of St. Louis* around the United States and into Mexico, Latin America, and the Caribbean paved the way for regularly scheduled airline service.[32]

Meanwhile, Lindbergh decided to retire the *Spirit of St. Louis* to the Smithsonian Institution, which had been working with him and his financial backers to add the aircraft to its collection. On April 30, 1928, Lindbergh flew the *Spirit of St. Louis* from St. Louis, where it had landed after the final leg of his Latin American tour, to Bolling Field in Washington, D.C., making a flight of 725 miles in just under five hours. Before its retirement, the aircraft had made 174 flights and had been in the air for 489 hours and twenty-eight minutes. In Washington, the aircraft was disassembled and transported through

In early 1930 Charles Lindbergh was pruning his new Lockheed Sirius for an attempt to break the cross-country record from Los Angeles to New York. NACA engineer Fred Weick is shown in the rear cockpit, Lindbergh in front, and Tom Hamilton standing. (NASA photo no. L90-3736)

the streets of the city to the North Hall of the Arts and Industries Building, where it was reassembled. On Sunday, May 13, 1928, the display of the aircraft was opened to the public. Lindbergh did place conditions on the display of the aircraft, insisting that it always remain on display and that it be exhibited "with a degree of dignity, prominence, and accessibility to the public equal to the setting and circumstance now surrounding its exhibition by the Smithsonian Institution."[33]

About a year after the aircraft had been placed on display, the financial arrangements of the donation were completed when Lindbergh and his backers sold the aircraft to the Smithsonian Institution for the sum of one dollar. In 1939, after the Smithsonian had received many requests to borrow the *Spirit*, Lindbergh wrote to Charles Greeley Abbot, secretary of the Smithsonian, and expressed the wish that the aircraft never leave the premises. The aircraft remained in the Arts and Industries Building until 1975. At that time it was moved to the Silver Hill restoration facility of the National Air and Space Museum. Several months before the museum's new building was scheduled to open in July 1976, the aircraft was placed on display in that building, where it has been on display suspended from the top of the Milestones of Flight Gallery for more than twenty-five years. During that time, the aircraft has been taken down from its place of honor twice: once in 1992 for a period of about two months for refurbishment, and in 2000 for about eight months, when it was relocated to the west end of the building to allow for repairs in the windows and skylights of the building. It now hangs in its accustomed place in the Milestones of Flight Gallery.[34]

Reassessing the Artifact

One might say that because it has become a "Milestone" aircraft in a chronological procession of landmark aircraft from the Wright Flyer to the X-15 and been enshrined in the National Air and Space Museum, the *Spirit of St. Louis* has suffered from a lack of historical interpretation and contextual analysis. As mentioned previously, no label script could do the aircraft justice, and most historical examination of it focuses on its flight across the Atlantic (and, consequently, Lindbergh's celebrity) to the detriment of its more important role in flying through the forty-eight states and Latin America. This lack of interpretation hearkens back to the question I posed earlier concerning the aircraft's symbolic and factual weight.

One would have to say that in its seventy-five-year history as one of the most important artifacts in the national aeronautical collection, the *Spirit of St. Louis*'s symbolic weight has outstripped its factual weight. It has in fact become worshiped, along with Lindbergh, the man with whom it is associated. This distortion of historical fact is what historian of technology Joseph Corn calls "object myth." The fallacy, Corn says, stems from the belief that "artifacts are 'primary documents,' that they can supplant libraries, or that like the Rosetta stone they can make mute objects speak." The result is the idealization of an object into a totem or symbol, in this case of the airplane as an icon of American technological progress. Wilcomb Washburn goes further, contending that "the object can become a fetish that, if we merely worship it, impedes our understanding of the object itself and its place in our society." Moreover, as Steven Lubar points out, it is in museums that the public learns or ideally should learn the history of technology. What the public too often finds in museums, Lubar says, "is the display of technological artifacts in chronological order, the great inventions arranged in proud procession across the exhibit floor." The implication here is that artifacts arranged in such a way—as are the artifacts in the Milestones of Flight gallery—present the false impression of an unimpeded march of technological progress.[35]

Why is this so? One reason for the *Spirit of St. Louis*'s lack of sufficient interpretation is a fundamental misunderstanding of its true importance in the history of aviation. This misinterpretation results from a capitulation on the part of the museum to popular demand to see the artifact as a product of Lindbergh's celebrity and heroism. Historian Daniel Boorstin has commented that Lindbergh "performed single-handed one of the heroic deeds of the century. His deed was heroic in the best epic mold. But he became degraded into a celebrity." Moreover, Boorstin says, "the biggest news about Lindbergh was that he was such big news. . . . Lindbergh's well-knownness was so sudden and so overwhelming. It was easy to make stories about what a big celebrity he was. . . . [T]here was little else anyone could say about him. Lindbergh's singularly impressive heroic deed was soon far overshadowed by his even more impressive publicity. If well-knownness made a celebrity, here was the greatest. Of course it was remarkable to fly the ocean by one-self, but far more remarkable thus to dominate the news. His stature as hero was nothing compared with his stature as celebrity. All the more because it had happened, literally, overnight." I would disagree with Boorstin in one particular: that while Lindbergh's heroism in regard to the *Spirit of St. Louis* is significant, it too is intricately bound up in Lindbergh's celebrity. It is the combination of Lindbergh's heroism and his celebrity that have clouded the issue of the aircraft's real significance to the history of aviation.[36]

How does one find one's way out of this all-too-familiar dilemma? One way is through periodic review of exhibits in museums to see whether new historical interpretations apply to the artifacts. Unfortunately, because of funding and emphasis on new exhibitions and other priorities, this reassess-ment does not take place as often as it should, and in some cases it does not take place at all. Another is to ensure that curatorial staff are profession-ally or academically trained, peer-reviewed, familiar with the literature, and capable of reinterpreting historical events in the broadest possible context. Another is to use monographs to interpret artifacts of national historical sig-nificance. Unfortunately, monographs on artifacts do not often receive the credit due them from peer review panels because such research and writing is deemed to be popular and not scholarly. This trend needs to be reversed if public history is to be beneficial to the public.

Concerning reinterpretation of an artifact in an exhibition milieu, one could take an example from the National Air and Space Museum's own ex-perience. The Wright Flyer is scheduled to be removed, at least temporarily, from the Milestones of Flight Gallery and to be reinterpreted as part of the commemoration of the hundredth anniversary of piloted, powered, and sus-tained flight in a heavier-than-air machine in December 2003. The proposed exhibition will make use of the museum's collection and others related to the initial decade of flight to explain how the aircraft worked, how it affected the aviation community at large, its potential use as a means of commercial transportation and as a weapon, and its effect on journalism, politics, and the arts.

Such a reinterpretation of the *Spirit of St. Louis* is also possible. An ideal scheme would be to make the aircraft the centerpiece of a new Golden Age of Flight Gallery. In such an exhibition, the artifact's factual weight could be balanced against its symbolic weight, and a public that is largely unfamiliar with the contextual story of Lindbergh's flights in the aircraft could see the aircraft in a new light. This is the true end of museum exhibition. But in today's climate, in which lack of funding for the preservation of our historical heritage and general indifference to the pursuit of new historical knowledge are prevalent, such a goal could prove to be elusive.

NOTES

1. Steven Lubar, "Learning from Technological Things," in W. David Kingery, ed., *Learning from Things: Method and Theory of Material Culture Studies* (Washington, D.C.: Smithsonian Institution Press, 1996), 32.

2. Perry D. Luckett, *Charles A. Lindbergh: A Bio-Bibliography* (Westport, Conn.: Greenwood Press, 1986), 129–30, quoted in A. Scott Berg, *Lindbergh* (New York: G. P. Putnam's Sons, 1998), 92.

3. Berg, *Lindbergh*, 92–93.

4. Ibid., 82, 93.

5. Ibid., 94.

6. Ibid., 95.

7. Ibid., 95, 96–97, 98.

8. Donald A. Hall, *Technical Preparation of the Airplane "Spirit of St. Louis,"* Technical Note No. 257, National Advisory Committee for Aeronautics, reprinted in Tom D. Crouch, ed., *Charles A. Lindbergh: An American Life* (Washington, D.C.: Smithsonian Institution, 1977), 84; Lindbergh quote is from his *Spirit of St. Louis*, 105; Berg, *Lindbergh*, 104.

9. Charles A. Lindbergh: *The Spirit of St. Louis* (New York: Charles Scribner's Sons, 1953), 148.

10. Museum of the City of New York, <http:www.mcny.org>; Kenneth S. Davis, *The Hero: Charles A. Lindbergh and the American Dream* (Garden City, N.Y.: Doubleday, 1959), 140–41.

11. Quoted in Davis, *Hero*, 144; Peter C. Allen, *The 91 before Lindbergh* (Shrewsbury, U.K.: Airlife, 1985), 15, 16.

12. There seems to be little or no consensus on how many persons crossed the Atlantic by air before Lindbergh made his solo flight in 1927. Peter C. Allen says there were ninety-one (*The 91 before Lindbergh*). R. E. G. Davies, *Charles Lindbergh: An Airman, His Aircraft, and His Great Flights* (McLean, Va.: Paladwr Press, 1997), 10, says there were 117. Dates for the transatlantic flights in 1919 and 1924 are from Eugene M. Emme, *Aeronautics and Astronautics: An American Chronology of Science and Technology in the Exploration of Space, 1915–1960* (Washington, D.C.: National Aeronautics and Space Administration, 1961), 10, 18.

13. Davis, *Hero*, 142; Allen, *91 before Lindbergh*, 160–61.

14. Joseph R. Hamlen, *Flight Fever* (Garden City, N.Y.: Doubleday, 1971), 89, 94–95; Robert J. Hoare, *Wings over the Atlantic* (Boston: Charles T. Branford, 1957), 51; Edward Jablonski, *Atlantic Fever* (New York: Macmillan, 1972), 103, 138. Jablonski (133–34) describes the soap opera atmosphere surrounding the Chamberlin-Levine attempt. The flight had to be delayed until June 4, but Chamberlin and Levine made it successfully from New York to Eisleben (Martin Luther's birthplace in the eastern foothills of the Harz Mountains in Germany), setting a world's record for distance in a straight line of 3,911 miles.

15. Lindbergh, *Spirit of St. Louis*, 65.

16. Luckett, *Charles A. Lindbergh*, 14.

17. Ibid., 14; Lindbergh, *Spirit of St. Louis*, 330.

18. Lindbergh, *Spirit of St. Louis*, 389. See also Susan M. Gray, *Charles A. Lindbergh and the American Dilemma: The Conflict of Technology and Human Values* (Bowling Green, Ohio: Bowling Green State University Popular Press, 1988), 83. Gray believes that when Lindbergh reported seeing apparitions in his book *The Spirit of St. Louis*, it was "an act fittingly symbolic of the changes which had occurred in his world view, the addition of spiritual elements to the primarily mechanical ones which had been there before."

19. Lindbergh, *Spirit of St. Louis*, 457, 459, 466, 477.

20. Ibid., 489, 492.

21. Ibid., 495.

22. "Lindbergh's Reception in France," *Current History*, July 1927, 522; Lindbergh, *Spirit of St. Louis*, 497, 499, 501; Luckett, *Charles A. Lindbergh*, 15.

23. Richard P. Hallion, *Legacy of Flight: The Guggenheim Contribution to American Aviation* (Seattle: University of Washington Press, 1977), 152–53.

24. Ibid., 154–56.

25. Walter S. Ross, *The Last Hero: Charles A. Lindbergh*, rev. ed. (New York: Harper and Row, 1976), 150–51.

26. Reginald M. Cleveland, *America Fledges Wings: The History of the Daniel Guggenheim Fund for the Promotion of Aeronautics*, foreword by Robert A. Millikan (New York: Pitman, 1942), 95, 99; Hallion, *Legacy of Flight*, 158. See also Donald E. Keyhoe, "Seeing America with Lindbergh: The Record of More than 20,000 Miles by Airplane through Forty-Eight States on Schedule Time," *National Geographic Magazine*, Jan. 1928, 1–46.

27. Hallion, *Legacy of Flight*, 158; Cleveland, *America Fledges Wings*, 99, 100.

28. Lindbergh's own account of his Latin American tour appears in "To Bogotá and Back by Air: The Narrative of a 9,500-Mile Flight from Washington, over Thirteen Latin-American Countries and Return, in the Single-Seater Airplane 'Spirit of St. Louis,'" *National Geographic Magazine* (May 1928): 529–602. See also "One Eagle Swoops Homeward on Last Leg of Triumphant Tour," *New York Herald Tribune*, Feb. 12, 1928, sec. 3, 1.

29. Wesley Phillips Newton, *The Perilous Sky: U.S. Aviation Diplomacy and Latin America, 1919–1931* (Coral Gables, Fla.: University of Miami Press, 1978), 126–28.

30. Ibid., 126–27, 128.

31. Ibid., 125; Marylin Bender and Selig Altschul, *The Chosen Instrument: Pan Am, Juan Trippe, the Rise and Fall of an American Entrepreneur* (New York: Simon and Schuster, 1982), 97.

32. Newton, *Perilous Sky*, 129; Berg, *Lindbergh*, 190, 191, 189, 205.

33. "Afterword: The *Spirit* and the Smithsonian Institution," in Dominick A. Pisano and F. Robert van der Linden, *Charles Lindbergh and the Spirit of St. Louis* (New York: Harry N. Abrams, 2002), 122.

34. Ibid., 126–31.

35. Joseph J. Corn, "Object Lessons/Object Myths? What Historians of Technology Learn from Things," in W. David Kingery, ed., *Learning from Things: Method and Theory of Material Culture Studies* (Washington, D.C.: Smithsonian Institution Press, 1996), 37; Wilcomb E. Washburn, "Collecting Information, Not Objects," *Museum News* (Feb. 1984), 15; Lubar, "Learning from Technological Things," 32.

36. Daniel J. Boorstin, *The Discoverers* (New York: Harry N. Abrams, 1991), 66, 68.

SELECTIVE ANNOTATED BIBLIOGRAPHY

Aeronautics and Space Report of the President. Washington, D.C.: U.S. Government Printing Office, 1959–2002. An annual series of reports, each of which discusses the federal government's activities in space flight during the previous year; includes key documents and tabular data.

Allard, Noel E., and Gerald N. Sandvick. *Minnesota Aviation History, 1857–1945.* Chaska, Minn.: MAHB Publishing, 1993. A history of aviation in Minnesota, commissioned by the Minnesota Office of Aeronautics as a fiftieth anniversary book.

Allen, Richard Sanders. *The Northrop Story, 1929–1939.* New York: Orion, 1990. An early history of the aerospace manufacturing giant Northrop.

———. *Revolution in the Sky: The Lockheeds of Aviation's Golden Age.* New York: Orion, 1988. A large-format illustrated history of the Lockheed aircraft from the Vega of the 1920s to the L-1011.

Anderson, John D., Jr. *A History of Aerodynamics and Its Impact on Flying Machines.* Cambridge, Eng.: Cambridge University Press, 1997. A thorough survey of aerodynamics written by an aeronautical engineer.

Andersson, Lennart. *Soviet Aircraft and Aviation, 1917–1941.* Annapolis, Md.: Naval Institute Press, 1994. A history of the Soviet experience in the air.

Anderton, David A. *Sixty Years of Aeronautical Research, 1917–1977.* Washington, D.C.: National Aeronautics and Space Administration, 1978. A general overview of research and development in the government laboratories of the National Advisory Committee for Aeronautics and its successor, NASA.

Andrews, Allen. *Back to the Drawing Board: The Evolution of Flying Machines.* Newton Abbot, U.K.: David & Charles, 1977. An analysis of the development of aircraft technology from ancient times to the present.

———. *The Flying Machine: Its Evolution through the Ages.* New York: G. P. Putnam's Sons, 1977. An American edition of the English *Back to the Drawing Board.*

———. *The History of the U.S. Air Force.* New York: Crescent, 1981. A handbook of U.S. Air Force history from 1908 to 1980 written for a popular audience.

Arbon, Lee. *They Also Flew: The Enlisted Pilot Legacy, 1912–1942.* Washington, D.C.: Smithsonian Institution Press, 1992. A history of the little-known program that allowed enlisted men to be trained as pilots and to fly in combat.

Astronautics and Aeronautics: A Chronology of Science, Technology, and Events. Washington, D.C.: National Aeronautics and Space Administration, 1962–1999. A multivolume series of chronologies covering the period 1915 to 1995.

Aymer, Brandt, ed. *Men in the Air: The Best Flight Stories of All Time from Greek Mythology to the Space Age—An Anthology of Fact and Fiction*. New York: Crown, 1990. A popular collection of stories about flight.

Baals, Donald D., and William R. Corliss. *Wind Tunnels of NASA*. NASA Special Publication 440. Washington, D.C.: U.S. Government Printing Office, 1981. An illustrated history of every tunnel built by the organization.

Baker, David. *Flight and Flying: A Chronology*. New York: Facts on File, 1993. A basic reference chronology of the progress of aviation.

Barlow, Jeffrey G. *Revolt of the Admirals: The Fight for Naval Aviation, 1945–1950*. Washington, D.C.: Naval Historical Center, 1994. A detailed account of the interservice rivalry in the Department of Defense during the Truman administration over which service would control air power.

Batchelor, John H., and Chris Chant. *Flight: The Story of Aviation*. New York: Gallery, 1990. A popular illustrated history of the progress of aviation.

Bauer, Eugene E. *Boeing: In Peace and War*. Enumclaw, Wash.: TAB, 1991. A corporate history recounting how one manufacturer took advantage of the business opportunities presented by the airplane and developed the largest aircraft company in the world.

Beaumont, Roland. *Testing Early Jets: Compressibility and the Supersonic Era*. Washington, D.C.: Smithsonian Institution Press, 1991. A history of flight testing with a discussion of the air compressibility problem near the speed of sound.

Becker, John V. *The High-Speed Frontier: Case Histories of Four NACA Programs, 1920–1950*. NASA Special Publication 445. Washington, D.C.: U.S. Government Printing Office, 1980. A history of four major research programs (high-speed airfoil program, transonic wind tunnel development, the high-speed propeller program, and high-speed cowlings, air inlets and outlets, and internal-flow systems) written by an NACA engineer.

Bednarek, Janet R. Daly. *America's Airports: Airfield Development, 1918–1947*. College Station: Texas A&M University Press, 2001. An account of the development of airports in the United States.

Bender, Marylin, and Selig Altschul. *The Chosen Instrument: Pan Am, Juan Trippe, the Rise and Fall of an American Entrepreneur*. New York: Simon and Schuster, 1982. A journalistic history of Pan American Airways, emphasizing its efforts to become the official air carrier of the United States in the international arena.

Berg, A. Scott. *Lindbergh*. New York: G. P. Putnam's Sons, 1998. A broad-ranging biography of the aviator Charles Lindbergh, a complex modern American hero, and his wife.

Berger, Carl. *B-29: The Superfortress*. New York: Ballantine, 1970. A history of the B-29 from its origins in World War II through its use as a long-range bomber in the 1950s.

Bernstein, Aaron. *Grounded: Frank Lorenzo and the Destruction of Eastern Airlines*. New York: Simon and Schuster, 1990. A recounting of the demise of Eastern Airlines, one of the top five commercial carriers in the nation until it collapsed in the deregulated environment of the industry in the mid-1980s.

Biddle, Tami Davis. *Rhetoric and Reality in Air Warfare: The Evolution of British and American Ideas about Strategic Bombing, 1914–1945*. Princeton, N.J.: Princeton University Press, 2002. An analysis of air-power doctrine through World War II.

Biddle, Wayne. *Barons of the Sky: From Earth Flight to Strategic Warfare—The Story of the American Aerospace Industry*. New York: Simon and Schuster, 1991. The story of four early aircraft manufacturers, Glenn L. Martin, Donald Douglas, Jack Northrop, and Allen and Malcolm Loughhead of Lockheed Aircraft Co.

Bilstein, Roger E. *The American Aerospace Industry: From Workshop to Global Enterprise*. New York: Twayne, 1996. Reprinted as *The Enterprise of Flight: The American Aviation and Aerospace Industry*. Smithsonian History of Aviation and Spaceflight Series. Washington, D.C.: Smithsonian Institution Press, 2001. A comprehensive overview of the history of the aerospace industry in the United States.

———. *Flight in America: From the Wrights to the Astronauts*. 3d ed. Baltimore, Md.: Johns Hopkins University Press, 2001. A synthesis of the origins and development of aerospace activities in America.

———. *Flight Patterns: Trends of Aeronautical Development in the United States, 1918–1929*. Athens: University of Georgia Press, 1983. A monograph analyzing developments in aviation in the pivotal decade of the 1920s.

———. *Orders of Magnitude: A History of the NACA and NASA, 1915–1990*. NASA Special Publication 4406. Washington, D.C.: U.S. Government Printing Office, 1989. A general, though scholarly, history of the National Aeronautics and Space Administration and its predecessor, the National Advisory Committee for Aeronautics.

Bilstein, Roger E., and Jay Miller. *Aviation in Texas*. Austin, Tex.: Texas Monthly Press, 1985. A well-researched account of aviation activity in Texas.

Blackman, Steven. *Planes and Flight*. New York: Franklin Watts, 1993. A general history, written and illustrated for the aeronautical-buff market.

Blatherwick, Francis John. *A History of Airlines in Canada*. Toronto, Ont.: Unitrade, 1989. A compendium of the airline history of Canada.

Boot, Roy. *From Spitfire to Eurofighter: 45 Years of Combat Aircraft Design*. Shrewsbury, U.K.: Airlife, 1990. An illustrated history of European military aircraft since World War II.

Borgiasz, William S. *The Strategic Air Command: Evolution and Consolidation of Nuclear Forces, 1945–1955*. Westport, Conn.: Praeger, 1996. A detailed account of the political machinations that created the U.S. strategic bomber force.

Borowski, Harry R. *A Hollow Threat: Strategic Air Power and Containment before Korea*. Westport, Conn.: Greenwood Press, 1982. An extended argument that with World War II demobilization the United States possessed no real power to deliver nuclear weapons to any enemy between 1945 and 1950.

Bowers, Peter M. *Boeing Aircraft since 1916*. New York: Funk and Wagnalls, 1968. A standard reference work.

Bowers, Peter M., and Ernest R. McDowell. *Triplanes: A Pictorial History of the World's Triplanes and Multiplanes*. Osceola, Wis.: Motorbooks International, 1993. An illustrated history of a design that gained popularity in World War I.

Bowman, Martin W. *The Men Who Flew the Mosquito: Compelling Accounts of the*

Wooden Wonder's Triumphant WWII Career. Osceola, Wis.: Motorbooks International, 1995. A popular history of an important military airplane.

Boyne, Walter J. *Beyond the Horizons: The Lockheed Story*. New York: St. Martin's Press, 1998. A history of the Lockheed corporation focusing on the period from World War II to the present.

———. *Silver Wings: A History of the United States Air Force*. New York: Simon and Schuster, 1993. A basic one-volume history of the U.S. Air Force.

———. *The Smithsonian Book of Flight*. Washington, D.C.: Smithsonian Books, 1987. Written for nonspecialists, a general illustrated history of flight.

Boyne, Walter J., and Donald S. Lopez. *The Jet Age: Forty Years of Jet Aviation*. Washington, D.C.: Smithsonian Institution Press, 1979. A general history of the civil and military uses of jet aircraft that examines aeronautical design and its evolution in detail.

Bradin, James W. *From Hot Air to Hellfire: History of Army Attack Aviation*. Boulder, Colo.: Presidio Press, 1994. An account of the evolution of attack aviation in the U.S. military from the use of balloons in the Civil War to the modern attack helicopters.

Briddon, Arnold E., and Ellmore Champie. *Federal Aviation Agency Historical Fact Book: A Chronology, 1926–1971*. Washington, D.C.: Department of Transportation, 1974. Although now somewhat dated, a reference with essential information on a myriad of aeronautical subjects.

Bright, Charles D. *The Jet Makers: The Aerospace Industry from 1945 to 1972*. Lawrence: Regents Press of Kansas, 1978. An analysis of the aircraft industry from the 1940s, when jets first came into use, until the mid-1970s.

———, ed. *Historical Dictionary of the U.S. Air Force*. Westport, Conn.: Greenwood Press, 1992. A reference work on the history of the U.S. Air Force, its aircraft, and its leaders.

Brooks-Pazmany, Kathleen. *United States Women in Aviation 1919–1929*. Washington, D.C.: Smithsonian Institution Press, 1991. Second work in the Smithsonian's series on the history of women in aviation in the United States.

Brown, Anthony E. *The Politics of Airline Deregulation*. Knoxville: University of Tennessee Press, 1987. A survey of the efforts since the late 1960s to deregulate the airline industry, the passage of the Airline Deregulation Act of 1978, and the consequences of that legislation.

Brown, Jerold E. *Where Eagles Land: Planning and Development of U.S. Army Airfields, 1910–1941*. Westport, Conn.: Greenwood Press, 1990. An analysis of the process of siting U.S. Army airfields in the United States and abroad before World War II.

Brown, Michael F. *Flying Blind: The Politics of the U.S. Strategic Bomber Program*. Ithaca, N.Y.: Cornell University Press, 1992. A detailed account of the politics of military aircraft acquisition from the B-36 to the B-2.

Bugos, Glenn E. *Engineering the F-4 Phantom II: Parts into Systems*. Annapolis, Md.: Naval Institute Press, 1996. A monograph on the development of the F-4, built by McDonnell Douglas; an examination of modern fighter aircraft that uses the social construction of technology approach.

Bungey, Lloyd M. *Pioneering Aviation in the West, As Told by the Pioneers*. Surrey, B.C.: Hancock House, 1992. A popular account of flight.

Cain, Anthony Christopher. *The Forgotten Air Force: French Air Doctrine in the 1930s*. Washington, D.C.: Smithsonian Institution Press, 2002. A study of French air power doctrine.

Campagna, Palmiro. *Storms of Controversy: The Secret Avro Arrow Files Revealed*. Toronto: Stoddart, 1992. An examination of the demise of the Canadian aerospace industry that focuses on the advanced fighter aircraft the Avro Arrow.

Clodfelter, Mark E. *The Limits of Air Power: The American Bombing of North Vietnam*. New York: Free Press, 1989. A historical analysis of why strategic bombardment did not subdue the enemy in the Vietnam conflict as promised by the doctrine of air power then used in the U.S. Air Force.

Coffey, Thomas M. *Hap: The Story of the U.S. Air Force and the Man Who Built It, General Henry H. "Hap" Arnold*. New York: Viking Press, 1982. A biography of the air officer who headed the U.S. Army Air Forces during World War II.

Constant, Edward W. II. *The Origins of the Turbojet Revolution*. Baltimore, Md.: Johns Hopkins University Press, 1980. An account of the development of the jet aircraft in World War II and after.

Cook, LeRoy. *American Aviation: An Illustrated History*. 2d ed. New York: McGraw-Hill, 1994. Another popular illustrated history of flight.

Cook, William H. *The Road to the 707: The Inside Story of Designing the 707*. Bellevue, Wash.: TYC, 1991. A basic discussion of the development of the 707 by a Boeing engineer.

Cooling, B. Franklin, ed. *Case Studies in the Achievement of Air Superiority*. Washington, D.C.: Center for Air Force History, 1994. A collection of essays in military history from World War II to the Israeli conflict in the Bekaa Valley with a focus on the issue of air superiority.

————. *Case Studies in the Development of Close Air Support*. Washington, D.C.: Office of Air Force History, 1990. A collection of essays by leading air power historians on close air support concentrating mostly on World War II, but with additional information on Korea, Southeast Asia, and the Israeli experience.

Copp, DeWitt S. *A Few Great Captains: The Men and Events That Shaped the Development of U.S. Air Power*. Garden City, N.Y.: Doubleday, 1980. A discussion of the history of the U.S. Army Air Corps between the two world wars and its struggle for independence.

Cordesman, Anthony H. *The Lessons and Non-Lessons of the Air and Missile Campaign in Kosovo*. Westport, Conn.: Praeger, 2001. This offers a detailed analysis of the air campaign in Kosovo.

Corn, Joseph J. *The Winged Gospel: America's Romance with Aviation, 1900–1950*. New York: Oxford University Press, 1983. A social history of the airplane and why Americans have been so attracted to it.

Crane, Conrad C. *American Airpower Strategies in Korea, 1950–1953*. Lawrence: University Press of Kansas, 2000. An examination of the development of a strategy for the use of air power in Korea, also focusing on the lessons learned and not learned from that conflict.

————. *Bombs, Cities, and Civilians: American Airpower Strategy in World War II*.

Lawrence: University Press of Kansas, 1993. An analysis of the development and explication of air power doctrine during World War II.

Craven, Wesley Frank, and James L. Cate, eds. *The United States Air Force in World War II*. 7 vols. Chicago: University of Chicago Press, 1948–1955. An official history of the U.S. Army Air Forces in World War II.

Crouch, Tom D. *The Bishop's Boys: A Life of Wilbur and Orville Wright*. New York: W. W. Norton, 1989. A comprehensive biography of the men who built the first successful powered flying machines.

———. *A Dream of Wings: Americans and the Airplane, 1875–1905*. Washington, D.C.: Smithsonian Institution Press, 1989. A look at the Gilded Age in America, when such pioneers in aviation as Octave Chanute and Samuel P. Langley, as well as the Wright Brothers, began their quest for successful powered flight; a social history of the ideal of flight.

Cunningham, Frank. *Sky Master: The Story of Donald Douglas and the Douglas Aircraft Company*. Philadelphia: Dorrance, 1943. An early biography of an early aeronautical engineer, aviation entrepreneur, and builder of the legendary DC-3.

Cunningham, William Glenn. *The Aircraft Industry: A Study in Industrial Location*. Los Angeles: Lorrin L. Morrison, 1951. An analysis of why the aircraft industry became a largely West Coast business.

Dale, Henry. *Early Flying Machines*. New York: Oxford University Press, 1993. An introduction to the development of aviation technology in the formative years before World War I.

Daley, Robert. *An American Saga: Juan Trippe and His Pan Am Empire*. New York: Random House, 1980. A biography of the leader who founded Pan Am in the 1920s and guided its operations until the 1960s.

Davies, R. E. G. *Airlines of the United States since 1914*. Rev. ed. Washington, D.C.: Smithsonian Institution Press, 1982. An encyclopedic history of U.S. airlines.

———. *Continental Airlines: The First Fifty Years*. The Woodlands, Tex.: Pioneer Publications, 1984. Focusing on an airline that dominated its routes for half a century, a history of Continental Airlines.

———. *Delta: The Illustrated History of a Major U.S. Airline and the People Who Made It*. Miami, Fla.: Paladwr Press, 1990. A popular history of the airline giant from its origins in the 1920s to the present.

———. *Fallacies and Fantasies of Air Transport History*. McLean, Va.: Paladwr Press, 1994. A colorfully written book on the general history of the airline industry.

———. *Pan Am: An Airline and Its Aircraft*. New York: Orion, 1987. With emphasis on photographs and aircraft, a basic history of the airline.

———. *Rebels and Reformers of the Airways*. Washington, D.C.: Smithsonian Institution Press, 1987. A collection of essays on colorful airline executives, many of whom fought with the regulators of the airline industry.

Davies, R. E. G., and I. E. Quastler. *Commuter Airlines of the United States*. Washington, D.C.: Smithsonian Institution Press, 1994. A basic source for information on commuter airlines.

Davis, Benjamin O., Jr. *Benjamin O. Davis Jr., American*. Washington, D.C.:

Smithsonian Institution Press, 1991. The autobiography of the first African American to serve and command in the U.S. Army Air Forces.

Davis, Richard G. *Carl A. Spaatz and the Air War in Europe*. Smithsonian History of Aviation Book Series. Washington, D.C.: Center for Air Force History, 1993. Also issued by Washington, D.C.: Smithsonian Institution Press, 1993. A study of the development of theories of air power and their application in World War II with particular reference to the personality of U.S. Army Air Forces commander Carl A. Spaatz.

Dawson, Virginia P. *Engines and Innovation: Lewis Laboratory and American Propulsion Technology*. NASA Special Publication 4306. Washington, D.C.: U.S. Government Printing Office, 1991. An institutional history of the Lewis Research Center from its creation in 1941 to the early 1990s.

de Syon, Guillaume. *Zeppelin! Germany and the Airship, 1900–1939*. Baltimore, Md.: Johns Hopkins University Press, 2002. A modern history of the German romance with the airship.

Dibbs, John, and Mike Jerram. *Hurricane: A Living Legend*. Osceola, Wis.: Motorbooks International, 1995. A "biography" of a World War II fighter.

Dierikx, Marc. *Fokker: A Transatlantic Biography*. Smithsonian History of Aviation Book Series. Washington, D.C.: Smithsonian Institution Press, 1997. The standard biography of the German aircraft manufacturer Fokker.

Dobson, Alan P. *Peaceful Air Warfare: The United States, Britain, and the Politics of International Aviation*. New York: Oxford University Press, 1991. A study of international commercial aviation and the power politics it engenders.

Doolittle, James H., with C. V. Glines. *I Could Never Be So Lucky Again*. New York: Bantam Books, 1991. The autobiography of one of the most influential airmen in the twentieth century.

Douglas, Deborah G. *United States Women in Aviation, 1940–1985*. Washington, D.C.: Smithsonian Institution Press, 1991. Fourth in the Smithsonian's series on women in aviation, an in-depth and broad-ranging examination of its topic.

Dunmore, Spencer, et al. *On Great White Wings: The Wright Brothers and the Race for Flight*. New York: Hyperion, 2001. A well-illustrated account of the Wright Brothers' efforts to fly.

Dwiggins, Don. *The Barnstormers: Flying Daredevils of the Roaring Twenties*. Blue Ridge Summit, Pa.: TAB Books, 1981. A popular account of aviation in the 1920s that focuses on the great Waldo Peppers, who traveled around the United States demonstrating the airplane after World War I.

Edgerton, David. *England and the Aeroplane: An Essay on a Militant and Technological Nation*. Basingstoke, U.K.: Macmillan, in association with the Centre for the History of Science Technology and Medicine, University of Manchester, 1991. An account of the fascination the English have for the airplane.

Emme, Eugene M. *The Impact of Air Power: National Security and World Politics*. Princeton, N. J.: D. Van Nostrand, 1959. An analysis of the use of the airplane for military purposes.

———, ed. *Two Hundred Years of Flight in America: A Bicentennial Survey*. San Diego, Calif.: Univelt, 1977. A collection of essays on flight in America in nearly all its permutations (general aviation, commercial airlines, and military applications).

Ermenc, Joseph J., ed. *Interviews with German Contributors to Aviation History*. Westport, Conn.: Meckler, 1990. A collection of oral histories.

Ethell, Jeffrey L. *Frontiers of Flight*. Washington, D.C.: Smithsonian Books, 1992. A popular Smithsonian book, well illustrated, discussing the development of aviation.

Fausel, Robert W. *Whatever Happened to Curtiss Wright? The Story of How a Very Successful Aircraft Company Took Itself Out of Business, 1945–1953, the Aviation View*. Manhattan, Kans.: Sunflower University Press, 1990. A chronicle of the demise of an aviation giant.

Fishbein, Samuel B. *Flight Management Systems: The Evolution of Avionics and Navigation Technology*. New York: Praeger, 1995. A detailed technical history of technological systems used in aircraft for control and navigation.

Flintham, Victor. *Air War and Aircraft: A Detailed Record of Air Combat, 1945 to the Present*. New York: Facts on File, 1990. A basic reference work on air combat since World War II.

Ford, Daniel. *Flying Tigers: Claire Chennault and the American Volunteer Group*. Smithsonian History of Aviation Book Series. Washington, D.C.: Smithsonian Institution Press, 1991. A history of the military organization that helped prevent the Japanese from overrunning China in 1941.

Francillon, René J. *McDonnell Douglas Aircraft since 1920*. 2 vols. Annapolis, Md.: Naval Institute Press, 1989–1990. A basic reference work.

Fredette, Raymond H. *The Sky on Fire: The First Battle of Britain, 1917–1918*. Repr. Smithsonian History of Aviation Book Series. Washington, D.C.: History of Aviation Book Series. Smithsonian Institution Press, 1991. An account of the first campaign of strategic bombardment in World War I.

Freudenthal, Elsbeth E. *The Aviation Business: From Kitty Hawk to Wall Street*. New York: Vanguard, 1940. A classic study of the rise of aviation as a business up to 1940.

Freydberg, Elizabeth A. H. *Bessie Coleman: The Brownskin Lady Bird*. New York: Garland, 1994. A biography of a pioneering African American aviator.

Fritzsche, Peter. *A Nation of Fliers: German Aviation and the Popular Imagination*. Cambridge, Mass.: Harvard University Press, 1992. A work of modern social history that describes Germany's romance with aviation from before World War I to the present.

Gandt, Robert L. *China Clipper: The Age of the Great Flying Boats*. Annapolis, Md.: Naval Institute Press, 1991. An account of the development of amphibious aircraft, as well as the development of the first intercontinental airlines that used them before World War II.

Gates, George D., ed. *From Jennies to Jets: A History of Aviation in the Pocatello Area*. Pocatello: Idaho State University Press, 1991. An account of aviation in an isolated section of the United States.

Geibert, Ron, and Patrick B. Nolan. *Kitty Hawk and Beyond: The Wright Brothers and the Early Years of Aviation: A Photographic History*. Dayton, Ohio: University Press of America for Wright State University, 1990. A pictorial history that focuses on the pioneer years of aviation.

Gero, David. *Aviation Disasters*. 2d ed. Osceola, Wis.: Motorbooks International, 1996. A recounting of major accidents in aviation history.

Gibbs-Smith, Charles Harvard. *The Aeroplane: An Historical Survey of Its Origins and Development*. London, U.K.: Her Majesty's Stationery Office, 1960. A reasoned analysis of the development of flight technology.

———. *Aviation: An Historical Survey from Its Origins to the End of World War II*. London, U.K.: Her Majesty's Stationery Office, 1970. A general history of flight in Europe, from the Wrights' first visits to 1945.

———. *A Brief History of Flying, from Myth to Space Travel*. London, U.K.: Her Majesty's Stationery Office, 1968. A short, general history.

———. *Flight through the Ages: A Complete, Illustrated Chronology from the Dreams of Early History to the Age of Space Exploration*. New York: Crowell, 1974. An expansion of *A Brief History of Flying*.

———. *The Rebirth of European Aviation, 1902–1908: A Study of the Wright Brothers' Influence*. London: Her Majesty's Stationery Office, 1974. A monograph that demonstrates that the Wright Brothers were more influential in Europe than in America during the first five years after Kitty Hawk.

Glines, Carroll V. *The Compact History of the United States Air Force*. New York: Hawthorn Books, 1963. A good short history of the subject.

———. *Roscoe Turner: Aviation's Master Showman*. Smithsonian History of Aviation Book Series. Washington, D.C.: Smithsonian Institution Press, 1995. A biography of an air racer and aviation ringmaster of the interwar period.

Goc, Michael J. *Forward in Flight: A History of Aviation in Wisconsin*. Friendship, Wis.: New Past Press, 1998. A celebratory history of aviation in Wisconsin commissioned by the Wisconsin Aviation Hall of Fame.

Gollin, Alfred. *No Longer an Island: Britain and the Wright Brothers, 1902–1909*. Stanford, Calif.: Stanford University Press, 1984. Another monograph that demonstrates that the Wright Brothers were more influential in Britain than in America during the 1900s.

Goodwin, Jacob. *Brotherhood of Arms: General Dynamics and the Business of Defending America*. New York: Times Books, 1985. A history of General Dynamics, one of the more important aerospace companies in the years after World War II.

Gorn, Michael H. *Expanding the Envelope: Flight Research at NACA and NASA*. Lexington: University Press of Kentucky, 2001. An in-depth exploration of a number of important research projects undertaken by the NACA and NASA.

———. *The Universal Man: Theodore von Kármán's Life in Aeronautics*. Washington, D.C.: Smithsonian Institution Press, 1992. A biography of the originator and first head of the organization that eventually became the Jet Propulsion Laboratory in Pasadena, California.

Gray, George W. *Frontiers of Flight: The Story of NACA Research*. New York: Alfred A. Knopf, 1948. A history of the National Advisory Committee for Aeronautics, the federal government's research and development laboratory dedicated to the development of aviation technology.

Green, William. *The Warplanes of the Third Reich*. Garden City, N. Y.: Doubleday, 1970. A reference work on German aircraft of World War II.

Gross, Charles J. *American Military Aviation: The Indispensable Arm*. College Station: Texas A&M University Press, 2002. A fine survey of a century of air power thinking and operations in the United States.

Gubert, Betty Kaplan, comp. *Invisible Wings: An Annotated Bibliography on Blacks in Aviation, 1916–1993*. Westport, Conn.: Greenwood Press, 1994. A basic reference work on African Americans in Aviation.

Gwynne-Jones, Terry. *Farther and Faster: Aviation's Adventuring Years, 1909–1939*. Washington, D.C.: Smithsonian Institution Press, 1991. An account of the development of flight during a formative thirty-year period.

Hallion, Richard P. *Legacy of Flight: The Guggenheim Contribution to American Aviation*. Seattle: University of Washington Press, 1977. The history of the philanthropic project that endowed numerous educational and research institutions with funding for flight research and operations from the 1920s to the late 1930s.

———. *On the Frontier: Flight Research at Dryden, 1946–1981*. NASA Special Publication 4303. Washington, D.C.: U.S. Government Printing Office, 1984. The institutional history of the Dryden Flight Research Center, the NASA facility in the Mojave Desert where hypersonic vehicles such as the X-15 were flown.

———. *Rise of the Fighter Aircraft, 1914–1918*. Baltimore, Md.: Nautical and Aviation Press, 1984. An account of the evolution of aeronautical technology during World War I.

———. *Storm over Iraq: Air Power and the Gulf War*. Washington, D.C.: Smithsonian Institution Press, 1992. A history of the U.S. Air Force in the Gulf War.

———. *Strike from the Sky: The History of Battlefield Air Attack, 1911–1945*. Washington, D.C.: Smithsonian Institution Press, 1989. A history of ground attack aviation.

———. *Test Pilots: The Frontiersmen of Flight*. Washington, D.C.: Smithsonian Institution Press, 1988. A scholarly approach to "the right stuff," eschewing flamboyance for analysis.

Hanle, Paul A. *Bringing Aerodynamics to America*. Cambridge, Mass.: MIT Press, 1982. An analysis of the development of flight technology that argues that many of the important ideas on which Americans capitalized in their airplanes were imported from the schools of Europe.

Hannah, Craig C. *Striving for Air Superiority: The Tactical Air Command in Vietnam*. College Station: Texas A&M University Press, 2002. An analysis of air power in Vietnam.

Hansen, James R. *Engineer in Charge: A History of the Langley Aeronautical Laboratory, 1917–1958*. NASA Special Publication 4305. Washington, D.C.: U.S. Government Printing Office, 1987. An institutional history of the facility that first became involved in space flight issues in the period after World War II and became the first home of Project Mercury, NASA's first human space flight initiative.

———. *Spaceflight Revolution: NASA Langley Research Center from Sputnik to Apollo*. Washington, D.C.: NASA SP-4308, 1995. A sequel to Hansen's *Engineer in Charge* that follows Langley through the end of the Apollo program.

Hardesty, Von D. *Red Phoenix: The Rise of Soviet Air Power, 1941–1945*. Washington, D.C.: Smithsonian Institution Press, 1991. A standard history of the Soviet air force in World War II.

———, ed. *Black Aviator: The Story of William J. Powell*. Smithsonian History of Aviation Book Series. Washington, D.C.: Smithsonian Institution Press, 1994. A reprint of the autobiography of a pioneering African American aviator with an introduction by Hardesty.

———. *Pacific War Diary*. Seattle: University of Washington Press, 1994. An account of aviation in the Pacific during World War II.

Hardesty, Von D., and Dominick A. Pisano. *Black Wings: The American Black in Aviation*. Washington, D.C.: National Air and Space Museum, Smithsonian Institution, 1983. A monograph on the contributions of African Americans to flight drawn from a Smithsonian exhibit.

Hardy, Michael. *World Civil Aircraft since 1945*. New York: Charles Scribner's Sons, 1979. A useful reference.

Harris, Sherwood. *The First to Fly: Aviation's Pioneer Days*. New York: Simon and Schuster, 1970. A basic history of the early years of flight.

Harrison, James P. *Mastering the Sky: A History of Aviation from Ancient Times to the Present*. New York: Sarpedon, 1996. Another basic history of flight.

Hart, Clive. *The Prehistory of Flight*. Berkeley: University of California Press, 1985. A history of flight before the Wright Brothers written for a popular audience.

Hartman, Edwin P. *Adventures in Research: A History of Ames Research Center, 1940–1965*. NASA Special Publication 4302. Washington, D.C.: U.S. Government Printing Office, 1970. The institutional history of the second aeronautical laboratory established by the National Advisory Committee for Aeronautics.

Harwood, William B. *Raise Heaven and Earth: The Story of Martin Marietta People and Their Pioneering Achievements*. New York: Simon and Schuster, 1993. A corporate history of the aerospace titan.

Hasimoto, Takehiko. *Theory, Experiment, and Design Practice: The Formation of Aeronautical Research, 1903–1930*. Baltimore, Md.: Johns Hopkins University Press, 1991. A study of the rise of aeronautical engineering as a discipline.

Haynes, Elmer E. *General Chennault's Secret Weapon: The B-24 in China*. Westport, Conn.: Praeger, 1992. A study of the employment of the B-24 Liberator bomber in the China-Burma-India theater of World War II.

Haynsworth, Leslie, and David Toomey. *Amelia Earhart's Daughters: The Wild and Glorious Story of American Women Aviators from World War II to the Dawn of the Space Age*. New York: Perennial, 1998. A book for a general audience on women aviators, including those who sought to become the first women astronauts.

Hennessy, Juliette A. *The United States Army Air Arm, April 1861 to April 1917*. Washington, D.C.: Office of Air Force History, 1985. A narrative of the use of flying machines, both balloons and airplanes, by the U.S. Army before U.S. entry into World War I.

Heppenheimer, T. A. *A Brief History of Flight: From Balloons to Mach 3 and*

Beyond. New York: John Wiley & Sons, 2001. Another popular account of the history of flight.

———. *Turbulent Skies: The History of Commercial Aviation*. New York: John Wiley & Sons, 1995. This work provides a history of the modern commercial aviation industry.

Heron, S. D. *History of the Aircraft Piston Engine: A Brief Outline*. Detroit, Mich.: Ethyl Corp., 1961. A historical discussion of the development of a critical aviation technology.

Hess, William N. *B-17 Flying Fortress*. Osceola, Wis.: Motorbooks International, 1994. Another "biography" of an airplane, this time the Boeing B-17.

———. *P-47 Thunderbolt: Combat and Development History of the Thunderbolt*. Osceola, Wis.: Motorbooks International, 1994. Another "biography" of a World War II aircraft.

Hess, William, and Thomas Ivie. *P-51 Mustang Aces*. Osceola, Wis.: Motorbooks International, 1992. A "biography" of the P-51.

Higham, Robin. *Air Power: A Concise History*. Manhattan, Kans.: Sunflower University Press, 1984. A history of the quest for air power with an international perspective.

Higham, Robin, and Jacob W. Kipp. *Soviet Aviation and Air Power: A Historical View*. Boulder, Colo.: Westview Press, 1978. A short history of Soviet military aviation.

Hinkle, Stacy C. *Wings over the Border: The Army Air Service Armed Patrol of the United States–Mexican Border, 1919–1921*. El Paso: Texas Western Press, 1970. A focus on one of the first efforts to use aircraft in combat by the U.S. military.

Holden, Henry M. *The Boeing 247: The First Modern Commercial Airplane*. Blue Ridge Summit, Pa.: TAB Books, 1991. A technical history of the first all-metal, monocoque, enclosed monoplane.

———. *The Douglas DC-3*. Blue Ridge Summit, Pa.: PNAero, 1991. A detailed history of what is arguably the most significant aircraft ever built.

———. *The Fabulous Ford Tri-Motors*. Blue Ridge Summit, Pa.: TAB Books, 1992. A tribute, but also a technical history, of the three motored Ford aircraft built for passenger use in the 1920s and 1930s.

Holden, Henry M., and Lori Griffith. *Ladybirds: The Untold Story of Women Pilots in America*. Rev. ed. Mt. Freedom, N. J.: Black Hawk, 1992. A popular history of women aviators from Katherine Stinson and Amelia Earhart to Jenna Yeager.

Holley, Irving B., Jr. *Buying Aircraft: Materiel Procurement for the Army Air Forces*. Washington, D.C.: Office of the Chief of Military History, 1964. An official history of the U.S. Army's efforts in World War II to equip airmen with the airplanes needed to help win the war.

———. *Ideas and Weapons: Exploitation of the Aerial Weapon by the United States during World War I: A Study in the Relationship of Technological Advance, Military Doctrine, and the Development of Weapons*. New Haven, Conn.: Yale University Press, 1953. A classic text for the argument that there is a complex interplay between the ideas that shape military strategy and the weapons that are developed to carry out that strategy.

Hopkins, George E. *The Airline Pilots: A Study in Elite Unionization*. Cambridge, Mass.: Harvard University Press, 1971. A survey of the history of the Air Line Pilots Association.

———. *Flying the Line: The First Half-Century of the Air Line Pilots Association*. Washington, D.C.: Air Line Pilots Association, 1982. A commissioned history of the organization.

———. *Pan Am Pioneer: A Manager's Memoir from Seaplane Clipper to Jumbo Jets*. Lubbock: Texas Tech University Press, 1995. The recollections of a lifetime spent as an official of Pan American Airways, the United States' largest international carrier until the 1980s.

Horgon, James J. *City of Flight: The History of Aviation in St. Louis*. Gerald, Mo.: Patrice Press, 1984. A history of aviation in one city important in its development.

Howard, Frank, and Bill Gunston. *The Conquest of the Air*. New York: Random House, 1972. An account of the development of flight with emphasis on the spectacular.

Howard, Fred. *Wilbur and Orville: A Biography of the Wright Brothers*. New York: Alfred A. Knopf, 1987. A biography of the Wright Brothers through World War II.

Hoyt, Edwin P. *The Airmen: The Story of American Fliers in World War II*. New York: McGraw-Hill, 1991. A social history of the men who flew American airplanes to victory around the world.

Hubler, Richard G. *Straight Up: The Story of Vertical Flight*. New York: Duell, Sloan, and Pearce, 1961. A history of the helicopter.

Hudson, James J. *Hostile Skies: A Combat History of the American Air Service in World War I*. Repr. Syracuse, N. Y.: Syracuse University Press, 1996. A classic history of U.S. air combat in World War I.

Hughes, Thomas Alexander. *Over Lord: General Pete Quesada and the Triumph of Tactical Air Power in World War II*. New York: Free Press, 1995. An example of the new military history, this one on the career of General Elwood Quesada, the wartime commander of the Ninth Air Force in Europe, the unit charged with American fighter operations in the theater.

Hurley, Alfred F. *Billy Mitchell: Crusader for Air Power*. New York: Franklin Watts, 1964. The standard biography of the air commander for the U.S. Army in World War I.

Ingalls, Douglas J. *The McDonnell Douglas Story*. Fallbrook, Calif.: Aero Publishers, 1979. A basic history of the St. Louis aerospace manufacturer.

Irving, Clive. *Wide-Body: The Triumph of the 747*. New York: Morrow, 1993. A basic history of the development of the wide-body airplanes built in the latter 1960s, the most successful of which was the Boeing 747.

Jakab, Peter L. *Visions of a Flying Machine: The Wright Brothers and the Process of Invention*. Smithsonian History of Aviation Book Series. Washington, D.C.: Smithsonian Institution Press, 1990. A detailed analysis of the methods by which the Wright Brothers built their aircraft and succeeded in controlled, powered flight when all beforehand had failed.

Jakeman, Robert J. *The Divided Skies: Establishing Segregated Flight Training at Tuskegee, Alabama, 1934–1942*. Tuscaloosa: University of Alabama Press, 1992.

A history of the development of the African American pilot training program attached to the Tuskegee Institute in the 1930s and 1940s.

James, John. *The Paladins: A Social History of the R. A. F. Up to the Outbreak of World War II*. London, U.K.: Macdonald, 1990. A study of the backgrounds and lifestyles of pilots in the Royal Air Force with special focus on Battle of Britain pilots in 1940.

Jamieson, Perry D. *Lucrative Targets: The U.S. Air Force in the Kuwaiti Theater of Operations*. Washington, D.C.: Air Force History and Museum Program, 2001. An official history of the U.S. Air Force in the Gulf War.

Jaros, Dean. *Heroes without Legacy: American Airwomen, 1912–1944*. Niwot: University Press of Colorado, 1993. An analytical study of why the participation of American women was so prominent up through World War II and declined afterward.

Johnson, Herbert A. *Wingless Eagle: U.S. Army Aviation through World War I*. Chapel Hill: University of North Carolina Press, 2001. A well-documented appraisal of very early U.S. Army aviation.

Jordan, William A. *Airline Regulation in America: Effects and Imperfections*. Baltimore, Md.: Johns Hopkins University Press, 1970. A history of the process of regulation, as well as a plea for reform.

Keil, Sally Van Wagenen. *Those Wonderful Women in Their Flying Machines: The Unknown Heroines of World War II*. New York: Four Directions Press, 1979. An early, popular history of the Women Airforce Service Pilots written shortly after their story was "rediscovered" in the mid-1970s.

Kelly, Charles J., Jr. *The Sky's the Limit: The History of the Airlines*. New York: Arno Press, 1963. A general history of airline transportation and the companies that provide it.

Kelsey. Benjamin S. *The Dragon's Teeth? The Creation of United States Air Power for World War II*. Washington, D.C.: Smithsonian Institution Press, 1982. A memoir by a military officer involved in airplane development in the 1930s.

Kennett, Lee. *The First Air War, 1914–1918*. New York: Free Press, 1991. A general history of air combat in World War I from an international perspective.

———. *A History of Strategic Bombing: From the First Hot-Air Balloons to Hiroshima and Nagasaki*. New York: Charles Scribner's Sons, 1982. A comprehensive overview of the subject from an international perspective.

Kent, Richard J., Jr. *Safe, Separated, and Soaring: A History of Federal Civil Aviation Policy, 1961–1972*. Washington, D.C.: U.S. Department of Transportation, 1974. The official history of the Federal Aviation Administration during the 1960s.

King, Peter. *Knights of the Air: The Life and Times of the Extraordinary Pioneers Who First Built British Aeroplanes*. Iowa City: University of Iowa Press, 1990. A biographical approach to understanding the early history of aviation in England.

Kirk, Stephen. *First in Flight: The Wright Brothers in North Carolina*. Raleigh, N.C.: John F. Blair, 1995. A study of the Wrights at Kitty Hawk between 1899 and 1905.

Knaack, Marcelle Size. *Post–World War II Bombers, 1945–1973*. Washington, D.C.:

Office of Air Force History, 1988. An encyclopedia with details about all U.S. Air Force bombers.

———. *Post-World War II Fighters, 1945-1973*. Washington, D.C.: Office of Air Force History, 1986. An encyclopedia with details about all U.S. Air Force fighters.

Komons, Nick A. *Bonfires to Beacons: Federal Civil Aviation Policy under the Air Commerce Act, 1926-1938*. Washington, D.C.: U.S. Government Printing Office, 1978. The first volume of the official history of the Federal Aviation Administration, which is much more than an institutional work, for it emphasizes the development of early aeronautical policy.

Launius, Roger D., ed. *Innovation and the Development of Flight*. College Station: Texas A&M University Press, 1999. A collection of essays on the nature of innovation and creativity in the development of flight systems.

Leary, William M. *Aerial Pioneers: The U.S. Air Mail Service, 1918-1927*. Washington, D.C.: Smithsonian Institution Press, 1985. The standard account of the rise of the air mail, the first commercial aviation activity in the United States.

———. *Perilous Missions: Civil Air Transport and CIA Covert Operations in Asia*. University: University of Alabama Press, 1984. The story of the CIA's secret air force in Southeast Asia.

———. *"We Freeze to Please": A History of NASA's Icing Research Tunnel and the Quest for Flight Safety*. NASA Special Publication 2002-4226. Washington, D.C.: U.S. Government Printing Office, 2002. A history of efforts to overcome one of the most significant safety hazards of aircraft—icing.

———, ed. *Aviation's Golden Age: Portraits from the 1920s and 1930s*. Iowa City: University of Iowa Press, 1989. A collection of biographical essays on figures in American history who furthered flight during the pivotal interwar era.

———. *Encyclopedia of American Business History and Biography: The Airline Industry*. New York: Facts on File, 1992. A collection of essays on people and companies and places associated with the airline industry.

———. *From Airships to Airbus: The History of Civil and Commercial Aviation*. Vol. 1: *Infrastructure and Environment*. Washington, D.C.: Smithsonian Institution Press, 1995. A fine collection of articles on the development of business operations in aviation from an international conference on the rise of the airplane.

Lebow, Eileen F. *Cal Rodgers and the Vin Fiz: The First Transcontinental Flight*. Washington, D.C.: Smithsonian Institution Press, 1989. A detailed story of Cal Rodgers's transcontinental flight in 1911.

Levine, Alan J. *The Strategic Bombing of Germany, 1940-1945*. Westport, Conn.: Praeger, 1992. A study of the effort to bring Germany down through the efforts of the Eighth Air Force in England and the daylight bombing campaign of World War II.

Lewis, W. David, ed. *Airline Executives and Federal Regulation: Case Studies in American Enterprise from the Airmail Era to the Dawn of the Jet Age*. Columbus: Ohio State University Press, 2000. A collection of essays on the entrepreneurs who built the American airline industry.

Lewis, W. David, and Wesley Phillips Newton. *Delta: The History of an Airline*.

Athens: University of Georgia Press, 1979. The history of one of the legendary airlines of the world, one that survived and even thrived in the post-1978 deregulated environment.

Lewis, W. David, and William F. Trimble. *The Airway to Everywhere: A History of All American Aviation, 1937–1953*. Pittsburgh, Pa.: University of Pittsburgh Press, 1988. A history of a small and innovative aviation company near Pittsburgh, Pennsylvania, that would become USAir.

Leyes, Richard A., and William A. Fleming. *The Story of North American Small Gas Turbine Engines*. Washington, D.C.: Smithsonian Institution Press and AIAA, 1999. A comprehensive examination of the subject.

Linnekin, Richard. *Eighty Knots to Mach 2: Forty-Five Years in the Cockpit*. Annapolis, Md.: Naval Institute Press, 1991. An interesting memoir.

Loftin, Laurence K., Jr. *Quest for Performance: The Evolution of the Subsonic Aircraft*. NASA Special Publication 463. Washington, D.C.: U.S. Government Printing Office, 1985. A discussion of the development of aviation technology from the standpoint of an engineer who spent a lifetime in aerodynamics.

Longyard, William H. *Who's Who in Aviation History: 500 Biographies*. Novato, Calif.: Presidio Press, 1995. A biographical reference work.

Lopez, Donald S. *Aviation: A Smithsonian Guide*. New York: Macmillan, 1995. An encyclopedic reference work, organized alphabetically.

Loving, Neal V. *Loving's Love: A Black American's Experience in Aviation*. Smithsonian History of Aviation Book Series. Washington, D.C.: Smithsonian Institution Press, 1994. A memoir of an African American pilot and aircraft designer covering much of the twentieth century.

Macklin, R. S., and H. A. Poitz. *Evolution of an Airline: Piedmont Airlines, 1948–1989*. Warrendale, Pa.: Society of Automotive Engineers, 1991. A history of a regional air carrier in the American southeast.

Mark, Eduard M. *Aerial Interdiction: Air Power and the Land Battle in Three American Wars*. Washington, D.C.: Center for Air Force History, 1994. A study of the use of air power to destroy enemy systems and materiel before it reaches the combat zone with special reference to World War II, Korea, and Vietnam.

Mason, Francis K. *British Fighters since 1912*. Annapolis, Md.: Naval Institute Press, 1992. An encyclopedic reference work.

Mason, Herbert Molloy, Jr. *Bold Men, Far Horizons: The Story of Great Pioneer Flights*. Philadelphia: Lippincott, 1966. A popular rendition of classic events in aviation history.

Masters, Charles J. *Glidermen of Neptune: The Neptune D-Day Glider Attack*. Carbondale: Southern Illinois University Press, 1995. An account of the little-known glider operations at D-Day, 6 June 1944, when the Allies invaded France.

Matthews, Birch. *Cobra! The Bell Aircraft Corporation, 1934–1946*. London, U.K.: Schiffer Publishing, 1996. A popular history of the aircraft company that developed the X-1, the first aircraft to fly faster than sound.

Maurer, Mauer. *Aviation in the U.S. Army, 1919–1939*. Washington, D.C.: Office of Air Force History, 1987. A narrative of military aviation between the two world wars.

McDougall, Walter A. . . . *The Heavens and the Earth: A Political History of the Space Age*. New York: Basic Books, 1985. Repr., Baltimore, Md.: Johns Hopkins University Press, 1997. An analysis of the race to the Moon in the 1960s arguing that Apollo prompted the space program to stress engineering over science, competition over cooperation, civilian over military management, and international prestige over practical applications.

McFarland, Stephen L. *America's Pursuit of Precision Bombing, 1910–1945*. Smithsonian History of Aviation Book Series. Washington, D.C.: Smithsonian Institution Press, 1995. A history of the development of aviation technology and procedures that made strategic bombardment a reality.

McFarland, Stephen L., and Wesley Phillips Newton. *To Command the Sky: The Battle for Air Superiority over Germany, 1942–1944*. Smithsonian History of Aviation Book Series. Washington, D.C.: Smithsonian Institution Press, 1991. A history of the quest for air superiority in Europe during World War II.

Mellberg, William F. *Famous Airliners: Seventy Years of Aviation & Transport Progress*. Boston: Plymouth Press, 1994. Primarily a popular reference work, with an aircraft-by-aircraft account of history.

Meyer, Henry Cord. *Airshipmen, Businessmen, and Politics, 1890–1940*. Smithsonian History of Aviation Book Series. Washington, D.C.: Smithsonian Institution Press, 1991. A history of the interplay of aviation and politics in the formative era.

Meyer, John R., and Clinton V. Oster Jr., eds. *Airline Deregulation: The Early Experience*. Boston: Auburn House, 1981. A collection of scholarly essays on the history of airline regulation and its evolution in the United States since the Kelley Act of 1926.

Mielinger, Phillip S. *Hoyt S. Vandenberg: The Life of a General*. Bloomington: Indiana University Press, 1989. A biography of an important pioneer of air power.

Mikesh, Robert C. *Zero: Japan's Legendary WWII Fighter*. Osceola, Wis.: Motorbooks International, 1994. A biography of the pathbreaking Japanese fighter of World War II.

Miller, Jay. *The X-Planes*. New York: Orion, 1988. Reference material on this class of experimental airplanes.

Miller, Roger G. *To Save a City: The Berlin Airlift, 1948–1949*. College Station: Texas A&M University Press, 2000. An account of the Berlin airlift.

————, ed. *Seeing Off the Bear: Anglo-American Air Power Cooperation during the Cold War*. Washington, D.C.: Air Force History and Museum Program, 1995. A collection of articles about the use of air power in the Anglo-American alliance.

Miller, Ronald, and David Sawers. *The Technical Development of Modern Aviation*. London, England: Routledge & Kegan Paul, 1968. A history of flight with a focus on well-known aircraft.

Miller, Samuel Duncan. *An Aerospace Bibliography*. Washington, D.C.: Office of Air Force History, 1978. A thorough general bibliography of aviation history.

Mills, Stephen E. *More Than Meets the Sky: A Pictorial History of the Founding and Growth of Northwest Airlines*. Seattle, Wash.: Superior Publishing, 1972. A popular history, well illustrated with photographs.

Mondey, David, comp. *Concise Guide to American Aircraft of World War II*. Novato, Calif.: Presidio Press, 1982. A good compendium.

———. *Concise Guide to British Aircraft of World War II*. Novato, Calif.: Presidio Press, 1982. Another good compendium.

Moody, Walton S. *Building a Strategic Force*. Washington, D.C.: Air Force History and Museums Program, 1996. An official history of the origins and development of the Strategic Air Command, the military command charged with nuclear warfare in the early Cold War era.

Morris, Lloyd R., and Kendall Smith. *Ceiling Unlimited: The Story of American Aviation from Kitty Hawk to Supersonics*. New York: Macmillan, 1953. A general history of the first half century of flight.

Morrow, John H., Jr. *German Air Power in World War I*. Lincoln: University of Nebraska Press, 1982. The standard work in English on its subject.

———. *The Great War in the Air: Military Aviation from 1909 to 1921*. Smithsonian History of Aviation Book Series. Washington, D.C.: Smithsonian Institution Press, 1993. A broad comprehensive history of the early development of aerial combat.

Mosley, Leonard. *Lindbergh: A Biography*. Garden City, N. Y.: Doubleday, 1976. A biography of the icon of American aviation.

Moy, Timothy. *War Machines: Transforming Technologies in the U.S. Military, 1920–1940*. College Station: Texas A&M University Press, 2001. A comparison of the technologies of the Air Corps's strategic bombardment doctrine with the technologies of the Marine Corps's amphibious assault doctrine.

Nalty, Bernard C., John F. Shiner, and George M. Watson. *With Courage: The U.S. Army Air Forces in World War II*. Washington, D.C.: Office of Air Force History, 1994. An official one-volume history of the U.S. Army Air Forces in World War II.

Nance, John J. *Blind Trust: The Human Crisis in Airline Safety*. New York: Morrow, 1986. A history of airline safety and the desire to enhance it, as well as a critique of the current situation.

———. *Splash of Colors: The Self-Destruction of Braniff International*. New York: Morrow, 1984. A journalistic account the demise of one of the premier air carriers in the United States in the early 1980s.

Newhouse, John. *The Sporty Game*. New York: Alfred A. Knopf, 1982. A financial history of aircraft development, production, and sales.

Newton, Wesley Phillips. *The Perilous Sky: U.S. Aviation Diplomacy and Latin America, 1919–1931*. Coral Gables, Fla.: University of Miami Press, 1978. An analysis of the role of aviation and air carriers in diplomacy with Latin America between World Wars I and II, with special reference to Pan American Airways' status as the nation's intercontinental carrier.

Nielson, Georgia Panter. *From Sky Girl to Flight Attendant: Women and the Making of a Union*. Ithaca, N. Y.: ILR Press, 1982. A history of flight attendants in the airline industry and their organizational activities to ensure equitable working conditions and salaries.

Nordeen, Lon O., and David Nicolle. *Phoenix over the Nile: A History of Egyptian Air Power, 1932–1994*. Washington, D.C.: Smithsonian Institution Press, 1996. A study of air power in a key nation within the troubled Middle East.

Oakes, Claudia M. *United States Women in Aviation, 1930–1939*. Washington, D.C.: Smithsonian Institution Press, 1991. The third in the Smithsonian's series on women in aviation; a brief account of the depression decade.

———. *United States Women in Aviation through World War I*. Washington, D.C.: Smithsonian Institution Press, 1978. The first in the Smithsonian's series on women in aviation, a short, informative account of a formative period.

O'Leary, Michael. *High VIZ: U.S. Military Aircraft, 1954–1964*. Osceola, Wis.: Motorbooks International, 1994. A basic airplane reference work.

Omissi, David E. *Air Power and Colonial Control: The Royal Air Force, 1919–1939*. Manchester, U.K.: Manchester University Press, 1990. An exploration of the British use of air power before World War II.

Ott, James, and Aram Gesar. *Jets: Airliners of the Golden Age*. Osceola, Wis.: Motorbooks International, 1993. A general history of the jet airliner.

Ottis, Sherri Green. *Silent Heroes: Downed Airmen and the French Underground*. Lexington: University Press of Kentucky, 2001. A discussion of rescues of various combat pilots.

Overy, R. J. *The Air War, 1939–1945*. New York: Stein and Day, 1980. Not a standard history of World War II in the air, but a broad international perspective on the interplay between air power and technology, the uses of air power to accomplish combat missions, and the results of air operations.

Owen, Kenneth. *Concorde and the Americans: International Politics of the Supersonic Transport*. Smithsonian History of Aviation Book Series. Washington, D.C.: Smithsonian Institution Press, 1997. A detailed analysis of the development and operations of the Concorde, the world's only supersonic commercial airliner.

Parramore, Thomas C. *First to Fly: North Carolina and the Beginnings of Aviation*. Chapel Hill: University of North Carolina Press, 2002. A history of early flight in North Carolina, including the Wright Brothers' experiments at Kitty Hawk.

———. *Triumph at Kitty Hawk: The Wright Brothers and Powered Flight*. Raleigh: North Carolina Division of Archives and History, 1993. A detailed account of the activities of the Wright Brothers on the Outer Banks of North Carolina in 1903.

Pattillo, Donald M. *A History in the Making: 80 Turbulent Years in the American General Aviation Industry*. New York: McGraw-Hill, 1998. A survey history of the general aviation industry that covers just about every airplane built for general aviation.

Pearcy, Arthur. *Sixty Glorious Years: A Tribute to the Douglas DC-3*. Osceola, Wis.: Motorbooks International, 1995. A discussion of the critically important DC-3 and the revolution in air transport it wrought.

———. *X-Plus: History of NACA/NASA Aircraft from 1915 to 1990*. Shrewsbury, U.K.: Airlife Publishing, 1992. An examination of the development and use of the NACA/NASA X series of aerospace craft that have pushed back the frontiers of flight.

Peebles, Curtis. *Dark Eagles: A History of Top Secret U.S. Aircraft Programs*. Novato, Colo.: Presidio Press, 1995. A popularly written account of U.S. Air Force "black programs," including reconnaissance aircraft such as the U-2 and SR-71 as well as the stealth technology in the F-117 and the B-2.

Perret, Geoffrey. *Winged Victory: The Army Air Forces in World War II.* New York: Random House, 1993. A popular history of American air power in the war.

Pisano, Dominick A. *To Fill the Skies with Pilots: A History of the Civilian Pilot Training Program.* Urbana: University of Illinois Press, 1993. The story of the little-known but critically important program run by the Civil Aeronautics Authority between 1939 and 1945 to train pilots at schools around the country.

Powell, William J. *Black Aviator: The Story of William J. Powell.* Washington, D.C.: Smithsonian Institution Press, 1994. A memoir by a pioneering African American pilot.

Powers, Sheryll Goecke. *Women in Flight Research at NASA Dryden Flight Research Center from 1946 to 1994.* Washington, D.C.: NASA History Office, Monographs in Aerospace History, No. 6, 1997. A brief but detailed description of the accomplishments of women at one U.S. laboratory and their community formation in the postwar and Cold War era.

Preston, Edmund. *Troubled Passage: The Federal Aviation Administration during the Nixon-Ford Term, 1973-1977.* Washington, D.C.: U.S. Department of Transportation, 1987. An official history of the FAA.

Rae, John B. *Climb to Greatness: The American Aircraft Industry, 1920-1960.* Cambridge, Mass.: MIT Press, 1968. A standard history of its subject.

Ragsdale, Kenneth Baxter. *Wings over the Mexican Border: Pioneer Military Aviation in the Big Bend.* Austin: University of Texas Press, 1984. The standard account of the first use of the airplane in American combat operations, during the 1916 punitive expedition against Pancho Villa.

Raines, Edgar F. *Eyes of Artillery: The Origins of Modern U.S. Army Aviation in World War II.* Washington, D.C.: Center of Military History, 1999. The official history of that part of U.S. Army aviation that remained in the U.S. Army after World War II.

Redding, Robert, and Bill Yenne. *Boeing: Planemaker to the World.* Greenwich, Conn.: Bison Books, 1983. A popular illustrated history of the largest commercial aircraft manufacturer in the world.

Reinhold, Ruth M. *Sky Pioneering: Arizona in Aviation History.* Tucson: University of Arizona Press, 1983. An examination of the importance of aviation to the Southwest.

Reynolds, Clark G. *Admiral John H. Tower: The Struggle for Naval Air Supremacy.* Annapolis, Md.: Naval Institute Press, 1991. A biography of one of the key figures in the development of naval aviation.

Rich, Ben R., and Leo Janos. *Skunk Works: A Personal Memoir of My Years at Lockheed.* Boston: Little, Brown, 1994. The story of the "Skunk Works," Lockheed's division for conducting highly classified research projects, and the efforts undertaken there.

Rich, Doris L. *Amelia Earhart: A Biography.* Washington, D.C.: Smithsonian Institution Press, 1996. The standard biography of the best-known American woman pilot, who was lost during her round-the-world flight in the late 1930s.

———. *Queen Bess: Daredevil Aviator.* Washington, D.C.: Smithsonian Institution

Press, 1993. A brief biography of a pioneering African American woman aviator.

Robie, William. *For the Greatest Achievement: A History of the Aero Club of America and the National Aeronautic Association.* Washington, D.C.: Smithsonian Institution Press, 1993. An institutional history of the organizations that have fostered aviation in America since 1911.

Robinson, Douglas H. *The Dangerous Sky: A History of Aviation Medicine.* Seattle: University of Washington Press, 1973. The only historical work of substance on a critically important aspect of flying.

Rochester, Stuart R. *Takeoff at Mid-Century: Federal Civil Aviation Policy in the Eisenhower Years, 1953–1961.* Washington, D.C.: U.S. Department of Transportation, 1976. The official history of the Federal Aviation Authority during the 1950s.

Rodgers, Eugene. *Flying High: The Story of Boeing and the Rise of the Jetliner Industry.* New York: Atlantic Monthly Press, 1996. A useful recent history of this aerospace firm.

Roessler, Walter, Leo Gomez, and Gail L. Greene. *Amelia Earhart—Case Closed?* New York: Markowski International Publishers, 1996. Another study of the aviator who disappeared during her round-the-world flight in the late 1930s.

Roland, Alex. *Model Research: The National Advisory Committee for Aeronautics, 1915–1958.* 2 vols. NASA Special Publication 4103. Washington, D.C.: U.S. Government Printing Office, 1985. An institutional study of NASA's predecessor, an organization that became involved in sounding rocket research and human space flight.

Roseberry, Cecil R. *The Challenging Skies: The Colorful Story of Aviation's Most Exciting Years, 1919–1939.* Garden City, N. Y.: Doubleday, 1966. A generic history of aviation in the period between World War I and World War II.

———. *Glen Curtiss: Pioneer of Flight.* Syracuse, N. Y.: Syracuse University Press, 1991. A reprint of a 1927 biography of one of the important early aviation entrepreneurs.

Ross, Walter S. *The Last Hero: Charles A. Lindbergh.* New York: Harper and Row, 1976. A standard biography of an American aviation icon.

Rotundo, Louis. *Into the Unknown: The X-1 Story.* Washington, D.C.: Smithsonian Institution Press, 1994. An account of the effort to exceed the speed of sound, finally achieved in 1947 with the Bell X-1 aircraft.

Rowe, Frank Joseph, and Craig Miner. *Borne on the South Wind: A Century of Kansas Aviation.* Wichita, Kans.: Wichita Eagle and Beacon Publishing, 1994. A well-illustrated history of flight in Kansas, the home of the light airplane industry.

Ruble, Kenneth D. *Flight to the Top.* New York: Viking Press, 1986. A popular history of Northwest Airlines.

Rummel, Robert W. *Howard Hughes and TWA.* Smithsonian History of Aviation Book Series. Washington, D.C.: Smithsonian Institution Press, 1991. The standard biography of a major early aeronautical entrepreneur and his airline.

Russo, Carolyn. *Women and Flight: Portraits of Contemporary Women Pilots.* Washington, D.C.: Smithsonian Institution Press, 1997. An exhibit catalog from the National Air and Space Museum with photographic and narrative portraits of women pilots of all ages and from all over the United States.

Sabach, Karl. *Twenty-first Century Jet: The Making and Marketing of the Boeing 777*. New York: Charles Scribner's Sons, 1996. A companion volume to a PBS documentary that traces the development of the Boeing 777, as well as the new design technologies that helped make the airplane possible.

Sandler, Stanley. *Segregated Skies: All-Black Combat Squadrons of World War II*. Washington, D.C.: Smithsonian Institution Press, 1992. A focus on the training and combat experience of African American military aviators in World War II.

Schaffel, Kenneth. *The Emerging Shield: The Air Force and the Evolution of Continental Air Defense, 1945–1990*. Washington, D.C.: Office of Air Force History, 1991. An official study of the U.S. Air Force's efforts to defend the United States from attack from the air.

Schaffer, Ronald. *Wings of Judgment: American Bombing in World War II*. New York: Oxford University Press, 1985. An analysis of American bombing in World War II.

Schatzberg, Eric. *Wings of Wood, Wings of Metal: Culture and Technology in American Airplane Material, 1914–1945*. Princeton, N. J.: Princeton University Press, 1999. An examination of the nontechnical factors behind the movement from wood to metal reflecting the latest scholarship on the sociology of technology.

Scheppler, Robert H. *Pacific Air Race*. Washington, D.C.: Smithsonian Institution Press, 1988. A history of the air races flown between southern California and Hawaii in the 1930s.

Schmid, S. H., and T. C. Weaver. *The Golden Age of Air Racing, Pre-1940*. Oshkosh, Wis.: EAA Aviation Foundation, 1991. A general study of the air race phenomenon that reigned among aviators before World War II.

Schoenfeld, Max. *Stalking the U-Boat: USAAF Offensive Antisubmarine Operations in World War II*. Smithsonian History of Aviation Book Series. Washington, D.C.: Smithsonian Institution Press, 1994. A study of the use of the airplane to hunt for German submarines.

Schoneberger, William A., and Paul Sonnenburg. *California Wings: A History of Aviation in the Golden State*. Woodland Hills, Calif.: Windsor Publications, 1984. An illustrated history of California airline companies and routes.

Schultz, James. *Winds of Change: Expanding the Frontiers of Flight, Langley Research Center's 75 Years of Accomplishment, 1917–1992*. Washington, D.C.: U.S. Government Printing Office, 1992. A well-illustrated popular history of the first government laboratory devoted exclusively to aeronautical research and development.

Scott, Phil. *The Shoulders of Giants: A History of Human Flight to 1919*. New York: Addison-Wesley, 1995. A popular history of the early efforts to fly around the globe.

Seamans, Robert C., Jr. *Aiming at Targets: The Autobiography of Robert C. Seamans Jr*. NASA Special Publication 4106. Washington, D.C.: U.S. Government Printing Office, 1996. A memoir by the deputy administrator of NASA during the 1960s and Nixon's secretary of the U.S. Air Force, 1969–1972.

Serling, Robert J. *Eagle: The Story of American Airlines*. New York: St. Martin's Press, 1985. A popular history of the company and the only work on the subject.

————. *From the Captain to the Colonel: An Informal History of Eastern Air Lines*. New York: Dial, 1980. A popular history of what once was one of the most important airlines in the nation.

————. *Howard Hughes' Airline: An Informal History of TWA*. New York: St. Martin's Press, 1983. A popular account of the history of Hughes and TWA with emphasis on Hughes's flamboyance.

————. *Legend and Legacy: The Story of Boeing and Its People*. New York: St. Martin's Press, 1992. A popular history of the aerospace company.

————. *Maverick: The Story of Robert Six and Continental Airlines*. Garden City, N. Y.: Doubleday, 1974. A history of a major air carrier and the man who created it.

————. *The Only Way to Fly: The Story of Western Airlines, America's Senior Air Carrier*. Garden City, N. Y.: Doubleday, 1976. A history of arguably the first passenger airline in the United States.

Sharader, Welman A. *Fifty Years of Flight: A Chronicle of the Aviation Industry in America, 1903–1953*. Cleveland, Ohio: Eaton Manufacturing, 1953. A privately published retrospective with important insights into the industry's history.

Sherry, Michael S. *The Rise of American Air Power: The Creation of Armageddon*. New Haven, Conn.: Yale University Press, 1987. An analysis of how the United States became wedded to the idea of the airplane as a means of national security.

Sherwood, John Darrell. *Officers in Flight Suits: The Story of American Air Force Fighter Pilots in the Korean War*. New York: New York University Press, 1996. A social history of fliers in the U.S. Air Force during the Korean conflict.

Smith, David J. *Britain's Military Airfields, 1939–1945*. Wellingborough, U.K.: Stephens, 1989. A basic guide to the English airfields, most of which date from World War II.

Smith, Dean C. *By the Seat of My Pants*. Boston: Little, Brown, 1961. A memoir of early flight in America by a seasoned aviator.

Smith, Frank Kingston. *Legacy of Wings: The Harold F. Pitcairn Story*. New York: Jason Aronsen, 1981. An in-depth look at both Harold Pitcairn and the aircraft, especially the autogiro, that made him famous.

Smith, Frank Kingston, and James P. Harrington. *Aviation and Pennsylvania*. Philadelphia: Franklin Institute Press, 1981. A brief narrative of aviation development in Pennsylvania.

Smith, G. Geoffrey. *Gas Turbines and Jet Propulsion*. London, U.K.: Iliffe & Sons, 1950. An early study of the history of the jet engine.

Smith, Henry Ladd. *Airways: The History of Commercial Aviation in the United States*. Smithsonian History of Aviation Book Series. Washington, D.C.: Smithsonian Institution Press, 1991. A reprint of a work first published in 1942 that remains an important work on the history of the airline industry.

————. *Airways Abroad: The Story of Commercial Aviation in the World*. Smithsonian History of Aviation Book Series. Washington, D.C.: Smithsonian

Institution Press, 1991. Another reprint of a classic history, originally published in 1950.

Smith, Herschel. *Aircraft Piston Engines: From the Manly Baltzer to the Continental Tiara*. New York: McGraw-Hill, 1981. A valuable survey of the engines used on general aviation aircraft.

Smithies, Edward, compiler. *War in the Air: Men and Women Who Built, Serviced, and Flew War Planes Remember the Second World War*. London, U.K.: Viking, 1990. A collection of oral histories.

Sochor, Eugene. *The Politics of International Aviation*. Iowa City: University of Iowa Press, 1991. A study of the International Civil Aviation Organization (ICAO), which negotiates international agreements on the use of airspace, procedures, etc., that places aviation politics in an international context.

Solberg, Carl. *Conquest of the Skies: A History of Commercial Aviation in America*. Boston: Little, Brown, 1979. A general history of the airline industry.

Sorenson, David S. *The Politics of Strategic Aircraft Modernization*. Westport, Conn.: Praeger, 1995. A scholarly monograph on the building of major weapons systems such as the B-52.

Spangenburg, Ray, and Diane Moser. *The Story of America's Air Transportation*. New York: Facts on File, 1992. A general history of the airline industry.

Spenser, Jay P. *Vertical Challenge: The Hiller Aircraft Story*. Seattle: University of Washington Press, 1992. A history of a company important in the history of the helicopter industry.

———. *Whirlybirds: A History of the U.S. Helicopter Pioneers*. Seattle: University of Washington Press and Museum of Flight, 1998. An extensive study of the pioneering men and machines in the U.S. helicopter industry.

Sterling, Christopher H. *Commercial Air Transport Books: An Annotated Bibliography of Airlines, Airliners, and the Air Transport Industry*. McLean, Va.: Paladwr Press, 1996. A valuable reference work.

Stevens, Robert W. *Alaskan Aviation History*. Vol. 1: *1897–1928*. Vol. 2: *1929–1930*. Des Moines, Iowa: Polynyas Press, 1990. A focus on the history of aviation in Alaska, where aviation is vital.

Stevenson, James Perry. *The Pentagon Paradox: The Development of the F-18 Hornet*. Annapolis, Md.: Naval Institute Press, 1993. A scholarly work on the politics of aircraft development.

Stillwell, W. H. *X-15 Research Results, with a Selected Bibliography*. NASA Special Publication 60. Washington, D.C.: U.S. Government Printing Office, 1965. A compendium of information on the X-15 aircraft development concept, flight research, aerodynamic characteristics of supersonic-hypersonic flight, hypersonic structure, and the flying laboratory, with a useful bibliography.

Stoff, Joshua. *From Airship to Spaceship: Long Island Aviation & Spaceflight*. New York: Heart of the Lakes Publishing, 1991. A study of the regional impact of aviation.

———. *The Thunder Factory: An Illustrated History of the Republic Aviation Corporation*. Osceola, Wis.: Motorbooks International, 1990. Another history of an aircraft manufacturer.

Stokesbury, James L. *A Short History of Airpower*. New York: Morrow, 1986.

A survey of military aviation from the first use of airplanes for combat in Libya in 1911 to the Falklands War of 1982.

Stroud, John. *Annals of British and Commonwealth Air Transport, 1919–1960*. London, U.K.: Putnam, 1962. A history of a British commercial airline.

Sturtivant, Ray. *British Naval Aviation: The Fleet Air Army, 1917–1990*. 2d ed. Annapolis, Md.: Naval Institute Press, 1990. A study of naval aviation in the United Kingdom.

―――. *British Research and Development Aircraft: Seventy Years at the Leading Edge*. Newbury Park, Calif.: Haynes, 1990. A study of aerospace research and development history with emphasis on the "X" aircraft used to test aerodynamic concepts.

Swanborough, Gordon, and Peter M. Bowers. *United States Navy Aircraft since 1911*. Annapolis, Md.: Naval Institute Press, 1990. The standard encyclopedia of naval aircraft.

Talay, T. A. *Introduction to the Aerodynamics of Flight*. NASA Special Publication 367. Washington, D.C.: U.S. Government Printing Office, 1975. A basic text on the atmosphere, fluid flow, subsonic flow effects, transonic flow, supersonic flow, aircraft performance, and stability and control.

Taylor, Frank J. *High Horizons: Daredevil Flying Postmen to Modern Magic Carpet—The United Air Lines Story*. New York: McGraw-Hill, 1964. A popular history of one of the most important of the nation's air carriers from the 1920s to the 1960s.

Taylor, Frank J., and Lawton Wright. *Democracy's Air Arsenal*. New York: Duell, Sloan, and Pearce, 1947. An important discussion of how the American aircraft industry was mobilized to build military bombers, fighters, and transports for World War II.

Taylor, John W. R. *A History of Aerial Warfare*. London, U.K.: Hamlyn, 1974. Another illustrated history.

―――, ed. *Jane's Pocket Book of Research and Experimental Aircraft*. New York: Collier, 1976. An encyclopedic summary of data on selected aircraft.

Taylor, John W. R., and David Mondey. *Milestones of Flight*. Boston: Van Nostrand Reinhold, 1983. A chronological approach to the history of flight.

Taylor, John W. R., and Kenneth Munson. *History of Aviation*. New York: Crown, 1972. Another popular illustrated history of flight.

Taylor, Michael. *History of Flight*. New York: Random House, 1991. Yet another popular history.

Tessendorf, K. C. *Barnstormers and Daredevils*. New York: Athenaeum, 1988. A popular account of the romantic age of flying featuring barnstorming pilots and impromptu air shows at county fairs in the period of the 1910s and 1920s.

Thayer, Frederick C., Jr. *Air Transport Policy and National Security: A Political, Economic, and Military Analysis*. Chapel Hill: University of North Carolina Press, 1965. An analysis of the history of airline policy at the national level, focusing on laws and process.

Thiele, Ray. *Kennedy's Hawaiian Air: Hawaii's Pioneer Airline*. Honolulu, Hawaii: Olomana Publishers, 1994. An account of the challenges in, opposition to, and early development of interisland service from creation of Hawaiian Air in the 1920s until its bankruptcy and restructuring in the 1980s.

Thompson, Milton O. *At the Edge of Space: The X-15 Flight Program*. Washington, D.C.: Smithsonian Institution Press, 1992. A description of the program from the perspective of one of the X-15 test pilots.

Tillett, Paul. *The Army Flies the Mails*. Tuscaloosa: University of Alabama Press, 1956. An account of the disastrous 1934 effort by the U.S. Army Air Corps to fly the mails.

Treadwell, Terry C. *Ironworks: A History of Grumman's Fighting Aeroplanes*. Shrewsbury, U.K.: Airlife, 1990. A handbook of all the fighters built by Grumman Aircraft from World War II on.

Trimble, William F. *Admiral William A. Moffett: Architect of Naval Aviation*. Washington, D.C.: Smithsonian Institution Press, 1993. A biography of the individual who more than any other built the navy carrier force that was so successful in World War II.

———. *High Frontier: A History of Aeronautics in Pennsylvania*. Pittsburgh, Pa.: University of Pittsburgh Press, 1982. A study of the impact of aviation within Pennsylvania.

———. *Jerome C. Hunsaker and the Rise of American Aeronautics*. Washington, D.C.: Smithsonian Institution Press, 2002. The first full-length biography of a major figure in the development of aeronautics in America.

———. *Wings for the Navy: A History of the Naval Aircraft Factory, 1917–1956*. Annapolis, Md.: Naval Institute Press, 1990. A history of an important navy laboratory.

———, ed. *From Airships to Airbus: The History of Civil and Commercial Aviation*. Vol. 2: *Pioneers and Operations*. Washington, D.C.: Smithsonian Institution Press, 1995. A collection of papers from an international symposium.

Tusa, Ann, and John Tusa. *The Berlin Airlift*. New York: Athenaeum, 1988. A detailed history of a pivotal event in the unfolding Cold War between the United States and the USSR.

Underwood, Jeffery S. *The Wings of Democracy: The Influence of Air Power on the Roosevelt Administration, 1933–1941*. College Station: Texas A&M University Press, 1991. An account of the U.S. Army Air Corps's building of its force during the years before World War II.

Underwood, John W. *The Stinsons: The Exciting Chronicle of a Flying Family and the Planes That Enhanced Their Fame*. Glendale, Calif.: Heritage Press, 1969. The story of a family of fliers from the 1910s who later founded an aircraft company.

Van der Linden, F. Robert. *Airlines and Air Mail: The Post Office and the Birth of Commercial Aviation Industry*. Lexington: University Press of Kentucky, 2002. A revisionist book, based on a doctoral dissertation, detailing the role of the U.S. Post Office in the development of commercial aviation in the United States.

———. *The Boeing 247: The First Modern Airliner*. Seattle: University of Washington Press, 1991. The standard history of its subject.

Vander Meulen, Jacob. *Building the B-29*. Smithsonian History of Aviation Book Series. Washington, D.C.: Smithsonian Institution Press, 1995. A history of the process and politics of developing an important bomber in World War II.

———. *The Politics of Aircraft: Building an American Military Industry*. Lawrence:

University Press of Kansas, 1991. A detailed history of how an industry developed in the decade before World War II and positioned itself to build thousands of aircraft to support American efforts in the war.

Verges, Marianne. *On Silver Wings: The Women Airforce Service Pilots of World War II, 1942–1944*. New York: Ballantine, 1991. Another popular history of the WASPs.

Villard, Henry Serrano. *Contact! The Story of the Early Birds*. Washington, D.C.: Smithsonian Institution Press, 1987. A history of the pioneers of early flight and the sometimes odd aircraft they built.

Vincenti, Walter G. *What Engineers Know and How They Know It: Analytical Studies from Aeronautical History*. Baltimore, Md.: Johns Hopkins University Press, 1990. One of the most influential works ever written, consisting of case studies on aeronautical research to illuminate the process of innovation and technological development in the aviation industry.

Wallace, Lane E. *Airborne Trailblazer: Two Decades with NASA Langley's Boeing 737 Flying Laboratory*. NASA Special Publication 4216. Washington, D.C.: U.S. Government Printing Office, 1994. A "biography" of a research aircraft that has been used by NASA to conduct all manner of experiments from wind shear research to avionics.

———. *Flights of Discovery: An Illustrated History of the Dryden Flight Research Center*. NASA Special Publication 4309. Washington, D.C.: U.S. Government Printing Office, 1996. An illustrated history of the Mojave facility.

Watson, George M., Jr. *The Office of the Secretary of the Air Force, 1947–1965*. Washington, D.C.: Center for Air Force History, 1993. An institutional history that concentrates on the founding years of the office, especially the pivotal tenure of Stuart Symington, Truman's secretary of the U.S. Air Force.

Wegg, John. *General Dynamics Aircraft and Their Predecessors*. Annapolis, Md.: Naval Institute Press, 1990. A basic compendium.

Weick, Fred C., and James L. Hansen. *From the Ground Up: The Autobiography of an Aeronautical Engineer*. Washington, D.C.: Smithsonian Institution Press, 1988. An autobiography of aviation pioneer Fred Weick that is a model for understanding the development of aviation between the 1930s and the 1970s.

Weiss, David A. *The Saga of the Tin Goose: The Story of the Ford Trimotor Airplane*. New York: Cumberland Enterprises, 1995. Another illustrated history.

Werrell, Kenneth P. *Blankets of Fire: U.S. Bombers over Japan during World War II*. Smithsonian History of Aviation Book Series. Washington, D.C.: Smithsonian Institution Press, 1996. An analysis of the strategic bombing campaign in the Pacific.

Weyl, A. R. *Fokker: The Creative Years*. London, U.K.: Putnam, 1965. A biography of the German aeronautical engineer and ingenious aircraft builder.

Wheatcroft, Stephen. *Aviation and Tourism Policies: Balancing the Benefit*. New York: Routledge, 1994. A survey of airline operations.

White, Robert P. *Mason Patrick and the Fight for Air Service Independence*. Washington, D.C.: Smithsonian Institution Press, 2001. A biography of the chief of the U.S. Army Air Service in the 1920s.

Whitnah, Donald R. *Safer Skyways: Federal Control of Aviation, 1926–1966*. Ames:

Iowa State University Press, 1966. A study of the national imperative to develop safe airways and aircraft from the passage of the Air Commerce Act of 1926 through the Federal Aviation Act of 1958.

Wilson, John R. W. *Turbulence Aloft: The Civil Aeronautics Administration amid Wars and Rumors of Wars, 1938–1953.* Washington, D.C.: Federal Aviation Administration, 1979. An official history of the FAA from the Kelley Act of 1938 through the end of the Truman administration.

Wilson, Stewart. *Legends of the Air: North American F-86 Sabre, MiG-15 and MiG-17, and the Hawker Hunter.* Osceola, Wis.: Motorbooks International, 1995. A collection of biographies of the fighters that engaged in the Korean Conflict, 1950–1953.

Wohl, Robert. *A Passion for Wings: Aviation and the Western Imagination, 1908–1918.* New Haven, Conn.: Yale University Press, 1994. A modern social history of the airplane; well-illustrated.

Wolfe, Martin. *Green Light! A Troop Carrier Squadron's War from Normandy to the Rhine.* Washington, D.C.: Center for Air Force History, 1993. A history of combat paratrooper operations in World War II.

Wooldridge, E. T., ed. *Carrier Warfare in the Pacific: An Oral History Collection.* Smithsonian History of Aviation Book Series. Washington, D.C.: Smithsonian Institution Press, 1993. A collection of oral histories from World War II veterans.

Wragg, David W. *Flight with Power: The First Ten Years.* New York: St. Martin's Press, 1978. Another history of the Wright Brothers and early aviation though 1914.

———. *Wings over the Sea.* New York: Arco Publishing, 1979. A history of naval aviation from an international perspective.

Wright, Monte D. *Most Probable Position: A History of Aerial Navigation to 1941.* Lawrence: University Press of Kansas, 1972. A history of technology on the ground and in aircraft that resolved the problems of aerial navigation.

Xu, Guangqiu. *War Wings: The United States and Chinese Military Aviation, 1929–1949.* Westport, Conn.: Greenwood Press, 2001. A study of international arms sales and bilateral military relations.

CONTRIBUTORS

Janet R. Daly Bednarek is chair of the history department at the University of Dayton. She is the author of several books and articles, including *The Changing Image of the City: Planning for Downtown Omaha, 1945–1974* (University of Nebraska Press, 1992) and *America's Airports: Airfield Development, 1918–1947* (Texas A&M University Press, 2001).

Tami Davis Biddle is professor of military history at the Army War College, Carlisle, Pennsylvania. She is the author of *Rhetoric and Reality in Air Warfare: The Evolution of British and American Ideas about Strategic Bombing, 1914–1945* (Princeton University Press, 2002).

Roger E. Bilstein is recently retired as professor of history at the University of Houston–Clear Lake. He is the author or editor of nine books and monographs, including *Flight in America: From the Wrights to the Astronauts* (Johns Hopkins University Press, 3d ed., 2000) and *Enterprise of Flight: The American Aviation and Aerospace Industry* (Smithsonian Institution Press, 2d ed., 2001). His honors and awards include recognition by the Aviation/Space Writers Association, the National Space Council, and the American Institute of Aeronautics and Astronautics. He has held the Lindbergh Chair of Aerospace History at the National Air and Space Museum (1992–93) and was visiting professor at the Air War College of the U.S. Air Force (1995–96).

Hans-Joachim Braun is professor of history at the Universität der Bundeswehr Hamburg. His previous publications in the United States are *The German Economy in the Twentieth Century: The German Reich and the Federal Republic* (Routledge, 1990) and *Music and Technology in the Twentieth Century* (Johns Hopkins University Press, 2002). His publications on aviation deal with the transfer of aviation technology, air armament, and aircraft production processes as well as with airports.

David T. Courtwright is professor of history at the University of North Florida, Jacksonville. He is the author of several books, including *Dark Paradise: A History of Opiate Addiction in America* (Harvard University Press, 2d ed. 2001), *Forces of Habit: Drugs and the Making of the Modern World* (Harvard University Press, 2001), and *Violent Land: Single Men and Social Disorder from the Frontier to the Inner City* (Harvard University Press, 1998). He is presently writing a history of the sky as frontier.

Anne Collins Goodyear is assistant curator of prints and drawings at the National Portrait Gallery, Smithsonian Institution, Washington, D.C. She is presently completing a study of the NASA Art Program.

Roger D. Launius is a member of the Division of Space History at the National Air and Space Museum, Smithsonian Institution, Washington, D.C. He has produced several books and articles on aerospace history, including *Imagining Space: Achievements, Possibilities, Predictions, 1950-2050* (Chronicle Books, 2001), *Reconsidering Sputnik: Forty Years since the Soviet Satellite* (Harwood Academic, 2000), *Innovation and the Development of Flight* (Texas A&M University Press, 1999), *NASA & the Exploration of Space* (Stewart, Tabori, & Chang, 1998), *Frontiers of Space Exploration* (Greenwood Press, 1998), *Organizing for the Use of Space: Historical Perspectives on a Persistent Issue* (Univelt, 1995), and *NASA: A History of the U.S. Civil Space Program* (Krieger, 1994).

William M. Leary, a former flight operations officer for KLM Royal Dutch Airlines at Gander, Newfoundland, is E. Merton Coulter Professor of History at the University of Georgia, Athens. He has written or edited numerous books and articles, including *Under Ice: Waldo Lyon and the Development of the Arctic Submarine* (Texas A&M University Press, 1999), *Project Coldfeet: Secret Mission to a Soviet Ice Station* (Naval Institute Press, 1996), *Aviation's Golden Age: Portraits from the 1920s and 1930s* (University of Iowa Press, 1989), *Aerial Pioneers: The U.S. Air Mail Service* (Smithsonian Institution Press, 1985), *Perilous Missions: Civil Air Transport and CIA Covert Operation in Asia* (University of Alabama Press, 1984), and *The Airline Industry*, a volume in the *Encyclopedia of American Business History and Biography* (Facts on File, 1990). He was the 1996–97 Guggenheim Fellow at the National Air and Space Museum, Smithsonian Institution.

David D. Lee is dean of the Potter College of Arts, Humanities, and Social Sciences at Western Kentucky University. He is the author of *Tennessee in Turmoil: Politics in the Volunteer State, 1920-1932* (Memphis State University Press, 1979) and *Sergeant York: An American Hero* (University Press of Kentucky, 1985).

W. David Lewis is Distinguished University Professor at Auburn University, Auburn, Alabama. He is the author or editor of many books on aviation history, including *Airline Executives and Federal Regulations: Case Studies in American Enterprise from the Airmail Era to the Dawn of the Jet Age* (Ohio State University Press, 2000), *The Airway to Everywhere: A History of All American Aviation, 1937-1953* (University of Pittsburgh Press, 1988), and *Delta: The History of an Airline* (University of Georgia Press, 1979). He is presently completing work on a biography of America's ace of aces, Eddie Rickenbacker.

John H. Morrow is Franklin Professor of History at the University of Georgia. He is the author of several books on aviation history, including *The Great War in the Air: Military Aviation from 1909 to 1921* (Smithsonian Institution Press, 1993) and *German Air Power in World War I* (University of Nebraska Press, 1982).

Dominick A. Pisano is chair of the aeronautics department at the National Air and Space Museum, Smithsonian Institution, Washington, D.C. He is the author of several books, including *Charles Lindbergh and the Spirit of St. Louis* (Harry N. Abrams, 2002), *To Fill the Skies with Pilots: The Civilian Pilot Training Program, 1939-46* (University of Illinois Press, 1993), and *Legend, Memory, and the Great War in the Air* (University of Washington Press, 1992).

A. Timothy Warnock is chief of the Organizational History Branch of the Air Force Historical Research Agency, Maxwell Air Force Base, Alabama. He has written *Short of War: Major United States Air Force Contingency Operations, 1947-1997* (Air Force History and Museum Program, 2000) and *Air Force Combat Medals, Streamers, and Campaigns* (Air Force History and Museum Program, 1991).

INDEX

stagnation of industry, 61–65; and strategic bombardment, 190–206; technology of, 15–49, 64–65; and turbojets, 75–78; U.S. government and, 2, 50–69; and World War I, 168–89, 191–94; and World War II, 3–4, 29–34, 59–61, 75–78, 196–203; and Wright brothers and U.S. army, 153–71

Aviation Corp., 100

Aviation Week and Space Technology, 23

B-17, 33, 196
B-29, 33
B-47, 36, 75, 214
B-52, 233
BAC 3-11, 80
Baeumker, Adolph, 59
Bairstow, Leonard, 74
Ball, Albert, 177, 247
Baxter, Warner, 27
Beechcraft Corp., 41
Bell Aircraft Co., 77
Bellanca, Giuseppe, 244, 249
Bender, Marylin, 255–56
Bendix Aviation, Inc., 121
Benjamin, Walter, 234
Bennett, Floyd, 249, 252
Benz, Karl, 16
Berlin Airlift, 36
Beuys, Joseph, 231
Bingham, Hiram, 93
Bingham, Ryan, 209–10
Bishop, William, 177, 247
Bixby, Harold, 244–45
Black, Archibald, 27
Black, Hugo, 106, 109, 110, 111, 112
Black Death, The, 194
Blériot, Louis, 16, 17, 171–72, 179
Boeing Aircraft Co., 25, 29, 32, 41–42, 63, 80, 220
Boeing 247, 25, 124, 210
Boeing 707, 214
Boeing 747, 40, 42, 214
Boeing 767, 80, 81

Boelke, Oswald, 175, 177, 179, 182
Bohnet, Frederick, 145
Boorstin, Daniel, 219, 258–59
Brancusi, Constantin, 229–30
Braniff International, Inc., 228
Braque, Georges, 223, 224
Braun, Wernher von, 237
Breech, Ernest R., 126
Breguet, Louis, 179
Bright, Charles D., 61, 62
British Advisory Committee for Aeronautics, 22
British Aircraft Corp., 79
British Airways, 79
Brock, William S., 140
Brookings Institution, 35
Brown, Arthur Whitton, 248
Brown, Walter Folger, 91, 95–108, 109, 110, 111, 112, 123
Buchnan, George, 17–18
Burge, Vernon, 159
Burgess, W. Starling, 165
Busemann, Adolph, 75
Butt, D. M., 196, 199
Byrd, Richard E., 248, 249, 252

Calder, Alexander, 225–28
California Institute of Technology (Caltech), 29, 74
Caniff, Milton, 27
Capper, J. E., 169–70
Caproni Aircraft Co., 186
Captain Easy, 27
Carroll, Thomas, 134, 135
Caudron, Gaston, 179
Caudron, René, 179
Cayley, Sir George, 16
Century Magazine, 18
Cessna, 41
CF-6 engine, 81
Challenger, 62
Chamberlin, Clarence, 245, 248, 249
Chambers, Reed, 119, 120
Chanute, Octave, 224
Chaplin, Charlie, 11